Improving the nutritional and nutraceutical properties of wheat and other cereals

It is widely recognised that agriculture is a significant contributor to global warming and climate change. Agriculture needs to reduce its environmental impact and adapt to current climate change whilst still feeding a growing population, i.e. become more 'climate-smart'. Burleigh Dodds Science Publishing is playing its part in achieving this by bringing together key research on making the production of the world's most important crops and livestock products more sustainable.

Based on extensive research, our publications specifically target the challenge of climate-smart agriculture. In this way we are using 'smart publishing' to help achieve climate-smart agriculture.

Burleigh Dodds Science Publishing is an independent and innovative publisher delivering high quality customer-focused agricultural science content in both print and online formats for the academic and research communities. Our aim is to build a foundation of knowledge on which researchers can build to meet the challenge of climate-smart agriculture.

For more information about Burleigh Dodds Science Publishing simply call us on +44 (0) 1223 839365, email info@bdspublishing.com or alternatively please visit our website at www.bdspublishing.com.

Related titles:

Achieving sustainable cultivation of wheat Volume 1: Breeding, quality traits, pests and diseases
Print (ISBN 978-1-78676-016-6); Online (ISBN 978-1-78676-018-0, 978-1-78676-019-7)

Achieving sustainable cultivation of wheat Volume 2: Cultivation techniques
Print (ISBN 978-1-78676-020-3); Online (ISBN 978-1-78676-022-7, 978-1-78676-023-4)

Achieving sustainable cultivation of sorghum Volume 1: Genetics, breeding and production techniques
Print (ISBN 978-1-78676-120-0); Online (ISBN 978-1-78676-122-4, 978-1-78676-123-1)

Achieving sustainable cultivation of sorghum Volume 2: Sorghum utilization around the world
Print (ISBN 978-1-78676-124-8); Online (ISBN 978-1-78676-126-2, 978-1-78676-127-9)

Advances in breeding techniques for cereal crops
Print (ISBN 978-1-78676-244-3); Online (ISBN 978-1-78676-246-7, 978-1-78676-247-4)

Chapters are available individually from our online bookshop:
https://shop.bdspublishing.com

climate•SMART•publishing
IN AGRICULTURAL SCIENCE

BURLEIGH DODDS SERIES IN AGRICULTURAL SCIENCE
NUMBER 81

Improving the nutritional and nutraceutical properties of wheat and other cereals

Edited by Professor Trust Beta, University of Manitoba, Canada

burleigh dodds
SCIENCE PUBLISHING

Published by Burleigh Dodds Science Publishing Limited
82 High Street, Sawston, Cambridge CB22 3HJ, UK
www.bdspublishing.com

Burleigh Dodds Science Publishing, 1518 Walnut Street, Suite 900, Philadelphia, PA 19102-3406, USA

First published 2021 by Burleigh Dodds Science Publishing Limited

Library of Congress Control Number: 2021931540

British Library Cataloguing in Publication Data
A catalogue record for this book is available from the British Library

ISBN 978-1-78676-479-9 (Print)
ISBN 978-1-78676-482-9 (PDF)
ISBN 978-1-78676-481-2 (ePub)
ISSN 2059-6936 (print)
ISSN 2059-6944 (online)

DOI 10.19103/AS.2021.0087

Typeset by Deanta Global Publishing Services, Dublin, Ireland

Contents

Series list

Title	Series number
Achieving sustainable cultivation of maize - Vol 1 From improved varieties to local applications Edited by: Dr Dave Watson, CGIAR Maize Research Program Manager, CIMMYT, Mexico	001
Achieving sustainable cultivation of maize - Vol 2 Cultivation techniques, pest and disease control Edited by: Dr Dave Watson, CGIAR Maize Research Program Manager, CIMMYT, Mexico	002
Achieving sustainable cultivation of rice - Vol 1 Breeding for higher yield and quality Edited by: Prof. Takuji Sasaki, Tokyo University of Agriculture, Japan	003
Achieving sustainable cultivation of rice - Vol 2 Cultivation, pest and disease management Edited by: Prof. Takuji Sasaki, Tokyo University of Agriculture, Japan	004
Achieving sustainable cultivation of wheat - Vol 1 Breeding, quality traits, pests and diseases Edited by: Prof. Peter Langridge, The University of Adelaide, Australia	005
Achieving sustainable cultivation of wheat - Vol 2 Cultivation techniques Edited by: Prof. Peter Langridge, The University of Adelaide, Australia	006
Achieving sustainable cultivation of tomatoes Edited by: Dr Autar Mattoo, USDA-ARS, USA; and Prof. Avtar Handa, Purdue University, USA	007
Achieving sustainable production of milk - Vol 1 Milk composition, genetics and breeding Edited by: Dr Nico van Belzen, International Dairy Federation (IDF), Belgium	008
Achieving sustainable production of milk - Vol 2 Safety, quality and sustainability Edited by: Dr Nico van Belzen, International Dairy Federation (IDF), Belgium	009
Achieving sustainable production of milk - Vol 3 Dairy herd management and welfare Edited by: Prof. John Webster, University of Bristol, UK	010
Ensuring safety and quality in the production of beef - Vol 1 Safety Edited by: Prof. Gary Acuff, Texas A&M University, USA; and Prof. James Dickson, Iowa State University, USA	011
Ensuring safety and quality in the production of beef - Vol 2 Quality Edited by: Prof. Michael Dikeman, Kansas State University, USA	012
Achieving sustainable production of poultry meat - Vol 1 Safety, quality and sustainability Edited by: Prof. Steven C. Ricke, University of Arkansas, USA	013
Achieving sustainable production of poultry meat - Vol 2 Breeding and nutrition Edited by: Prof. Todd Applegate, University of Georgia, USA	014

Acknowledgements

We wish to acknowledge the following for their help in reviewing a particular chapter:

- Chapter 2: Professor Fereidoon Shahidi, Memorial University of Newfoundland, Canada

Introduction

With more consumers moving away from traditional cereal-based foods due to concerns about health impacts, including wheat intolerance, the sector must develop next-generation nutritionally-enhanced cereal products to maximise market potential. This collection examines key research into the nutritional components of cereals and their role in preventing chronic diseases. The book is organised into two parts.

Part 1 focuses on the nutritional properties of cereals, with chapters covering aspects such as the nutritional value of starch, antioxidants and lipids in wheat. The first part of the book also addresses how cereal grains can prevent type 2 diabetes and cancer. Part 2 examines developing nutritionally enhanced cereal products. Chapters focus on areas such as understanding the genetics of the nutritional properties of cereals and optimising the effects of processing these properties. Chapters also discuss wheat flour fortification and human help, as well as reviewing the key steps in nutritional enhancement of cereal-based products, specifically emmer, einkorn spelt, sorghum and millet.

Part 1 Nutritional properties of cereals

The first part of the book opens with a chapter on advances in understanding the nutritional value of starch in wheat. Chapter 1 begins by providing an overview of wheat starch structure, focusing specifically on elements such as amylose and amylopectin. The chapter then goes on to discuss A-, B- and C-type wheat starch granules, before highlighting the role of wheat starch in food. It also examines the digestibility of diverse wheat starches, and highlights interactions of starch and other biomolecules such as proteins, lipids, and non-starch polysaccharides. The chapter concludes by emphasising the importance of rapidly digestible and slowly digestible starch in animal and human health.

Moving on from Chapter 1, Chapter 2 focuses on the nutritional value of antioxidants in wheat. The chapter provides a discussion of phytochemical antioxidants in wheat and how these can be extracted. It assesses the methods of analyses for phytochemical antioxidants in wheat grain, such as the analysis of total phenolic, anthocyanins, flavonoids, condensed tannins and carotenoid content. The chapter also discusses the analysis of phenolic compounds, carotenoids and tocochromanols in wheat grains. It then goes on to review the nutritional value of phytochemical antioxidants in wheat, focusing specifically on the effect of these compounds on the nutritional value of wheat food.

Chapter 3 addresses the nutritional value of lipids in wheat. Lipids are a small component in the wheat grain composition (2-4%) and roughly, two thirds

(66%) of them are contained in the germ, 15% are in the bran and particularly in the aleuronic layer, whereas about 20% are distributed in the endosperm, partly within the starch granules. They include a variety of different structural types, i.e. carboxylic acids (or fatty acids); mono, di and triacylglycerols (triglycerides or neutral fats); phospholipids; glycolipids; waxes; terpenes; steroids and alkylresorcinols. This chapter reviews the importance of lipids in wheat kernels and their significance in wheat technology and products. It also addresses the significance of wheat lipids and associated substances in human nutrition. A discussion on the wheat germ and how this wheat germ can be used to produce wheat oil is also included.

Moving on from the nutritional value of various compounds in wheat grains, Chapter 4 focuses on how dietary fibres can prevent the development of Type 2 diabetes mellitus. Dietary fibres are a heterogenous group of food compounds. The physicochemical properties of dietary fibres determine their effects on gastrointestinal and metabolic health, including effect on gastro-intestinal transit, glycaemic response, microbial composition and fermentative capacity. Most fibre rich foods contain insoluble, prebiotic and viscous fibres in varying ratios. The chapter begins by discussing underlying mechanisms of action of insoluble fibres. It then goes on to discuss the various dietary fibre intervention studies, specifically focusing on prebiotic fibres and soluble viscous fibres. The chapter concludes by highlighting the importance of consuming high levels dietary fibre and whole grains and their positive impact on metabolic health.

The final chapter of Part 1 examines fibre-associated wheat lignans and colorectal cancer prevention. Wheat, as a staple food, is largely consumed worldwide. In addition to nutritional values, whole grain including fibre-enriched wheat bran has been reported to provide many nutraceuticals such as wheat lignans. Chapter 5 reviews recent epidemiological and animal data on wheat lignans and their role in colorectal cancer prevention. It covers aspects of the lignan structure, biosynthesis, analysis, metabolism and potential health benefits with emphasis on anti-proliferative, antioxidant, anti-inflammation, antiestrogenic and cell cycle arrest mechanisms.

Part 2 Developing nutritionally-enhanced cereal products

Part 2 begins with an overview of the advances in understanding the genetics of the nutritional properties of cereals, specifically focusing on maize and oat proteins. Maize, after rice and wheat, is the most important cereal crop and is globally produced for food and feed. Similar to maize, oats are one of the six major cereal grains grown worldwide, with its production largely focused in the cooler parts of the world. Chapter 6 covers the genetics of maize and oat nutritional properties and the progress that has been made in terms of quality and the importance of these grains as macro- and micro-nutrients in humans. The chapter concludes by emphasising the importance of the biofortification of staple grains to help meet the nutritional demands of global populations.

Chapter 7 discusses developments in fractionation methods to improve extraction of aleurone from wheat grain. In wheat grain, the aleurone layer is located between the peripheral tissues and the starchy endosperm and is rich in soluble proteins, minerals, lipids, vitamins and micronutrients and contains several compounds with antioxidant activity. However, during grain fractionation it is mainly recovered in bran fractions, generally used to feed animals or for energy production. These last few years, the cereal scientist community and companies have developed research and new processing technologies (mainly protected with patents) to exploit its potential more deeply. This was mainly based on an enhanced knowledge of its composition and properties facilitated by a better monitoring of its behaviour along milling, de-branning and further isolation. The chapter summarises main strategies for aleurone layer isolation and pinpoints how its cell walls or cellular content may be of interest to obtain. It also highlights potential drawbacks, the synergistic effect of different compounds, question of bioavailability and possible future trends.

The next chapter focuses on wheat flour fortification and human health. Chapter 8 begins by providing a definition of wheat flour fortification and why its important to fortify wheat flour. It then goes on to discuss the status of wheat flour fortification worldwide and highlights the countries that currently mandate the process, allow voluntary fortification and have standards for wheat flour fortification. The chapter also examines how the impact of wheat flour fortification is measured in terms of human health and provides examples of health outcomes associated with the process. Additional considerations when assessing the health impact of wheat flour fortification are also included, such as the lack of a control group, the challenges of using birth defects registry data, selecting an outcome indicator that is only responsive to the nutrients added through fortification and allowing sufficient amount of time before measuring such outcomes. A section on health impact results observed from wheat flour fortification studies is also provided.

Chapter 9 addresses developing hulled wheat-based cereal products with enhanced nutritional properties, specifically emmer, einkorn and spelt. Hulled wheats have low yields but are suitable for organic and low-input agriculture under marginal or high-stress conditions. However, data on the composition of hulled wheats, often also called 'ancient wheats' is still scarce, especially on bioactive components such as vitamins. The chapter begins by discussing protein content and composition in hulled wheats compared to common and durum wheat as well as examining immunogenic and allergenic proteins in hulled wheats. Sections on starch and dietary fibre in hulled wheats are also provided, followed by an analysis of the various antioxidants and micronutrients that can be found as well. The chapter also provides case studies on yellow pigments in einkorn, before concluding with an overview of how hulled wheats have potential in terms of processing healthy and nutritious foods.

Moving on from Chapter 9, Chapter 10 focuses on the nutritional and nutraceutical properties of sorghum. Sorghum is a globally grown cereal. Many sorghum varieties contain high levels of polyphenolic compounds with potential health benefits. With a growing interest in using diet as a preventative measure against chronic diseases, the benefits of sorghum need to be examined. The chapter discusses current research on sorghum and its bioactive compounds, particularly the diversity of polyphenolic compounds present in sorghum. The effects of the phenolic compounds against cancer, their anti-inflammatory properties, anti-obesity effects and effects on gut microbiome are discussed. The chapter also discusses anti-nutritional effects of sorghum polyphenols as well as the effects of processing on bioactive compounds and bioavailability.

Chapter 11 concentrates on developing millet-based cereals products with enhanced nutritional properties. There is a growing awareness today about climate change and its expected ravaging effects on agricultural productivity, food production and ultimately, food and nutrition security. Against this backdrop, drought-tolerant hardy crops are gaining importance as significant sources of food and nutrition especially in the most vulnerable parts of the world mainly in sub-Saharan Africa and Asia. In this regard, the millets, which are major cereal staples in many parts of Africa and Asia are of importance for food and nutrition security in these regions. Millets are processed into various foods using traditional processing methods such as fermentation and malting and modern technologies such as extrusion cooking. This chapter focuses on how these processing methods can enhance the nutritional properties of millet-based foods. The simple practice of compositing millets with other plant foodstuffs for enhanced nutritional quality of millet-based foods is also explored.

Part 1

Nutritional properties of cereals

Chapter 1

Advances in understanding the nutritional value of starch in wheat

Senay Simsek and Jayani Kulathunga, North Dakota State University, USA; and Bahri Ozsisli, Kahramanmaras Sutcu Imam University, Turkey

1 Introduction

Starch is the major component in cereal grains; however, less attention has been given to its contribution to the health benefits of whole-grain foods (Luo and Zhang, 2018). Native starches are insoluble, semi-crystalline granules, and consist of two glucose homopolymers namely, amylose and amylopectin. They are highly variable in structure and functionality between and within plant species, growth conditions, and environment. Variability can also be observed in granule size, shape, amylose content, amylopectin chain length, distribution, and how amylose and amylopectin molecules are arranged into crystalline and amorphous regions within granules. The functional properties of starch, such as swelling, gelatinization retrogradation, pasting, and susceptibility to enzymatic digestion, are very important for food processing and human nutrition (Copeland et al., 2009; Wang et al., 2011).

Digestibility and glucose release are two main physiological properties of starch in terms of nutritional prospectives (Lehmann and Robin, 2007). Nutritionally important starch fractions are divided into rapidly digestible starch (RDS), slowly

http://dx.doi.org/10.19103/AS.2021.0087.01

digestible starch (SDS), and resistant starch (RS) based on the rate and degree of digestive enzymes hydrolysis (Englyst et al., 1992). Digestion and absorption of starch in the human body vary depending upon composition. Some starches are digested slowly and help in maintaining blood glucose levels and providing extended energy supply (Brennan et al., 2008; Woolnough et al., 2008). RDS can induce a rapid increase in blood glucose and insulin level and will cause a series of health complications (Jacobs and Gallaher, 2004; Lehmann and Robin, 2007). In contrast, SDS is digested slowly in the small intestine and it will have a lower glycemic response, which will help in controlling and preventing hyperglycemia-related diseases (Zhang and Hamaker, 2009). RS is not digested and absorbed in the small intestine, but it is fermented in the colon (Hoover and Zhou, 2003; Sajilata et al., 2006). The nutritional quality of different starch fractions can be described by the Glycemic index (GI) and it is a term used to describe the levels of postprandial glucose rise in blood as compared to a reference food or glucose (Jenkins et al., 1981). High glycemic index (GI>70) foods have rapid fluctuations in blood glucose levels (Ludwig et al., 1999) and long-term consumption of high-GI foods could lead to different complications such as the imbalance of glucose homeostasis and insulin resistance that are associated with several metabolic diseases such as obesity and type 2 diabetes (Lerer-Metzger et al., 1996).

This chapter aims to provide a nutritional overview of the wheat starch structure, function, digestibility of diverse wheat starches, and highlight interactions of starch and other biomolecules such as proteins, lipids, and non-starch polysaccharides.

2 Wheat starch structure

Wheat (*Triticum aestivum* L.) is an important food source for about 35% of the global population (Zhang et al., 2013). Starch is the main storage carbohydrate in wheat and comprises about 60–75% of grain and 70–80% of flour. It plays an important role in human nutrition as it contributes to >50% of caloric intake in the Western world and up to 90% in developing countries (Wang et al., 2015).

Wheat starch is made up of two polymers namely amylose and amylopectin. Amylose is a primarily linear polymer of (l→4)-linked α-D-glucopyranosyl units with some minor branching, whereas amylopectin is a high-molecular-weight-branched polymer synthesized through α (l→4) and α (l→6) linkages that create branch points (Wang et al., 2014).

2.1 Amylose

Amylose is a left-handed single helix structure with a relative molecular weight of about $1 \times 10^5 \sim 1 \times 10^6$ g/mol and the degree of polymerization of about 300–5000 (Buléon et al., 1998). Approximately 20% of the weight of wheat amylose is complexed with lysophospholipids (90% pure) (Seib, 1994). Amylose content varies for wheat starch granules with different granular sizes. The A-type granules have 4–10% higher amylose content than the B-type granules (Peng et al., 1999; Shinde et al., 2003; Geera et al., 2006; Ao and Jane, 2007; Liu et al., 2007). It could be due to the lower relative crystallinity of A-type granules compared to smaller B-type granules (Ao and Jane, 2007), as the amylose in the starch granules mainly remains in the amorphous lamellae. However, no significant differences in the amylose contents of A-type and B-type granules for certain wheat varieties have also been reported (Salman et al., 2009). In food systems, amylose molecules are largely responsible for the gelling and film-forming properties of cooked starch (Yoo and Jane, 2002; Chung et al., 2008). Amylose has a high tendency to retrograde and produce tough gels and strong films (Pérez and Bertoft, 2010 as cited in Hizukuri et al., 1970). Molecular weight (Mw) of amylose is reported to range from 1.88×10^5 (Mukerjea and Robyt, 2010) to 3.90×10^6 g/mol (Ong et al., 1994) using different wheat starches and measurement techniques. The starch molecules within A-type wheat starch granules seem to possess more of the longer chains (degree of polymerization (DP) > 25), and fewer of the short chains (DP <12) than B-type wheat starch granules (Liu et al., 2007).

2.2 Amylopectin

Wheat amylopectin has a cluster structure with a large number of branching structures on its molecular chains. Its relative molecular weight is about $1 \times 10^7 \sim 5 \times 10^8$ g/mol and its DP is more than 10^7 (Tester et al., 2004; Whistler and Daniel, 1984). But recent studies indicated that High Performance Size Exclusion Chromatography (HP-SEC) underestimates the size of amylopectin due to the shearing of that polysaccharide (Cave et al., 2009). The amylopectin molecules are associated with the crystallinity, gelatinization, and swelling of starch (Yoo and Jane, 2002; Chung et al., 2008). Amylopectin, dispersed in water, is more stable and produces soft gels and weak films (Pérez and Bertoft, 2010 as cited in Hizukuri et al., 1970).

3 Granules of wheat starch

Native starch occurs as semi-crystalline granules (Fig. 1) and has a very complex hierarchical structure. The starch granules are composed of an amorphous bulk core area surrounded by concentric semi-crystalline growth rings alternating with amorphous growth rings (Wang et al., 2015). The amorphous core contains amylose and amylopectin chains disordered at the reducing end (Wang and Copeland, 2012).

Starch is synthesized as discrete granules of varying sizes in amyloplasts. Different sizes and shapes of starch granules build up during the development

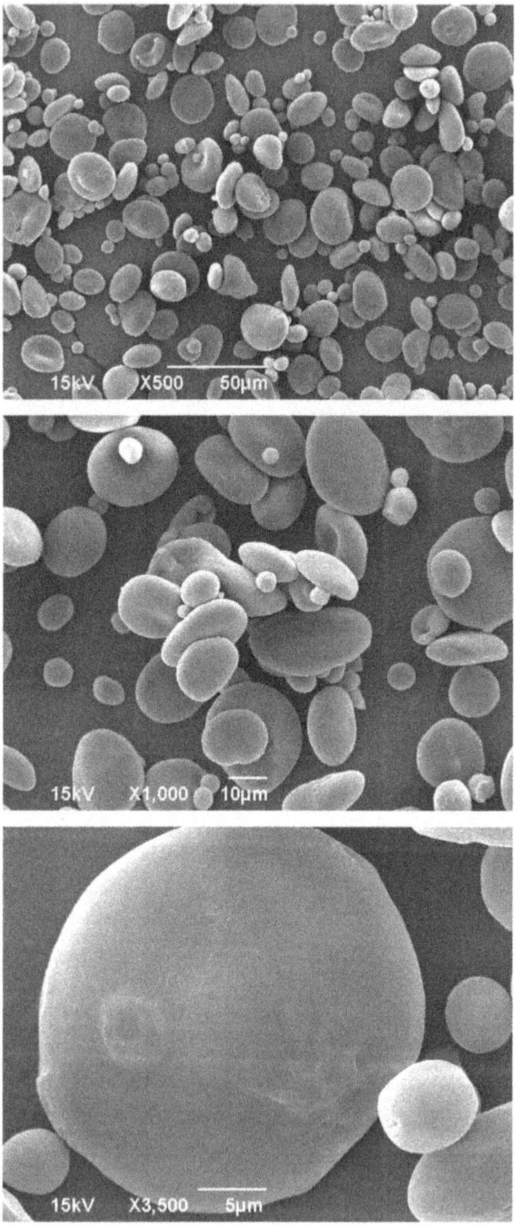

Figure 1 Scanning electron microscopy (SEM) images of wheat starch granules shown at X500, X1000, and X3500 magnification. (This material is based on work supported by the National Science Foundation under Grant Nos. 0619098, 0821655, 0923354, and 1229417. Any opinions, findings, and conclusions or recommendations expressed in this material are those of the author(s) and do not necessarily reflect the views of the National Science Foundation).

of grain. The starch granule distribution of different wheat starches is shown in Table 1. Wheat starch has a trimodal distribution of starch and consists of 3 different types of starch granules large (A), intermediate (B), and small (C) granules (Bancel et al., 2010; Cao et al., 2015).

3.1 A type granules

The A-type starch granules are disk or lenticular in shape with a diameter of >10 μm (Vermeylen et al., 2005; Ao and Jane, 2007; Kim and Huber, 2008; Wang et al., 2014). These granules contribute to more than 70% total weight of the starch in wheat (Bechtel et al., 1990; Peng et al., 1999; Shinde et al., 2003). A-type granules have higher gelatinization enthalpy, amylose content, pasting parameters such as peak, trough, breakdown, final and setback viscosities, and lower gelatinization onset and peak temperatures compared to B type granules (Sahlström et al., 2003; Geera et al., 2006; Soh et al., 2006; Kim and Huber, 2010; Yin et al., 2012). They also contain more amylose and less lipid than B-granules (Maningat et al., 2009; Tao et al., 2016). The proportions of A- and B-starch granules differ among genotypes by weight, volume, and number (Raeker et al., 1998; Li et al., 2001).

3.2 B-type granules

B-type starch granules are spherical or polygonal in shape and measure less than 10 μm in diameter (Vermeylen et al., 2005; Ao and Jane, 2007; Kim and Huber, 2008; Wang et al., 2014). Although they make up <30% of total starch weight, B-type granules comprise up to 90% of granules in number (Raeker et al., 1998). B-starch granule volume variation was wider in bread wheat, suggesting possibilities of genetic manipulation of the granule size distribution (Stoddard, 1999).

The B-granules occupied volumes in the range of 28.5–56.2% for hard red winter and hard red spring wheat (Park et al., 2009). The volume percentages of A- and B-type starch granules were 52.7–65.5% and 34.5–47.3% in seven Chinese wheat cultivars (Dai et al., 2009). In several studies, environmental stress such as temperature (Liu et al., 2011), water deficit (Dai et al., 2009; Zhang et al., 2010), nutrient supplementation (Ni et al., 2012; Li et al., 2013), and light intensity (Li et al., 2010) significantly changed starch granule size distribution and amylose content in wheat. B-type granules have lower gelatinization enthalpy, higher lipid-complexed amylose content higher swelling power, and broader gelatinization ranges compared to A-type granules (Sahlström et al., 2003; Geera et al., 2006; Soh et al., 2006; Kim and Huber, 2010; Yin et al., 2012). The proportion of B-granules is of particular importance in the commercial production of wheat starch because these granules owing to their smaller size are difficult to purify and recover than A-granules (Wei et al., 2010; Yu et al., 2015).

Table 1 Granule size distribution of wheat starches (Singh et al., 2008, 2009, 2010; Shevkani et al., 2011; Blazek and Copeland, 2008; Li et al., 2013; Kaur et al., 2016; Wilson et al., 2008 as cited in Shevkani et al., 2017).

Wheat species	Hard/Soft wheat	Type of granules	Distribution	% of A granules	% of B granules	% of C granules
Common			Trimodal	71.4–79.1	12.1–19.7	7.6–9.4
			Trimodal	45.6–73.2	14.0–37.0	10.5–17.5
			Trimodal	63.7–68.9	20.2–22.3	10.8–14.9
			Trimodal	71.4–79.1	12.1–19.7	7.6–9.4
			Bimodal	57.6	42.4	
	Hard		Bimodal	75.4	14.6	
	Soft		Bimodal	75.0	25.1	
	Hard	A	Bimodal	92.3	7.7	
	Soft	A	Bimodal	91.0	9.0	
	Hard	B	Bimodal	13.3	86.7	
	Soft	B	Bimodal	12.5	87.5	
Durum			Trimodal	69.9–85.4	9.4–20.1	4.7–10.0
Spelt			Trimodal	37.7–71.6	25.8–55.0	2.4–6.8

3.3 C type granules

C-granules are roughly spherical or oval and irregular or cuboidal in shape and exhibited an average size of 2–3 micrometers. However, B- and C-granules were sometimes considered together as small B-granules (Bancel et al., 2010; Cao et al., 2015).

A wide variation in granule size distribution exists amongst starches from different wheat type and cultivars/varieties (Table 1). The starches from hard wheat varieties were observed to contain a higher proportion of B-granules and a lower of A-granules than soft wheat (Li et al., 2008; Edwards et al., 2010; Singh et al., 2016). Normal and waxy starches have both A- and B-granules; however, these differed in relative granule size distribution and morphology. The starch from normal wheat contained more B-granules (Zhang et al., 2013), while that from waxy cultivars/varieties showed more spherical disc-like granule morphology (Zhou et al., 2014).

4 Role of wheat starch in food

Wheat starch has a significant role in food systems. Wheat starch can affect the appearance, texture, cooking characteristics, and eating quality of different foods. It can act as a tenderizer in cakes, doughnuts, cookies, and cheese analogs. It is also incorporated into batters, ice creams, soups, gravies to provide adhesion, structure, moisture control, and/or thickening (Maningat et al., 2009).

Wheat starch gelatinizes at 5–10°C below that of corn, rice, potato, and tapioca starches. The low gelatinization temperature of wheat starch results in rapid cooking in microwaveable soups, sauces, and gravies, but it has been reported that wheat starch generally has a lower paste viscosity compared with other commercial starches (Ral et al., 2008).

The wheat starch plays a significant role in bread making. It acts as a filler, provides extra sugars for fermentation. The sugars are provided by the degradation of damaged starch through amylase action (Glennie et al., 1987). Wheat starch provides a surface suitable for a strong union with the gluten adhesive and dilutes the gluten to a consistency desirable for dough processing during dough mixing (Hoseney et al., 1971; Hoseney, 1994). During baking, starch granules in the bread dough gelatinize, become flexible, and elongate as the gas cell walls expand, but the granules do not disintegrate (Lagrain et al., 2008). When granules swell, they also exude amylose that pools in the center of swollen granules, between granules, and at the starch-gluten interface (Hug-Iten et al., 2003). The crumb structure in fresh bread is set during cooling further by the gelling and crystallization processes of the pooled amylose and the stiffening of the swollen granules. Storage of bread over a period of time causes undesirable textural changes and loss of flavor. In bread staling, the crust becomes 'leathery' as moisture migrates from crumb to crust.

However, in some products such as noodles and pasta, retrogradation is promoted to improve textural and sensory properties as increased firmness, and reduced stickiness are desirable in such products. Additionally, retrogradation was also considered desirable in terms of the nutritional significance as retrograded starches were digested slowly and less completely than unretrograded starches, resulting in unstable postprandial blood glucose levels. Starch influenced cooking characteristics, appearance, and texture of cooked noodles (Ross et al., 1997; Zhao and Seib, 2005).

Wheat starch has been identified as an attractive choice to produce edible and/or biodegradable films due to low cost, high availability, renewability, and biodegradability (Lu et al., 2009).

5 Digestibility of wheat starch

Starch digestibility has gained substantial interest concerning the increasing incidence of non-communicable diseases (Wang et al., 2015). Nutritionally, starch can be categorized into 3 categories: rapidly digestible starch (RDS), slowly digestible starch (SDS), and resistant starch (RS) (Table 2).

5.1 Rapidly digestible starch

The starch that is digested rapidly or the amount of glucose release after 20 minutes (Englyst et al., 1992) in the upper gut causes a rapid elevation of glucose into the bloodstream. RDS can trigger physiological responses that over time are associated with increased risks for diabetes, cardiovascular disease, and cancer (Brand-Miller, 2003). It can produce large fluctuations in postprandial glycemia, which induces oxidative stress that may be a health concern (Monnier et al., 2006).

The postprandial physiological response was investigated to the ingestion of RDS in healthy subjects and type 2 diabetics (Ells et al., 2005, Seal et al., 2003). Significantly greater and more rapid changes in blood glucose and insulin were observed after the consumption of RDS compared to SDS. RDS differ from SDS in their ability to stimulate the secretion of gut incretin hormones. Glucagon-like peptide-1 (GLP-1) and glucose-dependent insulinotropic polypeptide (GIP) increased in the late postprandial phase (180–300 min) after SDS consumption. This could indicate beneficial effects of SDS in the late postprandial phase (Wachters-Hagedoorn et al., 2006).

Table 2 Nutritional classification of starch (Englyst et al., 1992).

Starch type	Starch structure	Digestion rate in the small intestine	Digestion timeline/place	Examples
RDS	Amorphous	Rapid	Within 20 minutes, the mouth, and small intestine	Freshly cooked starch, white bread
SDS	Amorphous/crystalline	Slow	20–120 minutes, small intestine	Most cereal starches
RS	Depending on the type mainly crystalline	Resistant	>120 minutes The main action in the colon	
RS1	Depending on the type	Resistant		Partially milled grain
RS2	Depending on the type	Resistant		Raw potato and banana starch
RS3	Depending on the type	Resistant		Cooled cooked potato
RS4	Depending on the type	Resistant		Cross-linking starch

*RDS – Rapidly Digestible Starch, SDS – Slowly digestible starch, RS – Resistant starch.

5.2 Slowly digestible starch

The amount of glucose released between 20 and 120 minutes, after hydrolysis of starch in the small intestine are called slowly digestible starches (SDS) (Englyst et al., 1992). SDS has a medium to low GI and thus reduces the glycemic load of a food product compared to rapidly digestible starch with a high GI (Ells et al., 2005; Englyst et al., 2003). Jenkins et al. (2002) stated that low GI diets are associated with decreased risk of diabetes and cardiovascular diseases. In addition, positive associations were found between dietary GI and risk of colon and breast cancer.

A few studies investigated the postprandial physiological responses to the ingestion of RDS and SDS in healthy subjects and type 2 diabetics (Ells et al., 2005 Seal et al., 2003). A reduction of potential risk factors for metabolic syndrome by the exchange of RDS by SDS was proposed (Ells et al., 2005). In obese, insulin-resistant subjects, Harbis et al. (2004) showed that the intake of slowly available glucose resulted in an improved metabolic profile, particularly in lower postprandial insulinemia, lower levels of circulating triacylglycerols and apolipoproteins B-100 and B-48 in the triacylglycerol-rich lipoproteins. Also, rapidly and slowly digestible starches differ in their ability to stimulate the secretion of gut incretin hormones. Glucagon-like peptide-1 (GLP-1) and glucose-dependent insulinotropic polypeptide (GIP) increased in the late postprandial phase (180–300 min) after SDS consumption. This could indicate beneficial effects of SDS in the late postprandial phase, for example, related to glucose homeostasis and energy storage (Wachters-Hagedoorn et al., 2006).

Epidemiological studies suggest that reduced postprandial glucose peaks, reduced episodes of hypoglycemia, improved lipid response, lower concentration of glycosylated hemoglobin and fructosamine, and greater insulin sensitivity are beneficial for diabetes management (Wolever et al., 2003). Slowly digestible starch intake results in a beneficial metabolic response for these conditions and is recommended for the prevention and management of diabetes (Axelsen et al., 1999; Ells et al., 2005; Seal et al., 2003). It was shown that SDS-containing foods at breakfast improved carbohydrate metabolism and reduced insulin requirement of insulin-treated type 2 diabetic patients (Golay et al., 1992).

Studies to date on the health benefits of SDS are limited. Furthermore, most studies do not make a precise distinction between starch fractions. The potential health benefits of SDS are linked to stable glucose metabolism, diabetes management, mental performance, and satiety.

6 Resistant starch

The portion of starch and starch products that resist digestion as they pass through the gastrointestinal tract are called resistant starch (Nugent, 2005). It

Table 3 Resistant starch content of wheat and wheat-based products

Wheat type or wheat product	Resistant starch content	Reference
Wheat grain	13.6*	(Lunn and Buttriss, 2007)
Wheat flour	1.7*	(Lunn and Buttriss, 2007)
White bread	1.9*	(Lunn and Buttriss, 2007)
Whole wheat cereal, flaked	1.0**	(Murphy et al., 2008)
Crackers, wheat thin	0.4**	(Murphy et al., 2008)
Whole wheat bread	1.0**	(Murphy et al., 2008)
Wheat rolls	0.1**	(Murphy et al., 2008)
Wheat germ bread	0.1**	(Murphy et al., 2008)
High fiber white bread	0.9**	(Murphy et al., 2008)

*-Grams per 100 g of dry matter.
**-Grams per 100 g of food.

is also defined as the fraction of starch that is not hydrolyzed to its monomeric units, D-glucose in the small intestine within 2 hours of being consumed but can be fermented in the colon, which does not lead to a significant increase in blood glucose concentration after consumption (Goni et al., 1997; Yamada et al., 2005). Resistant starch intake ranges approximately 30–40 g/day (Baghurst et al., 2001), from 3 g/day to 6 g/day (Dyssler and Hoffmann, 1994), and 5–7 g/day (Baghurst et al., 2001) in developing countries, the European Union, and Australia, respectively. Resistant starch content of wheat and wheat based products are shown in (Table 3).

Most of the studies have shown that resistant starch is a linear molecule of α-1,4-D-glucan derived from the retrograded amylose fraction and it has a relatively low molecular weight (1.2×10^5) (Tharanathan, 2002).

The reason for the resistance of digestion in the small intestine can depend upon several factors. Mostly, the starch granules are not accessible to digestive enzymes such as amylases due to the compact molecular structure (Haralampu, 2000). The resistance of starch to digestion is influenced by the nature of the association between starch polymers, with higher amylose levels in the starch being associated with slower digestibility rates.

Based on the structures contributing to the enzymatic resistance, RS can be categorized into five different types. These are discussed in the following sections.

6.1 Resistant Starch-1 (RS1)

The content of different types of resistant starch in wheat and wheat products are shown in Table 4. RS1 is the physically protected form of the starch found in whole grains, seeds, or tubers and physically inaccessible to digestion, due to the presence of intact cell walls in grains, seeds, or tubers (Hernández et al.,

Table 4 Different types of resistant starch content of wheat and wheat products

Wheat/wheat product	Type of resistant starch	Content (g/100g of total starch)	Reference
Wheat starch	Type II	0.2	(Themeier et al., 2005)
Wheat starch	Type II	0.3	(Themeier et al., 2005)
White flour	Type II	3.0	(BNF, 1990)
	Type III	Traces	(BNF, 1990)
Short bread	Type III	Traces	(BNF, 1990)
White bread	Type III	2.0	(BNF, 1990)
Whole wheat bread	Type III	2.0	(BNF, 1990)
White spaghetti	Type I	8.0	(BNF, 1990)
	Type III	1.0	(BNF, 1990)

2008). RS1 is heat stable in most normal cooking operations, which enables its use as an ingredient in a wide variety of conventional foods (Sajilata et al., 2006).

6.2 Resistant Starch-2

Resistant Starch-2 (RS2) are native, uncooked granules of starch, such as raw potato, banana starches, or high amylose maize starch, whose crystallinity makes them poorly susceptible to hydrolysis (Hernández et al., 2008; Wepner et al., 1999). In raw starch granules, starch is tightly packed in a radial pattern and is relatively dehydrated. This compact structure limits the accessibility of digestive enzymes and accounts for the resistant nature of RS2 such as ungelatinized starch (Sajilata et al., 2006).

6.3 Resistant Starch-3

Resistant Starch-3 (RS3) represents retrograded starch. In the formation of RS3, the starch granule is completely hydrated. Amylose leaches from the granules into the solution as a random coil polymer and during cooling, the polymer chains start to reassociate as double helices, stabilized by hydrogen bonds (Wu and Sarko, 1978). RS3 forms are generally formed during the retrogradation of starch granules (Wepner et al., 1999). RS3 can be formed in cooked foods that are kept at low or room temperature (Hernández et al., 2008). Most moist-heated foods contain some RS3 (Sajilata et al., 2006). RS3 is thermostable in cooking operations and therefore used in conventional foods (Haralampu, 2000). RS3 has a higher water-holding capacity than granular starch (Sanz et al., 2008). Retrograded amylose can generate RS3. Therefore, prolonged intake of an amylose-rich diet improves fasting triglyceride and cholesterol levels (Mikulíková et al., 2008).

6.4 Resistant Starch-4

Resistant Starch-4 (RS4) includes chemically modified or re-polymerized starches (chain linkage altered dextrins, ethers, or esters) used by food manufacturers to alter the functional characteristics of the starch (Ratnayake and Jackson, 2008; Sanz et al., 2009; Yao et al., 2009) and includes starches which have been etherized, esterified, or cross-banded with chemicals in such a manner as to decrease their digestibility (Nugent, 2005). RS4 can be produced by chemical modifications, such as conversion, substitution, or cross-linking, which can prevent its digestion by blocking enzyme access and forming linkages (Sajilata et al., 2006; Kim and Huber, 2008). The cross-linked RS4 wheat and waxy wheat starches were characterized by elevated phosphorus levels, low swelling powers (3 g/g) at 95°C, insolubilities (<1%) in 1M potassium hydroxide or 95% dimethyl sulfoxide, and increased temperatures and decreased enthalpies of gelatinization compared to native starches (Woo and Seib, 2002). Cross-linked resistant wheat starch (RS4) can be added in white pan bread with <15% loss of loaf volume. RS4 also can be used in sugar-snap cookies at levels up to 50% flour replacement without significantly changing the texture and the appearance of the cookies (Yeo and Seib, 2009).

6.5 Resistant Starch-5

Resistant Starch-5 (RS5) is amylose-lipid complexed starch (ALC) which is formed from high-amylose starches that require higher temperatures for gelatinization and are more susceptible to retrogradation (Cummings and Stephen, 2007).

7 Health benefits of resistant starch

7.1 Resistant starch as dietary fiber

Dietary fiber includes non-starch polysaccharides, resistant oligosaccharides, and analogous carbohydrates. It also includes resistant starch (Sharma et al., 2008). Currently, naturally occurring resistant starch found in whole grains, is considered dietary fiber under AACC's (American Association of Cereal Chemists) definition (Sajilata et al., 2006). Resistant starch behaves physiologically as a fiber and it should be detected in the total dietary fiber assay (Haralampu, 2000).

Resistant starch has gained interest as a new source of dietary fiber due to the increased awareness of consumers concerning the relationship between food, lifestyle, and health (Sanz et al., 2008). Resistant starch is not a cell wall component but is nutritionally more similar to non starch polysaccharides than to digestible starch (Sharma et al., 2008).

Resistant starch also has a positive impact on colonic health by increasing the crypt cell production rate or decreasing colonic epithelial atrophy in comparison to non-fiber diets. Some studies revealed resistant starch can influence tumorigenesis and reduces serum cholesterol and triglycerides. Resistant starch can increase fecal bulk and lower colonic pH (Slavin et al., 2009).

Resistant starch is a substrate/prebiotic for the colonic microbiota, forming metabolites including short-chain fatty acids (SCFA) such as acetic, propionic, and butyric acids. The production of SCFA causes a reduction in pathogenic bacteria by lowering the colonic pH.

Butyric acid is largely metabolized by the colonocyte and is the most important energy source for the cell. Butyric acid is also known to suppress colorectal cancer cell proliferation (Elmstahl, 2002).

A study conducted by Leu et al. (2002) combined resistant starch with insoluble dietary fiber for fermentation, and observed much higher SCFA levels, and butyrate were also observed in the feces. In rats, when RS was combined with a soluble fiber the site of RS fermentation was pushed more distally. As the distal colon is the site where most tumors arise, it may be of additional benefit for cancer protection if fermentation is further enhanced within the distal colon (Morita et al., 1999).

Liu and Xu (2008) showed that RS dose-dependently suppressed the formation of colonic aberrant crypt foci (ACF) only when it was present during the promotion phase to a genotoxic carcinogen in the middle and distal colon, suggesting that administration of RS may retard growth and/or the development of neoplastic lesions in the colon. Therefore, colon tumorigenesis may be highly sensitive to dietary intervention. Adults with preneoplastic lesions in their colon may therefore benefit from dietary RS. This suggests the usefulness of RS as a preventive agent for individuals at high risk for colon cancer development (Liu and Xu, 2008).

Resistant starch plays a role in functional foods as fermentable substrates for growth of lactobacilli and bifidobacteria, promoting beneficial physiological effects as dietary fiber, acting as encapsulation materials for probiotics to enhance their stability (Charalampopoulos et al., 2002).

7.2 Resistant starch and diabetes

Resistant starch consumption has also been related to reduce postprandial glycemic and insulinemic responses, which may have beneficial implications in the management of diabetes (Tharanathan and Mahadevamma, 2003). RS has assumed great importance in Type II diabetes owing to its slow rate of glucose release as well as the time it takes to metabolize (Englyst et al., 1992; Sui et al., 2011). The rate of digestion is slow in foods rich in resistant starch. The inaccessibility of starch to digestive enzymes (α-amylase, isoamylase, and pullulanase) in starchy products is responsible for the nutraceutical potential of starchy products as well as the reduction in postprandial blood glucose and insulin response. The long period it takes to metabolize resistant starch after consumption (5–7 hours) reduces postprandial glycemia and insulinemia and likely increases the period of satiety (Raben et al., 1994; Reader et al., 1997).

An RS3-containing bar decreased postprandial blood glucose and could play a role in providing improved metabolic control in type II diabetes (non-insulin-dependent) (Sajilata et al., 2006).

More recently, a study showed that RS reduces levels of glucose-dependent insulinotropic polypeptide mRNA along the jejunum and ileum in both normal and type 2 diabetes rats (Shimada et al., 2008).

Chemically modified starches (RS4) have also been found to generate different glucose responses. The effect of two test meals containing 1-2% acetylated potato starch and beta-cyclodextrin-enriched potato starch (2–3%), respectively, was studied in humans and only the latter was found to have lower body glucose levels. This may be due to the more distal absorption of beta-cyclodextrin in the intestine or due to delayed gastric emptying (Raben et al., 1997).

As RS has a low glycemic response, adding it as an ingredient to foods will help lower the overall GI value of the food (particularly if it is replacing existing readily absorbed forms of carbohydrate). RS is likely to become an increasingly attractive ingredient to many food manufacturers (particularly those of breads and cakes or similar products which traditionally may have had higher GI values) (Nugent, 2005).

Contrasting glycemic responses were also reported depending on the type of resistant starch. The cross-linked RS 4 was reported to reduce the glycemic response in the healthy subjects compared with the RS 2 (Haub et al., 2010).

7.3 Hypocholesterolemic effect of resistant starch

Some studies in humans revealed the beneficial effect of RS on fasting plasma triglyceride and cholesterol levels. According to several studies, RS ingestion may decrease the serum cholesterol level in rats fed a cholesterol-free diet (De-Deckere et al., 1993, Hashimoto et al., 2006). A pronounced reduction in total cholesterol and non-HDL-cholesterol was observed in RS consuming overweight and obese subjects compared to the fiber-consuming group (diet as recommended to patients with metabolic syndrome). However, a pronounced effect on glucoregulation was observed in the fiber group (Dodevska et al., 2016). In contrast, some studies indicated that RS consumption does not affect the measures of total cholesterol in humans. Therefore, more research studies are needed to understand the effects of RS on lipid metabolism in humans (Nugent, 2005).

8 Interactions of starch with other molecules

8.1 Starch-lipid interactions

Endogenous lipids and phospholipids in cereal starches have the ability to form complexes with amylose (Kwasniewska-Karolak et al., 2008) thereby it will reduce the susceptibility to amylolytic enzymes (Nebesny et al., 2002; Zhang et al., 2006a,b). *In vivo* and *in vitro* digestibility studies on the effect of these components have shown that they can considerably slow down the enzymatic digestibility (Singh et al., 2010).

Lysophosphatidylcholine (LPC) is a complexing agent that has shown high complexing ability with amylose, as indicated by Digital Scanning Colorimetry (DSC) (Ahmadi-Abhari et al., 2013; Cui and Oates, 1997). Also, the poor digestibility of the amylose-lipid complexes has been demonstrated (Seneviratne and Biliaderis, 1991). However, it is not clear if this complexation only affects the digestibility of complexed amylose chains or also the overall rate of digestion of starch granules to restrict amylolysis (Ye et al., 2019) as non-catalytic interactions between alpha-amylase and protein can lead to reduced alpha-amylase activity or availability. Adsorption of alpha-amylase on proteins is reported as a prime cause for a reduced rate of hydrolysis of starch in flour compared to isolated starch from wheat (Bhattarai et al., 2016) and barley (Yu et al., 2018)

8.2 Starch-protein interactions

The starch-gluten interactions are an important factor in reducing *in vitro* amylase digestion. Reduction in blood glucose response and decreased *in vitro* starch hydrolysis rate was reported upon consumption of white bread compared to gluten-free bread or gluten-free bread plus gluten and it was revealed that it could be due to starch-protein interactions in white bread. Furthermore, they found that the addition of gluten to starch did not inhibit amylase reaction (Jenkins et al., 1987).

8.3 Starch and non-starch polysaccharide interactions

The effect of dietary fibers such as alginate, β-glucan, gums, inulin, and pectin on *in vitro* digestion of starch was investigated and results revealed that fiber can bind to endogenous starch hydrolyzing enzymes in the GI track and inhibit their activity, thereby inhibiting starch hydrolysis. In addition, the authors reported that the presence of fibers in the diet can slow down glucose release and absorption (Singh et al., 2016).

In another study, postprandial serum glucose was lowered by dietary fiber, by inhibiting the activity of amylase and postponing the release of glucose from starch. When dietary fibers are incorporated into starch-based foods, they are expected to restrict the availability of water and to increase the viscosity of food products. As a result, the accessibility of starch granules to digestive enzymes is limited by dietary fibers in the human digestive system (Angioloni and Collar, 2011). *In vitro* starch digestibility in a wheat flour, gel system was affected by dietary fiber-enriched materials (FEMs) from apples (Bae et al., 2016). However, insoluble dietary fiber constituents such as cellulose and lignin have been shown to have minimal interactions with resistant starch compared to other constituents such as catechin (Escarpa et al., 1997).

8.4 Starch and phenolic compound interactions

Starch and phenolic compound interact to form either inclusion complex in the form of amylose single helices facilitated by hydrophobic effect, or complex with much weaker binding mostly through hydrogen bonds. The interactions between starch and phenolic compound can significantly contribute to the nutritional quality of food products. Berry extracts with phenolic compounds have decreased the insulin level and glycemic response in human after consumption of wheat breads (Coe et al., 2013; Törrönen et al., 2013). Tea extracts with phenolic acids were effective at inhibiting starch hydrolysis, and the effect was affected by the starch structure and tea type (Guzar et al., 2012).

9 Conclusion

Nutritionally important SDS and RS have several health benefits such as improved cholesterol metabolism, effective management of obesity, and reduced risk of type-2 diabetes and colon cancers. Interactions of starch with other molecules can affect the functional and nutritional properties of wheat starch such as gelatinization, swelling power, and digestibility. More nutritional information is available on wheat starch from common and durum wheat species; however, more studies are essential on ancient species of wheat, which is einkorn, emmer, and spelt. The interest in these species has recently increased; therefore, more studies are essential on the characterization and structure-function relationships of starches from these species. Moreover, the health benefits of starches should be further investigated by using animal and human models to gain the nutritional benefits associated with SDS and RS.

10 Where to look for further information

This book chapter summarizes the available information on nutritional value of wheat starch. Further information can be found via the following sources:

The health benefits of slowly digestible starch can be found at the link below:

http://onlinelibrary.wiley.com/doi/10.1002/9781118528723.ch06/summary.

British Nutrition Foundation (BNF) has published an in depth review in *Nutrition Bulletin* looking at the potential health benefits of resistant starch. The main points summarized in the review can be viewed at the link below:

http://onlinelibrary.wiley.com/doi/10.1111/nbu.12244/full.

11 References

Ahmadi-Abhari, S., Woortman, A. J. J., Hamer, R. J., Oudhuis, A. A. C. M. and Loos, K. (2013). Influence of lysophosphatidylcholine on the gelation of diluted wheat starch suspensions. *Carbohydrate Polymers* 93(1), 224–231.

Angioloni, A. and Collar, C. (2011). Physicochemical and nutritional properties of reduced-caloric density high-fibre breads. *LWT - Food Science and Technology* 44(3), 747–758.

Ao, Z. and Jane, J. L. (2007). Characterization and modeling of the A- and B-granule starches of wheat, triticale, and barley. *Carbohydrate Polymers* 67(1), 46–55.

Axelsen, M., Arvidsson Lenner, R., Lonnroth, P. and Smith, U. (1999). Breakfast glycaemic response in patients with type 2 diabetes: effects of bedtime dietary carbohydrates. *European Journal of Clinical Nutrition* 53(9), 706–710.

Bae, I. Y., Jun, Y., Lee, S. and Lee, H. G. (2016). Characterization of apple dietary fibers influencing the in vitro starch digestibility of wheat flour gel. *LWT - Food Science and Technology* 65, 158–163.

Baghurst, K., Baghurst, P. A. and Record, S. J. (2001). Dietary fiber, non-starch polysaccharide and resistant starch intakes in Australia. In: Spiller, G. A. (Ed). *CRC Handbook of Dietary Fiber in Human Health*. Boca Raton, FL: CRC Press Press. 583–591.

Bancel, E., Rogniaux, H., Debiton, C., Chambon, C. and Branlard, G. (2010). Extraction and proteome analysis of starch granule-associated proteins in mature wheat kernel (Triticum aestivum L.). *Journal of Proteome Research* 9(6), 3299–3310.

Bechtel, D. B., Zayas, I., Kaleikau, L. and Pomeranz, Y. (1990). Size-distribution of wheatstarch granules during endosperm development. *Cereal Chemistry* 67, 59–63.

Bhattarai, R. R., Dhital, S. and Gidley, M. J. (2016). Interactions among macronutrients in wheat flour determine their enzymic susceptibility. *Food Hydrocolloids* 61, 415–425.

Blazek, J. and Copeland, L. (2008). Pasting and swelling properties of wheat flour and starch in relation to amylose content. *Carbohydrate Polymers* 71(3), 380–387.

Brand-Miller, J. C. (2003). Glycemic load and chronic disease. *Nutrition Review* 61, S49–S55.

Brennan, C. S., Suter, M., Luethi, T., Matia-Merino, L. and Qvortrup, J. (2008). The relationship between wheat flour and starch pasting properties and starch hydrolysis: effect of non-starch polysaccharides in a starch gel system. *Starch* 60(1), 23–33.

British Nutrition Foundation [BNF] (1990). *Complex Carbohydrates in Foods: The Report of the British Nutrition Foundation's Task Force*. London: Chapman & Hall.

Buléon, A., Colonna, P., Planchot, V. and Ball, S. (1998). Starch granules: structure and biosynthesis. *International Journal of Biological Macromolecules* 23(2), 85–112.

Cao, H., Yan, X., Chen, G., Zhou, J., Li, X., Ma, W. and Yan, Y. (2015). Comparative proteome analysis of A-and B-type starch granule-associated proteins in bread wheat (*Triticum aestivum* L.) and Aegilops crassa. *Journal of Proteomics* 112, 95–112.

Cave, R. A., Seabrook, S. A., Gidley, M. J. and Gilbert, R. G. (2009). Characterization of starch by size-exclusion chromatography: the limitations imposed by shear scission. *Biomacromolecules* 10(8), 2245-2253.

Charalampopoulos, D., Wang, R., Pandiella, S. S. and Webb, C. (2002). Application of cereals and cereal components in functional foods: a review. *International Journal of Food Microbiology* 79(1-2), 131–141.

Chung, J. H., Han, J. A., Yoo, B., Seib, P. A. and Lim, S. T. (2008). Effects of molecular size and chain profile of waxy cereal amylopectins on paste rheology during retrogradation. *Carbohydrate Polymers* 71(3), 365–371.

Coe, S. A., Clegg, M., Armengol, M. and Ryan, L. (2013). The polyphenol-rich baobab fruit (Adansonia digitata L.) reduces starch digestion and glycemic response in humans. *Nutrition Research* 33(11), 888–896.

Copeland, L., Blazek, J., Salman, H. and Tang, M. C. M. (2009). Form and function of starch granules. *Food Hydrocolloids* 23(6), 1527–1534.

Cui, R. and Oates, C. G. (1997). The effect of amylose-lipid complex formation on enzyme susceptibility of sago starch. *Food Chemistry* 65, 417–425.

Cummings, J. H. and Stephen, A. M. (2007). Carbohydrate terminology and classification. *European Journal of Clinical Nutrition* 61(Suppl. 1), S5–18.

Dai, Z. M., Yin, Y. P. and Wang, Z. L. (2009). Starch granule size distribution from seven wheat cultivars under different water regimes. *Cereal Chemistry Journal* 86(1), 82–87.

De-Deckere, E. A. M., Kloots, W. J. and Van-Amelsvoort, J. M. M. (1993). Resistant starch decreases serum total cholesterol and triacylglycerol concentrations in rats. *Journal of Nutrition* 123(12), 2142–2151.

Dietary Fiber Definition Committee of the American Association of Cereal Chemists, AACC report, (2001). The definition of dietary fiber. *Cereal Foods World* 46, 112–126.

Dodevska, M. S., Sobajic, S. S., Djordjevic, P. B., Dimitrijevic-Sreckovic, V. S., Spasojevic-Kalimanovska, V. V. and Djordjevic, B. I. (2016). Effects of total fibre or resistant starch-rich diets within lifestyle intervention in obese prediabetic adults. *European Journal of Nutrition* 55(1), 127–137.

Dyssler, P. and Hoffmann, D. (1994). Estimation of resistant starch intake in Europe. In: Asp, N.-G., van Amelsvoort, J. M. M. and Hautvast, J. G. A. J. (Eds). *Proceedings of the Concluding Plenary Meeting of EURESTA*. Wageningen, The Netherlands: EURESTA. 84–86.

Edwards, M. A., Osborne, B. G. and Henry, R. J. (2010). Puroindoline genotype, starch granule size distribution and milling quality of wheat. *Journal of Cereal Science* 52(2), 314–320.

Ells, L. J., Seal,C. J., Kettlitz, B., Bal, W. and Mathers, J. C. (2005). Postprandial glycaemic, lipaemic and haemostatic responses to ingestion of rapidly and slowly digested starches in healthy young women. *British Journal of Nutrition* 94(6), 948–955.

Elmstahl, H. L. (2002). Resistant starch content in a selection of starchy foods on the Swedish market. *European Journal of Clinical Nutrition* 56(6), 500–505.

Englyst, H. N., Kingman, S. M. and Cummings, J. H. (1992). Classification, and measurement of nutritionally important starch fractions. *European Journal of Clinical Nutrition* 46(Suppl. 2), S33–S50.

Englyst, K. N., Vinoy, S., Englyst, H. N. and Lang, V. (2003). Glycaemic index of cereal products explained by their content of rapidly and slowly available glucose. *British Journal of Nutrition* 89(3), 329–340.

Escarpa, A., González, M. C., Morales, M. D. and Saura-Calixto, F. (1997).An approach to the influence of nutrients and other food constituents on resistant starch formation. *Food Chemistry* 60(4), 527–532.

Geera, B. P., Nelson, J. E., Souza, E. and Huber, K. C. (2006). Composition and properties of A-and B-type starch granules of wild-type, partial waxy, and waxy soft wheat. *Cereal Chemistry Journal* 83(5), 551–557.

Glennie, W. C., McDonald, A. M. L. and Roger Stark, J. (1987). Some observations on damage in commercial starch preparations. *Carbohydrate Research* 170(2), 263–268.

Golay, A., Koellreutter, B., Bloise, D., Assal, J. P. and Würsch, P. (1992).The effect of muesli or cornflakes at breakfast on carbohydratemetabolism in type 2 diabetic patients. *Diabetes Research and Clinical Practice* 15(2), 135–141.

Goni, I., Garcia-Alonso, A. and Saura-Calixto, F. (1997). A starch hydrolysis procedure to estimate glycemic index. *Nutrition Research* 17(3), 427–437.

Guzar, I., Ragaee, S. and Seetharaman, K. (2012). Mechanism of hydrolysis of native and cooked starches from different botanical sources in the presence of tea extracts. *Journal of Food Science* 77(11), C1192–C1196.

Haralampu, S. G. (2000). Resistant starch: a review of the physical properties and biological impact of RS3. *Carbohydrate Polymers* 41(3), 285–292.

Harbis, A., Perdreau, S., Vincent-Baudry, S., Charbonnier, M., Bernard, M. C., Raccah, D., Senft, M., Lorec, A. M., Defoort, C., Portugal, H., Vinoy, S., Lang, V. and Lairon, D. (2004). Glycemic and insulinemic meal responses modulate postprandial hepatic and intestinal lipoprotein accumulation in obese, insulin resistant subjects. *American Journal of Clinical Nutrition* 80(4), 896–902.

Hashimoto, N., Ito, Y., Han, K. H., Shimada, K., Sekikawa, M., Topping, D. L., Bird, A. R., Noda, T., Chiji, H. and Fukushima, M. (2006). Potato pulps lowered the serum cholesterol and triglyceride levels in rats. *Journal of Nutritional Science and Vitaminology* 52(6), 445–450.

Haub, M. D., Hubach, K. L., Al-Tamimi, E. K., Ornelas, S. and Seib, P. A. (2010). Different types of resistant starch elicit different glucose reponses in humans. *Journal of Nutrition and Metabolism* 2010, 230501.

Hernández, O., Emaldi, U. and Tovar, J. (2008). *In vitro* digestibility of edible films from various starch sources. *Carbohydrate Polymers* 71(4), 648–655.

Hizukuri, S., Tabata, S., Kagoshima and Nikuni, Z. (1970) Studies on starch phosphate Part 1. Estimation of glucose-6-phosphate residues in starch and the presence of other bound phosphate(s). *Starch - Stärke* 22(10), 338–343.

Hoover, R. and Zhou, Y. (2003). In vitro and in vivo hydrolysis of legume starches by α-amylase and resistant starch formation in legumes-a review. *Carbohydrate Polymers* 54(4), 401–417. https://doi.org/10.1016/S0144-8617(03)00180-2.

Hoseney, R. C. (1994). *Principles of Cereal Science and Technology* (2nd edn.). St. Paul, MN: AACC International.

Hoseney, R. C., Finney, K. F., Pomeranz, Y. and Shogren, M. D. (1971). Functional (breadmaking) and biochemical properties of wheat flour components. VIII. Starch. *Cereal Chemistry* 48, 191–201.

Hug-Iten, S., Escher, F. and Conde-Petit, B. (2003). Staling of bread: role of amylose and amylopectin and influence of starch-degrading enzymes. *Cereal Chemistry Journal* 80(6), 654–661.

Jacobs, D. R. and Gallaher, D. D. (2004). Whole grain intake and cardiovascular disease: a review. *Current Atherosclerosis Reports* 6(6), 415–423.

Jenkins, D. J. A., Kendall, C. W. C., Augustin, L. S. A., Franceschi, S., Hamidi, M., Marchie, A., Jenkins, A. L. and Axelsen, M. (2002). Glycemic index: overview of implications in health and disease. *American Journal of Clinical Nutrition* 76(1), 266S–273S.

Jenkins, D. J., Thorne, M. J., Wolever, T. M., Jenkins, A. L., Rao, A. V. and Thompson, L. U. (1987). The effect of starch-protein interaction in wheat on the glycemic response and rate of in vitro digestion. *The American Journal of Clinical Nutrition* 45(5), 946–951.

Jenkins, D. J., Wolever, T. M., Taylor, R. H., Barker, H., Fielden, H., Baldwin, J. M., Bowling, A. C., Newman, H. C., Jenkins, A. L. and Goff, D. V. (1981). Glycemic index of foods: a physiological basis for carbohydrate exchange. *American Journal of Clinical Nutrition* 34(3), 362–366.

Kaur, A., Shevkani, K., Katyal, M., Singh, N., Ahlawat, A. K. and Singh, A. M. (2016). Physicochemical and rheological properties of starch and flour from different durum wheat varieties and their relationships with noodle quality. *Journal of Food Science and Technology* 53(4), 2127–2138.

Kim, H. S. and Huber, K. C. (2008). Channels within soft wheat starch A-and B-type granules. *Journal of Cereal Science* 48(1), 159–172.

Kim, H. S. and Huber, K. C. (2010). Physicochemical properties and amylopectin fine structures of A- and B-type granules of waxy and normal soft wheat starch. *Journal of Cereal Science* 51(3), 256–264.

Kwasniewska-Karolak, I., Nebesny, E. and Rosicka-Kaczmarek, J. (2008). Characterization of amylose-lipid complexes derived from different wheat varieties and their susceptibility to enzymatic hydrolysis. *Food Science and Technology International* 14(1), 29–37.

Lagrain, B., Leman, P., Goesaert, H. and Delcour, J. A. (2008). Impact of thermostable amylases during breadmaking on wheat bread crumb structure and texture. *Food Research International* 41(8), 819–827.

Lehmann, U. and Robin, F. (2007). Slowly digestible starch-its structure and health implications: a review. *Trends in Food Science and Technology* 18(7), 346–355.

Lerer-Metzger, M., Rizkallal, S. W., Luo, J., Champ, M., Kabir, M., Bruzzo, F., Bornet, F. and Slama, G. (1996). Effects of long-term low-glycaemic index starchy food on plasma glucose and lipid concentrations and adipose tissue cellularity in normal and diabetic rats. *British Journal of Nutrition* 75(5), 723–732.

Leu, R. K. L., Hu, Y. and Young, G. P. (2002). Effects of resistant starch and non-starch polysaccharides on colonic luminal environment and genotoxin-induced apoptosis in the rat. *Carcinogenesis* 23(5), 713–719. https://doi.org/10.1093/carcin/23.5.713.

Li, J. H., Vasanthan, T., Rossnagel, B. and Hoover, R. (2001). Starch from hull-less barley: I. Granule morphology, composition, and amylopectin structure. *Food Chemistry* 74(4), 395–405.

Li, W. H., Shan, Y. L., Xiao, X. L. and Zheng, J. M. (2013). Effect of nitrogen and sulfur fertilization on accumulation characteristics and physiochemical properties of A- and B-wheat starch. *Journal of Agricultural and Food Chemistry* 61, 2418–2425.

Li, W. Y., Yan, S. H., Yin, Y. P., Li, Y., Liang, T., Gu, F., Dai, Z. and Wang, Z. (2008). Comparison of starch granule size distribution between hard and soft wheat cultivars in eastern china. *Agricultural Sciences in China* 7(8), 907–914.

Li, W., Yan, S., Yin, Y. and Wang, Z. (2010). Starch granules size distribution in wheat grain in relation to shading after anthesis. *Journal of Agricultural Science* 148(2), 183–189.

Liu, Q., Gu, Z., Donner, E., Tetlow, I. and Emes, M. (2007). Investigation of digestibility in-vitro and physicochemical properties of A- and B-type starch from soft and hard wheat flour. *Cereal Chemistry Journal* 84(1), 15–21.

Liu, P., Guo, W., Jiang, Z., Pu, H., Feng, C., Zhu, X., Peng, Y., Kuang, A. and Little, C. R. (2011). Effects of high temperature after anthesis on starch granules in grains of wheat (*Triticum aestivum* L.). *Journal of Agricultural Science* 149(2), 159–169.

Liu, R. and Xu, G. (2008). Effects of resistant starch on colonic preneoplastic aberrant crypt foci in rats. *Food and Chemical Toxicology: An International Journal Published for the British Industrial Biological Research Association* 46(8), 2672–2679.

Lu, D. R., Xiao, C. M. and Xu, S. J. (2009). Starch-based completely biodegradable polymer materials. *Express Polymer Letters* 3(6), 366–375.

Ludwig, D. S., Majzoub, J. A., Al-Zahrani, A., Dallal, G. E., Blanco, I. and Roberts, S. B. (1999). High glycemic index foods, overeating, and obesity. *Pediatrics* 103, e26.

Lunn, J. and Buttriss, J. L. (2007). Carbohydrates and dietary fibre. *Nutrition Bulletin* 32(1), 21–64.

Luo, K. and Zhang, G. (2018). Nutritional property of starch in a whole-grain-like structural form. *Journal of Cereal Science* 79, 113–117.

Maningat, C. C., Seib, P. A., Bassi, S. D., Woo, K. S. and Lasater, G. D. (2009). Wheat starch: production, properties, modification and uses. In: Bemiller, J. and Whistler, R. (Eds). *Starch: Chemistry and Technology* (3rd edn.). New York: Elsevier. 441–510.

Mikulíková, D., Masár, S. and Kraic, J. (2008). Biodiversity of legume health-promoting starch. *Starch* 60(8), 426–432.

Monnier, L., Mas, E., Ginet, C., Michel, F., Villon, L., Cristol, J. P. and Colette, C. (2006). Activation of oxidative stress by acute glucose fluctuations compared with sustained chronic hyperglycemia in patients with type 2 diabetes. *JAMA* 295(14), 1681–1687.

Morita, T., Kasaoka, S., Hase, K. and Kiriyama, S. (1999). Psyllium shifts the fermentation site of high-amylose corn starch toward the distal colon and increases fecal butyrate concentration in rats. *Journal of Nutrition* 129(11), 2081–2087.

Mukerjea, R. and Robyt, J. F. (2010). Isolation, structure, and characterization of the putative soluble amyloses from potato, wheat, and rice starches. *Carbohydrate Research* 345(3), 449–451.

Murphy, M. M., Douglass, J. S. and Birkett, A. (2008). Resistant starch intakes in the United States. *Journal of the American Dietetic Association* 108(1), 67–78.

Nebesny, E., Rosicka, J. and Tkaczyk, M. (2002). Effect of enzymatic hydrolysis of wheat starch on amylose-lipid complexes stability. *Starch* 54(12), 603–608.

Ni, Y., Wang, Z., Yin, Y. and Li, W. (2012). Starch granules size distribution in wheat grain in relation to phosphorus fertilization. *Journal of Agricultural Science* 150, 45–52.

Nugent, A. P. (2005). Health properties of resistant starch. *Nutrition Bulletin* 30(1), 27–54. doi: 10.1111/j.1467-3010.2005.00481.x.

Ong, M. H., Jumel, K., Tokarczuk, P. F., Blanshard, J. M. V. and Harding, S. E. (1994). Simultaneous determinations of the molecular weight distributions of amyloses and the fine structures of amylopectins of native starches. *Carbohydrate Research* 260(1), 99–117.

Park, S. H., Wilson, J. D. and Seabourn, B. W. (2009). Starch granule size distribution of hard red winter and hard red spring wheat: its effects on mixing and breadmaking quality. *Journal of Cereal Science* 49(1), 98–105.

Peng, M., Gao, M., Abdel-Aal, E.-S. M., Hucl, P. and Chibbar, R. N. (1999). Separation and characterization of A- and B-type starch granules in wheat endosperm. *Cereal Chemistry Journal* 76(3), 375–379.

Pérez, S. and Bertoft, E. (2010). The molecular structures of starch components and their contribution to the architecture of starch granules: a comprehensive review. *Starch* 62(8), 389–420.

Raben, A., Andersen, K., Karberg, M. A., Holst, J. J. and Astrup, A. (1997). Acetylation of or B-cyclodesetrin addition to potato starch: beneficial effect on glucose metabolism and appetite sensations. *American Journal of Clinical Nutrition* 66(2), 304–314.

Raben, A., Tagliabue, A., Christensen, N. J., Madsen, J., Holst, J. J. and Astrup, A. (1994). Resistant starch: the effect on postprandial glycemia, hormonal response, and satiety. *The American Journal of Clinical Nutrition* 60(4), 544–551.

Raeker, M. Ö, Gaines, C. S., Finney, P. L. and Donelson, T. (1998). Granule size distributions and chemical composition of starches from 12 soft wheat cultivars. *Cereal Chemistry Journal* 75(5), 721–728.

Ral, J. P., Cavanagh, C. R., Larroque, O., Regina, A. and Morell, M. K. (2008). Structural and molecular basis of starch viscosity in hexaploid wheat. *Journal of Agricultural and Food Chemistry* 56(11), 4188–4197.

Ratnayake, W. S. and Jackson, D. S. (2008). Thermal behavior of resistant starches RS 2, RS 3, and RS 4. *Journal of Food Science* 73(5), C356–C366.

Reader, D., Johnson, M. L., Hollander, P. and Franz, M. (1997). Response of resistant starch in a food bar vs. two commercially available bars in persons with type II diabetes mellitus. *Diabetes* 46(1), 254.

Ross, A. S., Quail, K. J. and Crosbie, G. B. (1997). Physicochemical properties of Australian flours influencing the texture of alkaline noodles. *Cereal Chemistry Journal* 74(6), 814–820.

Sahlström, S., Bævre, A. B. and Bråthen, E. (2003). Impact of starch properties on health bread characteristics. II. Purified A- and B-granule fractions. *Journal of Cereal Science* 37(3), 285–293.

Sajilata, M. G., Singhal, R. S. and Kulkarni, P. R. (2006). Resistant starch–a review. *Comprehensive Reviews in Food Science and Food Safety* 5(1), 1–17.

Salman, H., Blazek, J., Lopez-Rubio, A., Gilbert, E. P., Hanley, T. and Copeland, L. (2009). Structure-function relationships in A and B granules from wheat starches of similar amylose content. *Carbohydrate Polymers* 75(3), 420–427.

Sanz, T., Salvador, A., Baixauli, R. and Fiszman, S. M. (2009). Evaluation of four types of resistant starch in muffins. II. Effects in texture, colour and consumer response. *European Food Research and Technology* 229(2), 197–204.

Sanz, T., Salvador, A. and Fiszman, S. M. (2008). Resistant starch (RS) in battered fried products: functionality and high-fiber benefit. *Food Hydrocolloids* 22(4), 543–549.

Seal, C. J., Daly, M. E., Thomas, L. C., Bal, W., Birkett, A. M., Jeffcoat, R. and Mathers, J. C. (2003). Postprandial carbohydrate metabolism in healthy subjects and those with type 2 diabetes fed starches with slow and rapid hydrolysis rates determined in vitro. *British Journal of Nutrition* 90(5), 853–864.

Seib, P. A. (1994). Wheat starch: isolation, structure, and properties. *Oyo Toshitsu Kagaku* 41, 49–56.

Seneviratne, H. D. and Biliaderis, C. G. (1991). Action of -amylases on amylose-lipid complex superstructure. *Journal of Cereal Science* 13(2), 129–143.

Sharma, A., Yadav, B. S. and Ritika, B. (2008). Resistant starch: physiological roles and food applications. *Food Reviews International* 24(2), 193–234.

Shevkani, K., Singh, N., Bajaj, R. and Kaur, A. (2017). Wheat starch production, structure, functionality and applications—a review. *International Journal of Food Science and Technology* 52(1), 38–58.

Shevkani, K., Singh, N., Singh, S., Ahlawat, A. K. and Singh, A. M. (2011). Relationship between physicochemical and rheological properties of starches from Indian wheat lines. *International Journal of Food Science and Technology* 46(12), 2584–2590.

Shimada, M., Mochizuki, K. and Goda, T. (2008). Dietary resistant starch reduces levels of glucose-dependent insulinotropic polypeptide mRNA along the jejunum-ileum in both normal and Type 2 diabetic rats. *Bioscience, Biotechnology, and Biochemistry* 72(8), 2206–2209.

Shinde, S. V., Nelson, J. E. and Huber, K. C. (2003). Soft wheat starch pasting behavior in relation to A-and B-type granule content and composition. *Cereal Chemistry Journal* 80(1), 91–98.

Singh, J., McCarthy, O. J., Singh, H. and Moughan, P. J. (2008). Low temperature post-harvest storage of New Zealand Taewa (Maori potato): effects on starch physico-chemical and functional characteristics. *Food Chemistry* 106, 583–596.

Singh, J., Kaur, L. and McCarthy, O. J. (2009). Potato starch and its modification J. Singh, L. Kaur (Eds.), Advances in Potato Chemistry and Technology, Academic Press, New York, 273–318.

Singh, J., Dartois, A. and Kaur, L. (2010). Starch digestibility in food matrix: a review. *Trends in Food Science and Technology* 21(4), 168–180.

Singh, N., Kaur, N., Katyal, M., Kaur, A. and Shevkani, K. (2016). Characteristics of starch separated from coarse and fine flour fractions obtained from hard, medium-hard and soft Indian wheat cultivars. *Starch* 69(1-2). doi: 10.1002/star.201600012.

Slavin, J., Stewart, M., Timm, D. and Hospattankar, A. (2009). Fermentation patterns and short chain-fatty acid (scfa) profiles of wheat dextrin and other functional fibers. In: *Proceedings of the 4th International Dietary Fibre Conference*. Vienna, Austria: International Association for Cereal Science and Technology (ICC). 35.

Soh, H. N., Sissons, M. J. and Turner, M. A. (2006). Effect of starch granule size distribution and elevated amylose content on durum dough rheology and spaghetti cooking quality. *Cereal Chemistry Journal* 83(5), 513–519.

Stoddard, F. L. (1999). Survey of starch particle-size distribution in wheat and related species. *Cereal Chemistry Journal* 76(1), 145–149.

Sui, Z., Shah, A. and BeMiller, J. N. (2011). Cross-linked and stabilized in-kernel heat-moisture-treated and temperature-cycled normal maize starch and effects of reaction conditions on starch properties. *Carbohydrate Polymers* 86(4), 1461–1467.

Tao, H., Wang, P., Wu, F., Jin, Z. and Xu, X. (2016). Particle size distribution of wheat starch granules in relation to baking properties of frozen dough. *Carbohydrate Polymers* 137, 147–153.

Tester, R. F., Karkalas, J. and Qi, X. (2004). Starch–composition, fine structure and architecture. *Journal of Cereal Science* 39(2), 151–165.

Tharanathan, R. N. (2002). Food-derived carbohydrates: structural complexity and functional diversity. *Critical Reviews in Biotechnology* 22(1), 65–84.

Tharanathan, R. N. and Mahadevamma, S. (2003). Grain legumes: a boon to human nutrition. *Trends in Food Science and Technology* 14(12), 507–518.

Themeier, H., Hollmann, J., Neese, U. and Lindhauer, M. G. (2005). Structural and morphological factors influencing the quantification of resistant starch II in starches of different botanical origin. *Carbohydrate Polymers* 61(1), 72–79.

Törrönen, R., Kolehmainen, M., Sarkkinen, E., Poutanen, K., Mykkänen, H. and Niskanen, L. (2013). Berries reduce postprandial insulin responses to wheat and rye breads in healthy women. *Journal of Nutrition* 143(4), 430–436.

Vermeylen, R., Goderis, B., Reynaers, H. and Delcour, J. A. (2005). Gelatinization related structural aspects of small and large wheat starch granules. *Carbohydrate Polymers* 62(2), 170–181.

Wachters-Hagedoorn, R. E., Priebe, M. G., Heimweg, J. A. J., Heiner, A. M., Englyst, K. N., Holst, J. J., Stellaard, F. and Vonk, R. J. (2006). The rate of intestinal glucose absorption is correlated with plasma glucose-dependent insulinotropic polypeptide concentrations in healthy men. *Journal of Nutrition* 136(6), 1511–1516.

Wang, S., Li, C., Copeland, L., Niu, Q. and Wang, S. (2015). Starch retrogradation: a comprehensive review. *Comprehensive Reviews in Food Science and Food Safety* 14(5), 568–585.

Wang, S., Luo, H., Zhang, J., Zhang, Y., He, Z. and Wang, S. (2014). Alkali-induced changes in functional properties and in vitro digestibility of wheat starch: the role of surface proteins and lipids. *Journal of Agricultural and Food Chemistry* 62(16), 3636–3643.

Wang, S. J., and Copeland, L. (2012). New insights on the mechanism of acid degradation of pea starch. *Carbohydrate Polymers* 87(3), 1941– 1949.

Wang, S. J., Sharp, P. and Copeland, L. (2011). Structural and functional properties of starches from field peas. *Food Chemistry* 126(4), 1546–1552.

Wei, C., Zhang, J., Chen, Y., Zhou, W., Xu, B., Wang, Y. and Chen, J. (2010). Physicochemical properties and development of wheat large and small starch granules during endosperm development. *Acta Physiologiae Plantarum* 32(5), 905–916.

Wepner, B., Berghofer, E., Miesenberger, E., Tiefenbacher, K. and N. K. Ng, P. (1999). Citrate starch: application as resistant starch in different food systems. *Starch* 51(10), 354–361.

Whistler, R. L. and Daniel, J. R. (1984). Chapter vi. Molecular structure of starch. In: Whistler, R. L. and James, N. (Eds). *Be Miller, Eugene F. Paschall, in Food Science and Technology, Starch: Chemistry and Technology* (2nd edn.), Academic Press, San Diego, CA, 153–182.

Wilson, J. D., Bechtel, D. B., Wilson, G. W. T. and Seib, P. A. (2008). Bread quality of spelt wheat and its starch. *Cereal Chemistry Journal* 85(5), 629–638.

Wolever, T. M., Vorster, H. H., Björck, I., Brand-Miller, J., Brighenti, F., Mann, J. I., Ramdath, D. D., Granfeldt, Y., Holt, S., Perry, T. L., Venter, C. and Wu, X. (2003). Determination of the glycaemic index of foods: interlaboratory study. *European Journal of Clinical Nutrition* 57(3), 475–482.

Woo, K. S. and Seib, P. A. (2002). Cross-linked resistant starch: preparation and properties. *Cereal Chemistry Journal* 79(6), 819-825.

Woolnough, J. W., Monro, J. A., Brennan, C. S. and Bird, A. R. (2008). Simulating human carbohydrate digestion in vitro: a review of methods and the need for standardization. *International Journal of Food Science and Technology* 43(12), 2245-2256.

Wu, H. W. and Sarko, A. (1978). The double-helical molecular structure of crystalline A-amylose. *Carbohydrate Research* 61(1), 27-40.

Yamada, Y., Hosoya, S., Nishimura, S., Tanaka, T., Kajimoto, Y., Nishimura, A. and Kajimoto, O. (2005). Effect of bread containing resistant starch on postprandial blood glucose levels in humans. *Bioscience, Biotechnology, and Biochemistry* 69(3), 559-566.

Yao, N., Paez, A. V. and White, P. J. (2009). Structure and function of starch and resistant starch from corn with different doses of mutant amylose-extender and Floury-1 alleles. *Journal of Agricultural and Food Chemistry* 57(5), 2040-2048.

Ye, X., Zhang, Y., Qiu, C., Corke, H. and Sui, Z. (2019). Extraction and characterization of starch granule-associated proteins from rice that affect in vitro starch digestibility. *Food Chemistry* 276, 754-760.

Yeo, L. L. and Seib, P. A. (2009). White pan bread and sugar-snap cookies containing wheat starch phosphate, A cross-linked resistant starch. *Cereal Chemistry Journal* 86(2), 210-220.

Yin, Y. A., Qi, J. C., Li, W. H., Cao, L., Wang, Z. and Wang, Z. B. (2012). Formation and developmental characteristics of A- and B-type starch granules in wheat endosperm. *Journal of Integrative Agriculture* 11(1), 73-81.

Yoo, S. H. and Jane, J. L. (2002). Structural and physical characteristics of waxy and other wheat starches. *Carbohydrate Polymers* 49(3), 297-305.

Yu, H., Cao, Y., Fang, Q. and Liu, Z. (2015). Effects of treatment temperature on properties of starch-based adhesives. *BioResources* 10(2), 3520-3530.

Yu, W., Zou, W., Dhital, S., Wu, P., Gidley, M. J., Fox, G. P. and Gilbert, R. G. (2018). The adsorption of α-amylase on barley proteins affects the in vitro digestion of starch in barley flour. *Food Chemistry* 241, 493-501.

Zhang, G., Ao, Z. and Hamaker, B. R. (2006a). Slow digestion property of native cereal starches. *Biomacromolecules* 7(11), 3252-3258.

Zhang, G. and Hamaker, B. R. (2009). Slowly digestible starch: concept, mechanism, and proposed extended glycemic index. *Critical Reviews in Food Science and Nutrition* 49(10), 852-867.

Zhang, G., Venkatachalam, M. and Hamaker, B. R. (2006b). Structural basis for the slow digestion property of native cereal starches. *Biomacromolecules* 7(11), 3259-3266.

Zhang, T., Wang, Z., Yin, Y., Cai, R., Yan, S. and Li, W. (2010). Starch content and granule size distribution in grains of wheat in relation to post-anthesis water deficits. *Journal of Agronomy and Crop Science* 196(1), 1-8.

Zhang, H., Zhang, W., Xu, C. and Zhou, X. (2013). Morphological features and physicochemical properties of waxy wheat starch. *International Journal of Biological Macromolecules* 62, 304-309.

Zhao, L. F. and Seib, P. A. (2005). Alkaline-carbonate noodles from hard winter wheat flours varying in protein, swelling power and polyphenol oxidase activity. *Cereal Chemistry Journal* 82(5), 504-516.

Zhou, Z., Zhang, Y., Chen, X., Zhang, M. and Wang, Z. (2014). Multi-scale structural and digestion properties of wheat starches with different amylose contents. *International Journal of Food Science and Technology* 49(12), 2619-2627.

Chapter 2

Advances in understanding the nutritional value of antioxidants in wheat

Franklin Brian Apea-Bah and Trust Beta, University of Manitoba, Canada

1 Introduction

Wheat is the most important staple crop in the temperate regions of the world and its demand is increasing in countries undergoing rapid urbanization and industrialization (Shewry and Hey, 2015). World production of wheat in 2019 stood at 767 million tons, and among cereals, was second only to corn (FAO, 2019). They are widely cultivated around the world with Asia (334 million tons), Europe (266 million tons) and North America (84 million tons) being the leading wheat-producing continents (FAO, 2019). Wheat grains find food applications in bread (Yu et al., 2013), muffins (Rupasinghe et al., 2008), cookies (Heredia-Sandoval et al., 2019), pasta (Feillet et al., 2000), wheat flour noodles (Fu, 2008) and flat breads (e.g. tortilla, chapati, pita, parotta, yufka, and tandoori roti) (Gocmen et al., 2009).

Wheat plants are classified by farmers and breeders based on the following: (1) growing season (winter wheat or spring wheat); (2) protein content ranging from 10% in soft wheat to 15% in hard wheat; (3) gluten quality, yielding bread wheat with strong, elastic gluten for dough formation, and durum wheat with non-elastic gluten for pasta preparation; (4) grain color, comprising red, amber or yellow, white and other intensely colored cultivars (purple, blue and black wheat) (Beta et al., 2019).

http://dx.doi.org/10.19103/AS.2021.0087.04

Nutritionally, wheat grains are a good source of carbohydrates (particularly starch), protein, dietary fiber and vitamins (especially the B-vitamins) (Shewry and Hey, 2015). In addition to these nutritional components, they also contain bioactive phytochemicals which include phenolic compounds (Yu and Beta, 2015), carotenoids (Ndolo and Beta, 2013) and tocochromanols (which comprise tocopherols and tocotrienols) (Labuschagne et al., 2017). Most of these phytochemicals have antioxidant capacity and other health-promoting properties.

This chapter presents an overview of the antioxidant bioactive compounds present in wheat and their relevance to the nutritional value of wheat.

2 Phytochemical antioxidants of wheat

2.1 Phenolic compounds

Phenolic compounds are the most diverse group of phytochemicals present in wheat. Most health benefits associated with wheat consumption have been attributed to these compounds, but most significant are their antioxidant properties. Wheat phenolic compounds may be categorized into the following subclasses: phenolic acids (Fig. 1), flavonoids (Fig. 2), anthocyanins (Fig. 3), proanthocyanidins (Fig. 4) and alkylresorcinols (Fig. 5).

2.1.1 Phenolic acids

Two classes of phenolic acids are reportedly present in wheat grains and these are the hydroxybenzoic acids and the hydroxycinnamic acids. The hydroxybenzoic acids include *p*-hydroxybenzoic, protocatechuic, 2,4-dihydroxybenzoic, vanillic and syringic acids, as well as syringaldehyde which is the aldehyde form of syringic acid (Li et al., 2008; Yu and Beta, 2015). The hydroxycinnamic acids include *o*-coumaric, *p*-coumaric, caffeic, ferulic and sinapic acids (Lu et al., 2014), as well as diferulic acids, which are composed of two ferulic acids covalently bound together by C-C or C-O-C bonds (Li et al., 2008; Liu et al., 2010). In addition to these compounds, Ma et al. (2015) reported the presence of chlorogenic acid in winter wheat. Figure 1 shows the structure of phenolic acids identified in wheat. Moore and Hao (2012) reviewed the antioxidant and health-promoting properties of wheat and provided information about their constituent phenolic compounds and the effects of processing methods on their bioaccessibility and bioavailability. This current chapter adds to the existing information, recent developments in the area of wheat antioxidants and their nutritional value.

Phenolic acids are usually concentrated in the bran of wheat grains and exist in the soluble-free, soluble-esterified and insoluble-bound forms (Liyana-Pathirana and Shahidi, 2006). Several researchers have identified the previously

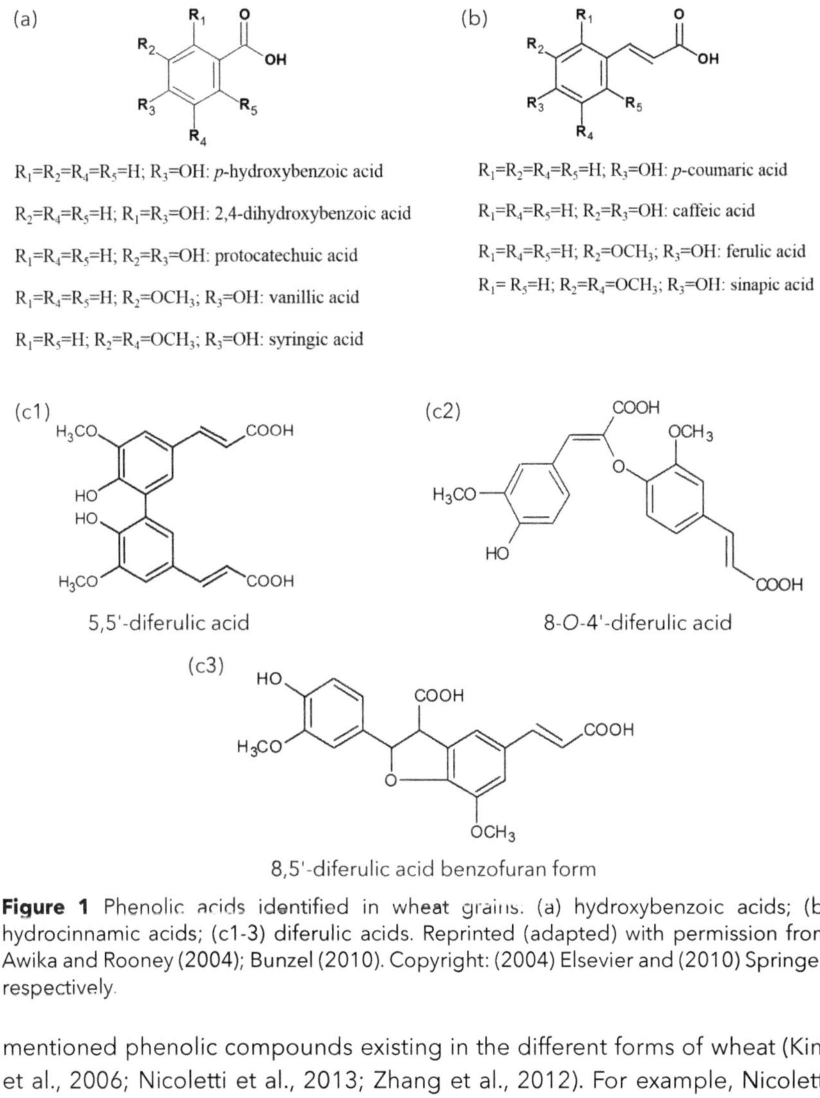

Figure 1 Phenolic acids identified in wheat grains. (a) hydroxybenzoic acids; (b) hydrocinnamic acids; (c1-3) diferulic acids. Reprinted (adapted) with permission from Awika and Rooney (2004); Bunzel (2010). Copyright: (2004) Elsevier and (2010) Springer, respectively.

mentioned phenolic compounds existing in the different forms of wheat (Kim et al., 2006; Nicoletti et al., 2013; Zhang et al., 2012). For example, Nicoletti et al. (2013) identified *p*-hydroxybenzoic, vanillic, syringic, *p*-coumaric, ferulic and sinapic acids in the soluble-free, soluble-esterified and insoluble-bound forms in durum whole meal, coarse bran, semolina and pasta. Vanillic acid was the major hydroxybenzoic acid derivative and was dominant in the soluble-esterified form, compared to the soluble-free and insoluble-bound forms. While ferulic acid was overall, the dominant compound in the soluble-free and insoluble-bound forms, sinapic acid was the dominant compound in the soluble-esterified form (Nicoletti et al., 2013).

There are more bound phenolic acids in wheat followed by the esterified forms while the free phenolic acids are the least abundant in the grains (Shewry et al., 2008). The diferulic acids are mostly bound to cell wall components such as arabinoxylans to form feruloylated arabinoxylans (Malunga and Beta, 2016). Arabinoxylans are the predominant components of dietary fiber in wheat (Malunga et al., 2017).

Within the bran of wheat grains, the concentration of phenolic acids varies among the different layers or fractions. Ndolo and Beta (2014) studied the distribution of phenolic acids (caffeic, *p*-coumaric, ferulic, iso-ferulic, sinapic, *p*-hydroxybenzoic, vanillic and syringic acids) in the different fractions of cereal grains. They observed the phenolic acids to be most abundant in the pericarp of wheat, followed by the aleurone and then the germ, while the endosperm had the least amount of phenolic acids. Only *p*-coumaric and ferulic acids were identified in the endosperm. Ferulic acid was found to be the dominant phenolic acid in the wheat varieties followed by vanillic acid, while syringic acid was present in the least amount (Ndolo and Beta, 2014). Table 1 shows a summary of phenolic acid contents reported in the wheat species *Triticum monococcum* (Einkorn), *T. turgidum* (Khorasan, Emmer, Durum) and *T. aestivum* (winter wheat, spring wheat, soft wheat, hard wheat, purple wheat, blue wheat and spelt wheat).

2.1.2 Flavonoids

Flavonoids are arguably the most abundant polyphenols in the plant kingdom. Structurally, they are composed of two phenol rings (A and B) joined together by a 3-carbon chain which cyclizes into a third ring (C) (Fig. 2). Like phenolic acids, flavonoids are mostly concentrated in the bran of whole grains such as wheat (Dykes and Rooney, 2007). Six main flavonoid classes are known and these are: the flavan-3-ols (e.g. (+)-catechin, (−)-epicatechin and their glycosides); flavan-4-ols (e.g. apiforol, luteoforol and their glycosides); flavonols (e.g. kaempferol, quercetin, myricetin and their glycosides); flavones (e.g. apigenin, luteolin, chrysoeriol and their glycosides); flavanones (e.g. naringenin, eriodictyol and their glycosides) and flavanonols (e.g. taxifolin and its glycosides) (Pietta, 2000). The general structures of the flavonoids are shown in Fig. 2.

Of the six flavonoid classes mentioned above, only two have been reported in wheat. These are the flavones and flavonols. Structurally, the two classes bear marked resemblance except that flavonols have a 3-OH substituent on the C-ring, but this is lacking in flavones. Liu et al. (2010) identified 20 flavonoids from 7 wheat (3 purple, 2 red, 1 yellow and 1 white) varieties comprising 11 *C*-conjugated glycosides, 7 *O*-glycosides and 2 acylated *C*-glycosides. Seventeen of the compounds identified were flavones and the remaining three were flavonols. Thirteen of the flavones were either *C*-glycosides or their acyl derivatives, while the remaining four were *O*-glycosides. On the contrary, all the flavonols were *O*-glycosides.

Table 1 Pooled means of phenolic acids present in *Triticum monococcum*, *Triticum turgidum* and *Triticum aestivum* and their sources

Wheat type	*p*-HBA	VNA	SYA	SYD	p-CA	DHBA	CFA	FRA	SNA	2-HCNA	Total	References
						Soluble free						
Triticum monococcum	-	-	-	-	-	-	-	390.0±125.9	-	-	390.0±125.9	Abdel-Aal and Rabalski (2008)
Triticum monococcum	-	0.4±0.4	0.2±0.2	0.8±0.5	0.3±0.2	1.2±0.6	0.8±0.8	4.0±0.4	3.6±2.2	0.1±0.1	11.4±1.5	Li et al. (2008)
Triticum turgidum	-	-	-	-	-	-	-	337.3±72.6	-	-	337.3±72.6	Abdel-Aal and Rabalski (2008)
Triticum turgidum	-	1.8±0.1	1.8±0.8	-	1.0±0.2	1.0±0.2	0.6±0.6	3.1±0.2	3.6±1.5	0.3±0.1	13.0±1.2	Li et al. (2008)
Triticum aestivum	-	-	-	-	-	-	-	472.7±43.1	-	-	472.7±43.1	Abdel-Aal and Rabalski (2008)
Triticum aestivum	-	1.1±0.4	0.7±0.2	-	0.2±0.0	-	-	1.8±0.5	-	-	3.7±1.2	Moore et al. (2005)
Triticum aestivum	0.1±0.0	1.8±0.5	1.7±0.9	0.1±0.2	0.6±0.0	0.4±0.2	0.4±0.5	3.2±0.5	2.2±0.3	0.1±0.0	10.6±1.1	Li et al. (2008)
Triticum aestivum (Öelands hvede)	-	-	-	-	-	-	-	2.50 ± 0.20	-	-	2.50 ± 0.20	Yu and Beta (2015)

(*Continued*)

Table 1 (*Continued*)

Wheat type	*p*-HBA	VNA	SYA	SYD	p-CA	DHBA	CFA	FRA	SNA	2-HCNA	Total	References
Triticum aestivum (Purple Indigo)	-	-	-	-	-	-	-	2.02 ± 0.07	-	-	2.02 ± 0.07	Yu and Beta (2015)
Triticum aestivum (Purple Konini)	-	-	-	-	-	-	-	2.25 ± 0.01	-	-		Yu and Beta (2015)
						Soluble conjugated						
Triticum monococcum	4.6±0.4	9.6±1.2	3.8±0.4	-	5.0±0.6	86.8±1.6	-	56.6±5.3	92.8±9.0	1.8±0.3	261.0±39.6	Li et al. (2008)
Triticum turgidum	9.4±2.2	13.3±2.8	5.8±0.1	-	5.4±1.1	83.6±35.1	-	35.6±13.2	75.0±16.4	2.5±0.0	230.4±32.9	Li et al. (2008)
Triticum aestivum	6.2±0.8	13.2±1.5	10.1±1.7	0.4±0.1	-	48.8±9.4	-	29.6±1.3	52.0±4.1	2.2±0.3	167.4±20.0	Li et al. (2008)
Triticum aestivum	-	5.3±0.7	8.4±2.9	-	1.1±0.2	-	-	38.3±5.2	-	-	53.1±9.0	Moore et al. (2005)
						Total soluble						
Triticum aestivum	-	6.4±1.1	9.2±3.1	-	1.3±0.3	-	-	40.0±0.7	-	-	56.9±10.2	Moore et al. (2005)
Triticum aestivum	-	80.5±13.4	23.7±5.5	-	50.7±6.5	-	-	674.7±75.0	50.7±7.2	-	880.3±279.5	Ndolo and Beta (2014)

Triticum aestivum (purple wheat)	-	121.0±4.5	55.0±1.2	-	63.0±0.9	-	-	731.0±3.9	114.0±2.2	-	1084.0±289.0	Ndolo and Beta (2014)
Insoluble bound												
Triticum monococcum	0.7±0.2	2.2±0.2	1.1±0.2	-	2.9±0.0	55.8±9.0	-	241.4±34.9	25.3±1.3	2.7±0.1	332.1±83.0	Li et al. (2008)
Triticum turgidum	3.0±1.3	3.0±0.4	1.1±0.1	0.1±0.1	3.3±0.2	66.5±60.2	-	402.4±66.3	24.5±0.9	4.9±0.3	508.7±131.5	Li et al. (2008)
Triticum aestivum	2.3±0.5	3.7±0.9	3.6±1.2	0.4±0.1	6.4±6.4	49.7±25.9	-	358.3±21.8	20.4±2.7	4.3±0.3	449.0±116.7	Li et al. (2008)
Triticum aestivum	-	4.0±0.8	3.6±0.8	-	10.9±1.3	-	-	518.7.0±54.5	-	-	537.1±57.3	Moore et al. (2005)
Triticum aestivum (Öelands hvede)	22.7±1.3	9.3±0.5	9.0±0.6	-	26.1±1.2	-	4.5±0.2	532.2±30.0	69.4±2.7	-		Yu and Beta (2015)
Triticum aestivum (Indigo)	25.7±1.2	24.8±0.0	10.0±0.2	-	19.1±1.5	-	6.5±0.2	447.0±24.1	36.0±2.8	-		Yu and Beta (2015)
Triticum aestivum (Konini)	24.6±1.2	19.2±1.0	9.1±0.1	-	21.6± 0.7	-	7.1±0.3	466.7±0.3	52.5±2.2	-		Yu and Beta (2015)

Key: *p*-HBA - *p*-hydroxybenzoic acid; VNA - vanillic acid; SYA - syringic acid; *p*-CA - *p*-coumaric acid; DBHA - 2,4-dihydroxybenzoic acid; CFA - caffeic acid; FRA - ferulic acid; SNA - sinapic acid; 2-HCNA - 2-hydroxycinnamic acid. *Öelands hvede* is common wheat while *Indigo* and *Konini* are purple wheat cultivars. *Triticum monococcum* represents Einkorn. *Triticum turgidum* includes Emmer, Khorasan and Durum. *Triticum aestivum* includes spelt, soft wheat, hard wheat and common wheat.

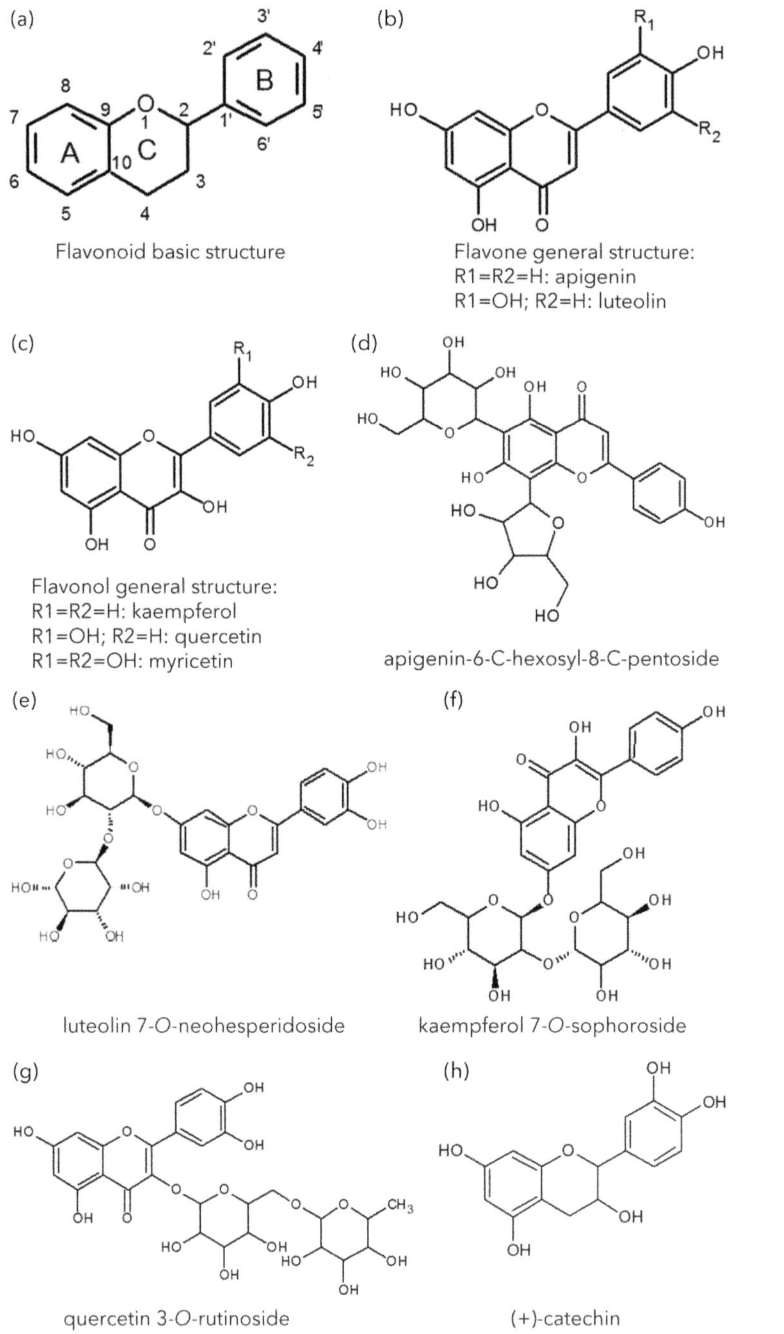

Figure 2 Flavonoids identified in wheat grains. Reprinted (adapted) with permission from Pietta (2000). Copyright: (2000) American Chemical Society.

The flavone *C*-glycosides included: two isomers of apigenin-6,8-di-*C*-hexoside, chrysoeriol-6,8-di-*C*-pentoside, three isomers of apigenin-6-*C*-pentoside-8-*C*-hexoside, three isomers of chrysoeriol-6-*C*-pentoside-8-*C*-hexoside, two isomers of apigenin-6,8-di-*C*-pentoside, apigenin-6-*C*-sinapoylpentoside-8-*C*-hexoside and apigenin-8-*C*-sinapoylpentoside-6-*C*-hexoside. The two latter compounds are flavone acyl glycosides. The flavone *O*-glycosides included: luteolin-7-*O*-hexoside, luteolin-7-*O*-neohesperidoside, apigenin-7-*O*-neohesperidoside and chrysoeriol-7-*O*-neosperidoside. The flavonols, on the other hand, included: kaempferol-7-*O*-sopheroside, kaempferol-7-*O*-neohesperidoside and quercetin-3-*O*-rutinoside (Liu et al., 2010).

Earlier work by Feng et al. (1988) identified two flavones: apigenin-6-*C*-arabinoside-8-*C*-hexoside and apigenin-6-*C*-hexoside-8-*C*-pentoside, in the bran of hard red spring wheat. Ioset et al. (2007) also identified the glycosides of the flavones isoorientin and tricin in wheat.

The following are apparent from the above studies: (1) flavones are the dominant flavonoid class in wheat grains, while flavonols are present to a lesser extent; (2) wheat flavones exist mainly as *C*-glycosides with few of them present in acylated forms, while the flavonols exist mostly as *O*-glycosides; (3) *C*-glycosylation occurs preferentially at the *C*-6 and *C*-8 positions on the flavonoid A-ring, while *O*-glycosylation occurs mainly at positions *C*-7 (on the A-ring) and *C*-3 (on the C-ring); (4) the *C*-glycosides comprised mainly monosaccharides (one or two pentoses or hexoses attached to the *C*-6 or *C*-8 positions), while the *O*-glycosides comprised mainly disaccharides (rutinose, neohesperidose or sopherose) attached to the *C*-7 or *C*-3 positions. While there is some information about flavonoid composition of some wheat cultivars, there appears to be a dearth of information about the content of flavonoids measured by chromatography.

2.1.3 Anthocyanins

Anthocyanins are a unique class of polyphenols in that their flavylium basic chemical structure has a net positive charge on the O-1 atom. This distinguishes them from other flavonoid classes. Structurally, anthocyanins are similar to flavan-3-ols in that they have a 3-hydroxyl (OH) group but lack a 4-keto group (Fig. 3) (Pietta, 2000). Anthocyanins are water-soluble pigment compounds and are responsible for the purple, blue and black colors of specialty (colored) wheat cultivars. Like flavonoids, they exist mostly as glycosides in the bran of colored wheat grains. Anthocyanins therefore consist of an aglycone (anthocyanidin) component and a glycosyl moiety. The most common anthocyanidins reported in colored wheat include cyanidin, delphinidin, malvidin, pelargonidin, peonidin and petunidin. These aglycones are usually covalently bonded to

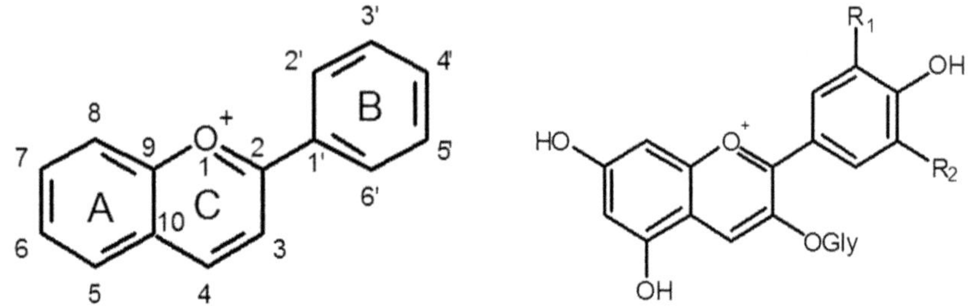

$R_1 = R_2 = H$; pelargonidin
$R_1 = OH, R_2 = H$; cyanidin
$R_1 = R_2 = OH$; delphinidin
$R_1 = OCH_3, R_2 = H$; peonidin
$R_1 = OH, R_2 = OCH_3$; petunidin
$R_1 = R_2 = OCH_3$; malvidin

Figure 3 Anthocyanins identified in wheat grains. Reprinted (adapted) with permission from Hosseinian et al. (2008). Copyright: (2008) Elsevier.

pentoses, hexoses or disaccharides of hexose units which are the glycosyl moieties (Fig. 3).

Abdel-Aal et al. (2006) studied the anthocyanin composition of colored wheat varieties and observed differences in the composition of these compounds in blue wheat and purple wheat. For blue wheat, they identified delphinidin 3-*O*-glucoside as the major compound, followed by delphinidin 3-*O*-rutinoside, cyanidin 3-*O*-glucoside and cyanidin 3-*O*-rutinoside, all of which had concentrations in the range 17–56 μg/g. Other anthocyanins identified in blue wheat to lesser amounts included petunidin 3-*O*-glucoside, petunidin 3-*O*-rutinoside, peonidin rutinoside and malvidin rutinoside, all of which had concentrations between 1 μg/g and 5 μg/g. Purple wheat, on the other hand, contained cyanidin 3-*O*-glucoside as the dominant compound, followed by peonidin 3-*O*-glucoside, cyanidin malonylglucoside and cyanidin succinylglucoside, all of which had concentrations ranging between 1 μg/g and 4 μg/g. Other anthocyanins identified in purple wheat included peonidin malonylglucoside, peonidin succinylglucoside and peonidin malonylsuccinylglucoside, all of which had concentrations <1 μg/g (Abdel-Aal et al., 2006). Table 2 shows anthocyanin contents of some colored (purple and blue) wheat cultivars. It is worth noting that anthocyanins are present only in colored grains.

Similarly, Hosseinian et al. (2008) identified the following anthocyanins in purple wheat: cyanidin 3-*O*-glucoside, cyanidin 3-*O*-galactoside, cyanidin 3-*O*-arabinoside, cyanidin aglycone, delphinidin 3-*O*-galactoside, delphinidin 3-*O*-arabinoside, malvidin 3-*O*-glucoside, pelargonidin 3-*O*-glucoside,

Table 2 Comparison of anthocyanin composition (μg/g) of normal purple wheat, heat-stressed purple wheat and normal blue wheat

Anthocyanins	Normal purple wheat by HPLC method[1]	Normal purple wheat by pH differential method	Heat-stressed purple wheat by HPLC method[1]	Heat-stressed purple wheat by pH differential method	Normal purple wheat by HPLC method	Normal blue wheat by HPLC method	Purple wheat (*Indigo*) differential method[2]	Purple wheat (*Konini*) differential method[2]
Delphinidin 3-galactoside	38.3 ± 0.6	35.4 ± 0.3	37.7 ± 0.4	39.4 ± 0.1	-	-	-	-
Delphinidin 3-glucoside	-	-	-	-	-	56.5 ± 4.6	-	-
Delphinidin 3-rutinoside	-	-	-	-	-	49.6 ± 2.4	-	-
Delphinidin 3-arabinoside	16.7 ± 0.4	15.1 ± 0.9	16.0 ± 0.8	15.3 ± 0.5	-	-	-	-
Cyanidin 3-arabinoside	25.1 ± 0.1	23.2 ± 0.3	26.5 ± 0.8	25.5 ± 0.6	-	-	-	-
Petunidin 3-glucoside	40.4 ± 0.3	38.3 ± 0.6	40.7 ± 0.3	40.1 ± 0.3	-	2.2 ± 0.1	-	-
Petunidin 3-rutinoside	-	-	-	-	-	4.5 ± 0.2	-	-
Cyanidin 3- galactoside	72.0 ± 0.10	76.1 ± 0.7	66.0 ± 0.3	71.3 ± 0.1	4.0 ± 0.1	20.3 ± 1.5	-	-
Cyanidin 3-glucoside	103.0 ± 0.4	109.2 ± 0.1	104.9 ± 1.4	110.8 ± 0.4	-	-	104.1 ± 3.9	53.1 ± 0.2
Cyanidin 3-rutinoside	-	-	-	-	-	16.8 ± 0.8	-	-
Cyanidin malonylglucoside	-	-	-	-	1.2 ± 0.1	-	-	-
Pelargonidin 3-glucoside	28.8 ± 0.0	29.3 ± 0.0	27.3 ± 0.0	32.2 ± 0.1	-	-	-	-
Peonidin 3-glucoside	2.7 ± 0.1	3.3 ± 0.4	3.5 ± 0.0	6.0 ± 0.2	2.1 ± 0.1	-	-	-
Peonidin rutinoside	-	-	-	-	-	1.2 ± 0.1	-	-
Cyanidin succinylglucoside	-	-	-	-	0.6 ± 0.0	-	-	-
Malvidin 3-glucoside	51.6 ± 0.7	60.3 ± 0.3	85.7 ± 0.2	90.4 ± 0.6	-	-	-	-
Malvidin rutinoside	-	-	-	-	-	2.0 ± 0.1	-	-

(Continued)

Table 2 (*Continued*)

Anthocyanins	Normal purple wheat by HPLC method[1]	Normal purple wheat by pH differential method	Heat-stressed purple wheat by HPLC method[1]	Heat-stressed purple wheat by pH differential method	Normal purple wheat by HPLC method	Normal blue wheat by HPLC method	Purple wheat (*Indigo*) differential method[2]	Purple wheat (*Konini*) differential method[2]
Cyanidin chloride	5.2 ± 0.1	6.8 ± 0.2	9.8 ± 0.0	9.2 ± 0.1	–	–	–	–
Pelargonidin 3-galactoside	26.1 ± 0.5	22.1 ± 0.7	25.6 ± 0.2	21.5 ± 0.1	–	–	–	–
Pelargonidin 3-arabinoside	9.3 ± 0.3	11.3 ± 0.3	10.3 ± 0.6	10.2 ± 0.5	–	–	–	–
Peonidin 3-arabinoside	28.4 ± 0.4	29.3 ± 0.8	27.7 ± 0.8	26.1 ± 0.2	–	–	–	–
Cyanidin succinylglucoside	–	–	–	–	1.1 ± 0.1	–	–	–
Peonidin malonylglucoside	–	–	–	–	0.6 ± 0.2	–	–	–
Cyanidin succinylglucoside	–	–	–	–	1.2 ± 0.1	–	–	–
Peonidin succinylglucoside	–	–	–	–	0.9 ± 0.1	–	–	–
Peonidin succinylglucoside	–	–	–	–	0.6 ± 0.1	–	–	–
Peonidin malonyl succinylglucoside	–	–	–	–	0.5 ± 0.1	–	–	–
Total	447.6 ± 1.0	460.7 ± 0.9	481.7 ± 0.3	499.1 ± 1.2	12.8	153.1	–	–
References	Hosseinian et al. (2008)	Hosseinian et al. (2008)	Hosseinian et al. (2008)	Hosseinian et al. (2008)	Abdel-Aal et al. (2006)	Abdel-Aal et al. (2006)	Yu and Beta (2015)	Yu and Beta (2015)

Reprinted (adapted) with permission from (Abdel-Aal et al., 2006; Hosseinian et al., 2008; Yu and Beta, 2015). Copyright (2006) ACS, (2008) Elsevier and (2015) MDPI (Creative Commons Attribution 4.0).

pelargonidin 3-*O*-galactoside, pelargonidin 3-*O*-arabinoside, peonidin 3-*O*-glucoside, peonidin 3-*O*-arabinoside and petunidin 3-*O*-glucoside. Cyanidin 3-*O*-glucoside was the predominant compound. The compounds Hosseinian et al. (2008) identified in purple wheat were more diverse, and their concentrations were also higher than their counterparts reported by Abdel-Aal et al. (2006).

Liu et al. (2010) also identified cyanidin-3-*O*-glucoside, cyanidin-3-*O*-rutinoside, delphinidin-3-*O*-glucoside, pelargonidin-3-*O*-glucoside, peonidin-3-*O*-glucoside and petunidin-3-*O*-rutinoside in purple wheat, with cyanidin-3-*O*-glucoside being the predominant compound. It may be inferred from these studies that genotype and environment may affect the anthocyanin composition of colored wheat. It is clear from the foregoing that the C-3 of the flavylium ion is the preferred position for glycosylation in anthocyanins. Figure 3 shows the basic flavylium structures, and samples of anthocyanins identified in colored wheat grains.

2.1.4 Proanthocyanidins

Proanthocyanidins are colorless oligomeric or polymeric compounds made up of two or more flavan-3-ol units as building blocks, joined together by covalent bonds. Upon acid hydrolysis, they yield anthocyanidins (McCallum, 1989). The building blocks are linked together covalently through C4→C8 bonds and occasionally C4→C6 bonds which result in branching of the polymer. As expected of flavan-3-ols, the A-ring usually has OH-groups at positions C5 and C7, while the B-ring has either two OH-groups at positions C3′ and C4′, resulting in procyanidins, or three OH-groups at positions C3′, C4′ and C5′ resulting in

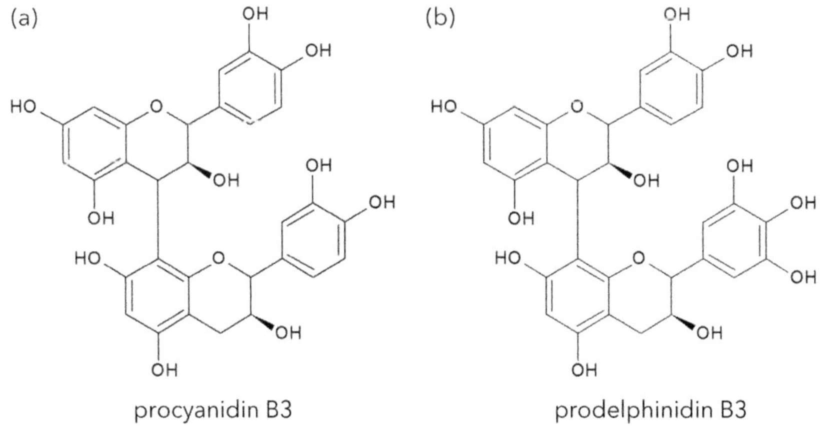

Figure 4 Proanthocyanidins identified in wheat grains. Reprinted (adapted) with permission from McCallum and Walker (1990). Not copyrightable: (1990) American Association of Cereal Chemists Inc.

prodelphinidins (McCallum, 1989). The proanthocyanidins identified in wheat bran include propelargonidin units, procyanidin B3 (catechin-[4α→8]-catechin) and prodelphinidin B3 (gallocatechin-[4α→8]-catechin) (McCallum and Walker, 1990) (Fig. 4). Due to the many OH-groups in each molecule, proanthocyanidins are highly susceptible to enzymatic or air oxidation leading to the formation of phlobaphenes which impart color to the bran of the grain (McCallum, 1989).

2.1.5 Alkylresorcinols

Alkylresorcinols are amphiphilic phenolic lipids comprising a polar (hydrophilic) head and a non-polar (hydrophobic) tail. The hydrophilic head is a 1,3-dibenzene moiety while the hydrophobic tail is usually an odd-numbered hydrocarbon

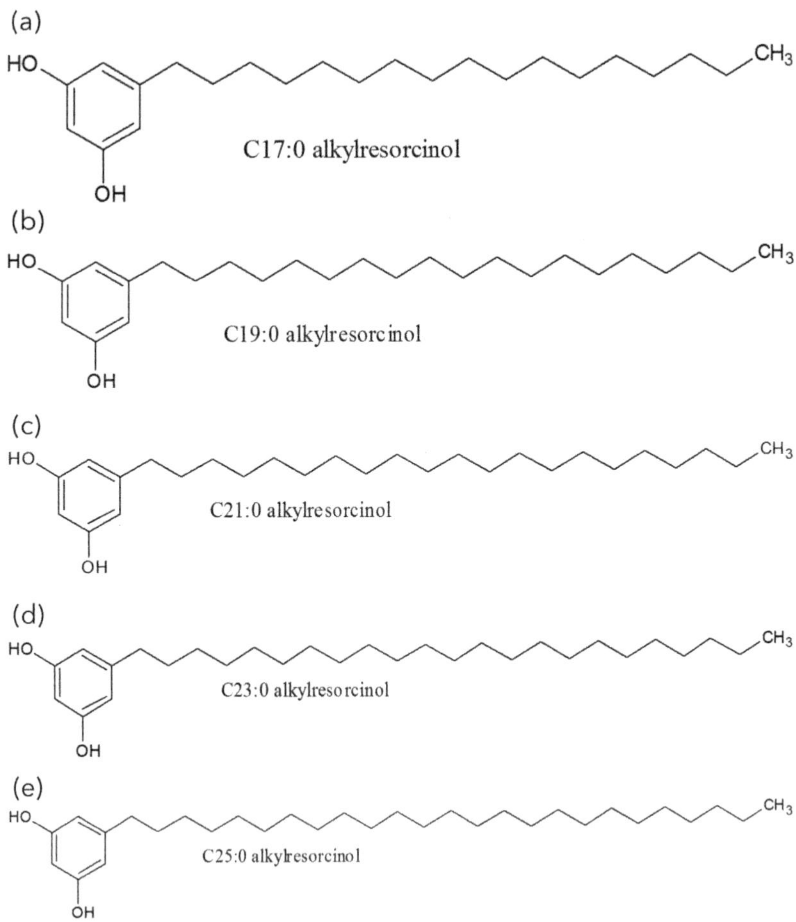

Figure 5 Alkylresorcinols identified in wheat grains. Reprinted (adapted) with permission from Kozubek and Tyman (1999). Copyright: (2007) American Chemical Society.

structure which may be saturated or unsaturated. This hydrocarbon tail is usually attached to position 5 of the 1,3-dibenzene ring (Piironen et al., 2009). The most common alkylresorcinol homologues in wheat have alkyl chain lengths C17:0, C19:0, C21:0, C23:0 and C25:0 (Fig. 5), and therefore are mostly saturated (Hengtrakul et al., 1991; Ross, 2012). Unsaturated alkenylresorcinols, such as C21:1, are rare and exist at very low concentrations in wheat (Hengtrakul et al., 1991).

Alkylresorcinols in wheat are concentrated in an intermediate layer of the caryopsis, comprising the hyaline layer, testa and inner pericarp of the kernel (Piironen et al., 2009). Specifically, they are located in the outer cuticle of testa or inner cuticle of the pericarp, but are absent from the endosperm and germ of wheat grains (Landberg et al., 2008). Alkylresorcinols are therefore considered as suitable markers for whole-grain wheat consumption (Landberg et al., 2008).

Alkylresorcinol composition and concentrations vary widely among different types of wheat. For example, Andersson et al. (2008) studied the alkylresorcinol composition and concentrations among 131 winter wheats, 20 spring wheats, 10 durum wheats, 5 spelt wheats and 10 early cultivated types (5 diploid einkorn and 5 tetraploid emmer). They reported the following concentration ranges (on dry weight basis) for the different wheat types: winter (220–652 μg/g); spring (254–537 μg/g); durum (194–531 μg/g); spelt (490–741 μg/g); einkorn (545–654 μg/g); emmer (531–714 μg/g). They also observed variation in the relative homologue composition among the different types of wheat, with a C17:0 to C21:0 ratio of 0.1 for winter, spring and spelt wheats, 0.04 for einkorn and emmer wheat and 0.01 for durum wheat. Relative homologue compositions reported by different laboratories for the same type of wheat, however, appear comparable (Table 3). This suggests that alkylresorcinols may not be good markers for distinguishing among the different types of wheat.

2.2 Carotenoids present in wheat

Carotenoids are a group of yellow, orange or red natural pigments found in wheat and other grains (Ndolo and Beta, 2013). Structurally, they are polyisoprenoid hydrocarbons comprising conjugated double bonds whose arrangements and number determine the amount of light they can absorb, giving rise to their distinctive colors (yellow, orange and red). At the end of their hydrocarbon chains, wheat carotenoids cyclize into ring structures. Depending on whether the cyclic molecules bear OH-groups or not, carotenoids may be classified as either carotenes (tetraterpenoid hydrocarbons with no OH-groups) or xanthophylls (they bear OH-groups) (Fig. 6) (Hidalgo et al., 2006).

Although carotenoids are lipophilic due to their hydrocarbon backbone, the presence of OH-groups in the ring structures of xanthophylls makes them

Table 3 Alkylresorcinols content (μg/g dry weight basis) and relative homologue composition (%) of different types of wheat

	Relative homologue composition, % (mean ± SD)								Total	
Wheat type	C17:0	C19:0	C19:1	C21:0	C21:1	C23:0	C25:0	Ratio of C17:0 to C21:0	μg/g dry basis	Reference
Common winter	5±1	38±3	-	46±3	-	9±1	3±1	0.10±0.02	410±93	Andersson et al. (2008)
Common wheat	4	33	-	50	-	10	3	0.08	-	Hengtrakul et al. (1991)
Spring	4±1	34±3	-	48±3	-	10±2	4±1	0.09±0.02	416±73	Andersson et al. (2008)
Durum	0.4±0.01	12±2	-	61±4	-	21±2	6±2	0.01±0.001	399±109	Andersson et al. (2008)
Durum	1	16	1	52	4	21	6	0.02	-	Hengtrakul et al. (1991)
Spelt	4±0.5	32±3	-	50±2	-	11±1	4±1	0.09±0.01	605±103	Andersson et al. (2008)
Einkorn	2±1	14±1	-	47±2	-	27±1	10±1	0.04±0.01	595±50	Andersson et al. (2008)
Emmer	2±1	16±8	-	53±2	-	21±6	7±1	0.05±0.02	581±76	Andersson et al. (2008)

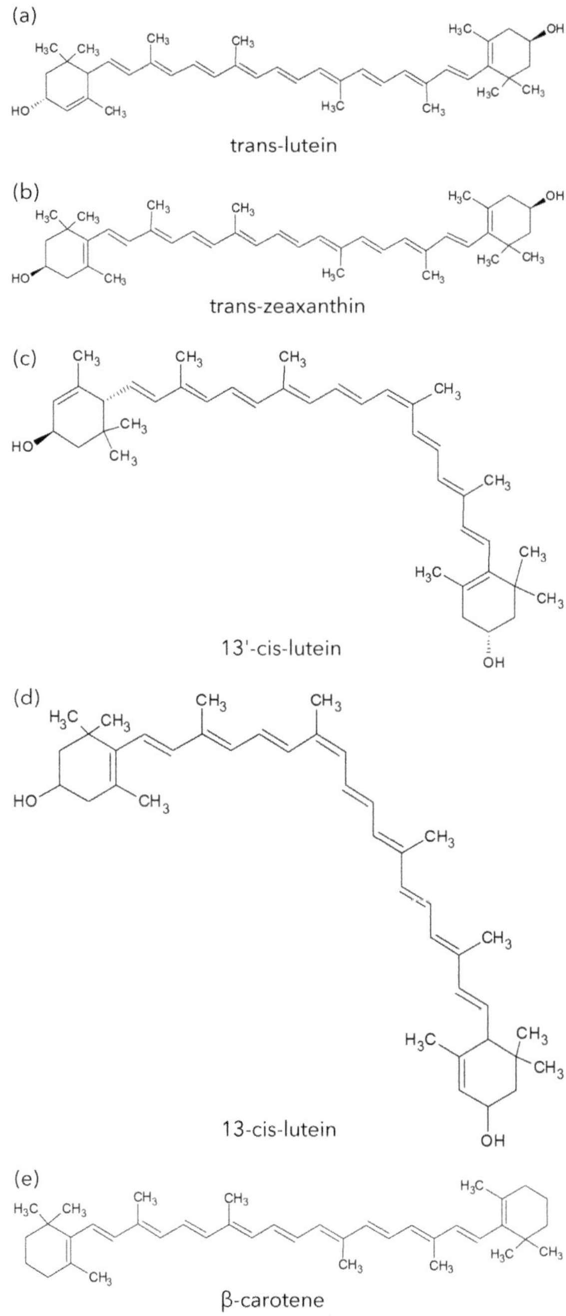

Figure 6 Carotenoids identified in wheat grains. Reprinted (adapted) with permission from Abdel-Aal et al. (2007). Copyright: (2007) American Chemical Society.

Table 4 Pooled means of carotenoids (μg/g) present in *Triticum monococcum, Triticum turgidum* and *Triticum aestivum* and their sources

Sample	All trans-lutein	All trans-zeaxanthin	15-cis-lutein	13-cis-lutein	13-cis-lutein	9-cis-lutein	9-cis-lutein	9-cis-zeaxanthin	All trans-β-carotene	Total	References
Triticum monococcum (Einkorn)	7.41 ± 0.19	0.94 ± 0.05	0.31 ± 0.02	0.37 ± 0.02	0.35 ± 0.03	nd	0.11 ± 0.01	nd	0.13 ± 0.02	9.62±2.40	Abdel-Aal et al. (2007)
Triticum monococcum (Einkorn)	7.69 ± 1.63	-	-	-	-	-	-	-	0.19±0.23	7.88±5.30	Hidalgo et al. (2006)
Triticum monococcum (Einkorn)	7.80±0.50	-	-	-	-	-	-	-	-	7.8±0.5	Abdel-Aal and Rabalski (2008)
Triticum turgidum	5.47 ± 0.08	0.60 ± 0.16	nd	0.12 ± 0.01	0.12 ± 0.02	nd	nd	nd	0.09 ± 0.00	6.40 ± 1.79	Abdel-Aal et al. (2007)
Triticum turgidum	4.50±0.60	-	-	-	-	-	-	-	-	4.5±0.6	Abdel-Aal and Rabalski (2008)
Triticum turgidum	3.00 ± 1.17	-	-	-	-	-	-	-	0.03±0.07	3.03±2.10	Hidalgo et al. (2006)
Triticum aestivum	1.00 ± 0.11	0.29 ± 0.06	-	-	-	-	-	-	0.16 ± 0.04	1.45 ± 0.03	Moore et al. (2005)
Triticum aestivum	1.89 ± 0.09	nd	nd	nd	nd	nd	nd	nd	0.05 ± 0.00	1.94 ± 0.63	Abdel-Aal et al. (2007)
Triticum aestivum	1.5±0.4	-	-	-	-	-	-	-	-	1.5±0.4	Abdel-Aal and Rabalski (2008)
Triticum aestivum	0.89±0.15	0.40±0.04	-	-	-	-	-	-	-	1.30±0.35	Ndolo and Beta (2013)
Triticum aestivum	1.90±0.47	-	-	-	-	-	-	-	0.11 ± 0.12	2.01 ± 1.27	Hidalgo et al. (2006)
Triticum aestivum (purple wheat)	0.60	0.54	-	-	-	-	-	-	-	1.14±0.04	Ndolo and Beta (2013)

Key: *Triticum monococcum* represents Einkorn. *Triticum turgidum* includes Emmer, Khorasan and Durum. *Triticum aestivum* includes spelt, soft wheat, hard wheat and common wheat.

slightly polar. The carotenoids identified in wheat are mostly xanthophylls, and they include trans-lutein, trans-zeaxanthin and their cis-isomers. Lutein is the predominant carotenoid in wheat, followed by trans-zeaxanthin and the cis-isomers (e.g. 15-cis-lutein, 13-cis-lutein, 9-cis-lutein, 9′-cis-lutein) (Abdel-Aal et al., 2007; Ndolo and Beta, 2013). All-trans-β-carotene is the only carotene reported in wheat, but is present at low concentrations (Abdel-Aal et al., 2007). Table 4 shows carotenoid contents of some wheat cultivars.

2.3 Tocopherols and tocotrienols

Tocopherols and tocotrienols (collectively called tocochromanols or tocols) are a group of phytochemicals that constitute the vitamin E family. Structurally, they are composed of a chroman head and a phytyl tail. The chroman head consists of a phenol ring joined to an O-containing 6-membered heterocyclic ring while the phytyl tail is a C-13 hydrocarbon chain bearing three methyl substituents. Tocols exist in four isomeric forms: alpha (α), beta (β), gamma (γ) and delta (δ). These isomers differ from each other based on the number and positions of attachment of methyl substituents on the chroman head. Tocopherols differ from tocotrienols in that, while tocopherols have a saturated phytyl chain, tocotrienols have polyunsaturated phytyl chain bearing three double bonds (Fig. 7) (Kamal-Eldin and Appelqvist, 1996).

Wheat grains contain all the isomeric forms of tocols. However, their composition and content vary among genotypes and are also affected by the environment. Hidalgo et al. (2006) compared the composition and contents of tocols of 54 domesticated Einkorn wheat (*Triticum monococcum* ssp. *monococcum*) accessions with 15 non-Einkorn wheat cultivars (comprising eight *Triticum turgidum* and seven *Triticum aestivum*). They observed that the Einkorn wheat had higher total tocols content than the non-Einkorn wheat (*T. turgidum* and *Triticum aestivum*) cultivars. The trend for the mean values of the total tocols content was: Einkorn wheat (*Triticum monococcum*; 61.5–115.2 μg/g dry matter) > *Triticum aestivum* (53.2–75.0 μg/g dry matter) > *Triticum turgidum* (38.9–67.9 μg/g dry matter) (Hidalgo et al., 2006). Among the four tocols identified (α-tocopherol, β-tocopherol, α-tocotrienol and β-tocotrienol), β-tocotrienol (18.7–66.8 μg/g dry matter) was the most abundant for all the species studied. Among the Einkorn wheat types, the trend for the tocols was: β-tocotrienol > α-tocotrienol ≈ α-tocopherol > β-tocopherol. Among *T. turgidum* cultivars, the tocols concentration were in the order: β-tocotrienol > α-tocopherol > α-tocotrienol ≈ β-tocopherol while for the *T. aestivum* cutlivars, the trend was: β-tocotrienol > α-tocopherol > β-tocopherol > α-tocotrienol (Hidalgo et al., 2006). Table 5 shows the content of tocochromanols of some wheat cultivars.

Moore et al. (2005), on the other hand, identified α-tocopherol (3.4–10.1 μg/g) as the only tocol in soft wheat, while Zhou et al. (2005) identified

(a)

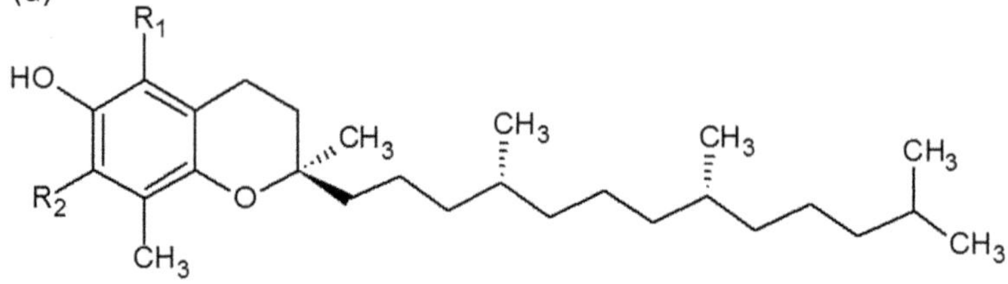

Name	R_1	R_2
α-tocopherol	CH_3	CH_3
β- tocopherol	CH_3	H
γ- tocopherol	H	CH_3
δ- tocopherol	H	H

(b)

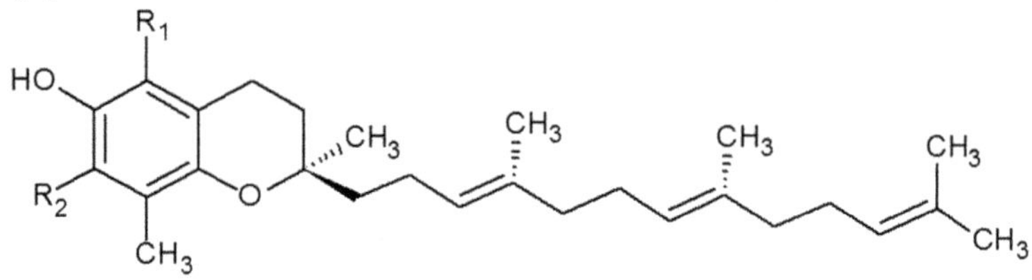

Name	R_1	R_2
α-tocotrienol	CH_3	CH_3
β- tocotrienol	CH_3	H
γ- tocotrienol	H	CH_3
δ- tocotrienol	H	H

Figure 7 Tocochromanols identified in wheat grains: (a) tocophenols; (b) tocotrienols. Reprinted (adapted) with permission from Colombo (2010). Copyright: (2010) MDPI, Creative Common Attribution 4.0.

α-tocopherol, γ-tocopherol and δ-tocopherol in the bran of hard red winter wheat, with α-tocopherol being the most abundant. Therefore, depending on the sample type and geographical location, different tocols may be present as dominant compounds in wheat grains. It is important to note that the tocol content and composition of wheat cultivars are significantly affected by genetic and environmental factors (Lampi et al., 2010) as well their interaction (Labuschagne et al., 2017). Also, some wheat genotypes are more sensitive to environmental effects than others (Lampi et al., 2010). These are important

Table 5 Pooled means of tocochromanols (μg/g) present in *Triticum monococcum, Triticum turgidum* and *Triticum aestivum* and their sources

Sample	α-T	β-T	α-T3	β-T3	δ-T3	T3/T	Total tocols	References
Triticum monococcum (Einkorn)	7.1±0.2	2.2±0.2	6.3±1.6	18.8±1.8	-	2.7	34.4±7.1	Abdel-Aal and Rabalski (2008)
Triticum monococcum (Einkorn)	9.1±2.0	2.8±1.2	9.8±1.8	35.3±9.5	-	3.8	57.0±14.4	Lampi et al. (2008)
Triticum monococcum (Einkorn)	12.2±0.8	5.0±1.1	12.8±2.4	48.9±6.3	-	3.6	78.8±19.8	Hidalgo et al. (2006)
Triticum turgidum	6.7±1.0	3.4±1.7	2.9±0.6	15.0±6.5	-	1.8	28.0±5.6	Abdel-Aal and Rabalski (2008)
Triticum turgidum	9.2±2.1	3.9±1.1	5.4±1.0	23.8±4.1	-	2.2	42.3±9.1	Lampi et al. (2008)
Triticum aestivum	9.0±1.5	4.5±0.9	3.4±0.5	14.6±2.0	-	1.3	31.5±5.1	Abdel-Aal and Rabalski (2008)
Triticum aestivum (bread wheat)	14.4±1.7	6.4±1.3	6.0±0.9	31.1±7.7	-	1.8	57.9±11.7	Lampi et al. (2010)
Triticum aestivum (bread wheat)	12.8	6.3±0.1	4.7±0.1	24.8±0.6	-	1.6	48.6±9.1	Lampi et al. (2008)
Triticum aestivum (whole flour)	13.3	6.4	4.9	24.0	0.6	2.8	49.4	Labuschagne et al. (2017)
Triticum aestivum (white flour)	3.3	2.0	1.2	12.7	0.5	1.5	19.7	Labuschagne et al. (2017)
Triticum aestivum (white flour)	0.8±0.3	0.9±0.3	0.3±0.1	6.0±1.7	-	3.7	8.0±2.4	Labuschagne et al. (2014)

Key: T - tocopherol; T3 - tocotrienol.

points to consider in determining the nutritional value of tocols as wheat antioxidants.

Tocols are important constituents of wheat since they are vitamers (isomeric forms) of the fat-soluble vitamin E. Vitamin E has very important biological properties including antioxidant capacity. Although α-tocopherol is the most credited with vitamin E activity, all the other vitamers have potent antioxidant and other biological activities (Peh et al., 2016). For example, tocotrienols reportedly have neuroprotective, anticancer, cholesterol-lowering and cardio-protective functions (Sen et al., 2007).

3 Extraction of phytochemical antioxidants in wheat

3.1 Extraction of phenolic antioxidants

Methods for extracting phenolic compounds from wheat grains are based on the pioneering work of Krygier et al. (1982), who extracted soluble-free, soluble-esterified and insoluble-bound phenolic acids from rapeseed, using a combination of methanol, acetone and water (70% aqueous methanol-70% aqueous acetone, 1:1) as solvents. Liyana-Pathirana and Shahidi (2005) performed experiments to optimize the three conditions of solvent system (v/v; water with methanol, ethanol or acetone), temperature (15–95°C) and duration (15–105 min) for soluble-free phenolic extraction from wheat, using total antioxidant activity as a measurable indicator. They found aqueous ethanol to be the most effective solvent system for extracting phenolic compounds from whole grain and bran of both soft and hard wheat. Consequently, they reported the optimum phenolic extraction conditions from whole grain and bran of soft wheat and whole grain and bran of hard wheat as: 54%, 61°C, 64 min; 49%, 64°C, 60 min; 54%, 61°C, 65 min and 50%, 62°C, 60 min, respectively (Liyana-Pathirana and Shahidi, 2005).

Other phenolic extraction conditions have been reported. For example, phenolic acids have been extracted with 80% (v/v) aqueous ethanol (Li et al., 2008) or methanol (Yu and Beta, 2015) due to their polar nature, since they bear a carboxylic acid functional group and a phenol ring. The crude phenolic extract contains the soluble-free and soluble-esterified phenolic acids. After concentrating the crude phenolic extract, it is acidified to pH 2 using 6 M HCl, followed by centrifugation to separate the soluble-free phenolic acids from the soluble-esterified phenolic acids (Krygier et al., 1982). A mixture of equal volumes of ethyl acetate-diethyl ether is used to extract the soluble-free phenolic acids from the supernatant while the cloudy precipitate is alkaline (4 M NaOH) hydrolyzed, followed by acidification and extraction with ethyl acetate-diethyl ether to obtain the soluble-esterified phenolic acids. Hexane is used to remove fatty acids and other lipid impurities from the soluble-free and esterified phenolic acids (Krygier et al., 1982).

The residue obtained from crude phenolic extraction contains the insoluble-bound phenolic acids which are covalently bonded to cell wall components, such as arabinoxylans. The insoluble-bound phenolic acids are first released by alkaline hydrolysis using 4 M NaOH under nitrogen gas to prevent oxidation (Li et al., 2008). The sodium phenolate salts so formed are then acidified to pH 2.0 with 6 M HCl to release the free phenolic acids, which are then extracted from the aqueous solution using solvents such as ethyl acetate (Liu et al., 2010) or a mixture of equal volumes of ethyl acetate and diethyl ether (Lu et al., 2014; Moore et al., 2005). The extract is then evaporated to dryness and re-dissolved in aqueous methanol for analysis (Yu and Beta, 2015).

Flavonoids and proanthocyanidins in grains are usually extracted with acetone/water (70:30 v/v) (Liu et al., 2010) or acetone/water/acetic acid (70:29:1 v/v) (Qiu et al., 2009). This is because they are less polar than phenolic acids due to their larger polyphenol structure and lack of carboxylic acid functional group. Extraction of anthocyanins is done using methanol that is acidified with 1 M HCl at a ratio of 85:15 v/v (methanol-HCl, pH < 2) (Liu et al., 2010; Yu and Beta, 2015). Acidification of anthocyanins during extraction stabilizes the flavylium cationic structure (Fig. 3), which is more extractable in highly acidic medium (Liu et al., 2010).

Alkylresorcinols can be extracted using acetone, ethyl acetate or methanol depending on the method of analysis (Ross et al., 2004). For gas chromatographic analysis, for example, ethyl acetate will be preferred due to its volatile nature. In such an instance, a known weight of the powdered sample will be extracted with a predetermined volume of the solvent for a predetermined period of time. The extract will then be evaporated to dryness and re-dissolved in a much smaller volume of the solvent such as 200 μL for analysis (Andersson et al., 2008; Ross et al., 2001).

3.2 Extraction of carotenoid antioxidants

A number of extraction solvents have been reported for wheat carotenoids, including hexane (Leenhardt et al., 2006), a mixture of equal volumes of methanol and tetrahydrofuran (Burkhardt and Böhm, 2007; Hentschel et al., 2002; Moore et al., 2005; Zhou et al., 2005), water-saturated 1-butanol (Abdel-Aal and Rabalski, 2008; Hidalgo et al., 2014; Ndolo and Beta, 2013; Trono et al., 1999) and aqueous ethanol (Werner and Böhm, 2011). The choice of solvent system is important since it determines which compounds will be extracted. While the predominant carotenoids in wheat are xanthophylls which bear an OH-group on the cyclic molecule, they also have small amounts of carotenes. The extraction methods therefore explore solvent polarities that can extract these two classes of compounds.

Abdel-Aal et al. (2007) compared the extraction efficiency of 80% aqueous ethanol, 80% aqueous methanol, water-saturated butanol, methyl-*t*-butyl ether and tetrahydrofuran for wheat carotenoids. They found the extraction efficiency for lutein (the predominant carotenoid in wheat) to be highest when extracted using 80% aqueous ethanol or water-saturated 1-butanol (Abdel-Aal et al., 2007). Similarly, Hidalgo et al. (2006) compared the extraction efficiency of three solvent systems: water-saturated 1-butanol, tetrahydrofuran and hot saponification with KOH followed by hexane-ethyl acetate (9:1, v/v). They also found water-saturated 1-butanol to be a more efficient solvent for extracting lutein, compared to the other two methods.

Carotenoid extraction is a rather simple process. A measured volume of the extraction solvent is added to a known weight of the finely divided sample and agitated under dim light or in the dark to prevent photo-oxidation. A chemical antioxidant such as butylated hydroxytoluene (BHT) is also added while atmospheric oxygen is removed by covering the extraction mixture with nitrogen gas, all to prevent oxidation. Internal standards such as β-apo-8′-carotenal may also be added to ensure quantitative extraction (Hentschel et al., 2002; Werner and Böhm, 2011). After extraction, the supernatant is removed by filtration or centrifugation, and the residue may be re-extracted using the same solvent a few more times. The pooled supernatants are evaporated under vacuum at low temperature (30–45°C) and dried under nitrogen (N_2) gas. The dried extract is then re-dissolved in a small amount of a solvent of choice, such as the extraction solvent or mobile phase for chromatographic (HPLC) analysis, and the compounds analyzed (Abdel-Aal et al., 2007; Abdel-Aal and Rabalski, 2008; Burkhardt and Böhm, 2007; Hentschel et al., 2002; Hidalgo et al., 2014; Moore et al., 2005; Ndolo and Beta, 2013; Zhou et al., 2005).

3.3 Extraction of tocochromanol antioxidants

Tocols are extracted by first performing hot saponification which releases the tocols that are tightly bound to matrix components, followed by either solvent extraction or supercritical fluid (CO_2) extraction (SFE). Without the hot saponification step, only soluble-free and soluble-esterified tocols will be extracted (Fratianni et al., 2002). During the hot saponification, a known weight of the powdered wheat sample is dispersed in ethanol and reacted with potassium hydroxide and aqueous sodium chloride, in the presence of an antioxidant such as pyrogallol or BHT, and under a stream of N_2 gas to prevent oxidation. The reaction occurs in the dark or under subdued light to prevent photo-oxidation, and at elevated temperatures typically between 70°C and 100°C, with periodic agitation to facilitate mixing of the reactants.

After saponification, the released tocols are separated from the residue by either filtration or centrifugation, and extracted from the aqueous mixture using an appropriate solvent system such as hexane-ethyl acetate (9:1, v/v) (Fratianni et al., 2002; Hidalgo and Brandolini, 2010), heptane-ethyl acetate (8:2, v/v) (Lampi et al., 2008) or stepwise extraction with ethanol, methyl-*t*-butyl ether (MTBE) and petroleum ether (Werner and Böhm, 2011). The extract is then evaporated to dryness and its residue re-dissolved in a smaller volume of a non-polar solvent such as 1% (v/v) isopropanol in hexane (Fratianni et al., 2002; Panfili et al., 2003; Panfili et al., 1994), 2% (v/v) MTBE in hexane (Werner and Böhm, 2011) or heptane (Lampi et al., 2008) for analysis. It is worth noting that, just like for carotenoid extraction, an internal standard such as α-tocopheryl acetate may be used here as well, to estimate losses during extraction (Werner and Böhm, 2011).

Supercritical CO_2 extraction requires specialized equipment with pumps and pressure vessels. Extraction is usually achieved above the critical temperature of 31°C and critical pressure of 74 Bar. Fratianni et al. (2002) observed that when supercritical CO_2 was compared to conventional methods for extracting tocols from cereals, the supercritical CO_2 did not show any significant selectivity for any tocol. The extraction efficiency was also comparable to conventional methods such as Soxhlet and Folch methods. However, supercritical CO_2 provided the advantage of requiring less solvent and time for extraction (Fratianni et al., 2002).

4 Methods of analyses for phytochemical antioxidants in wheat grains

4.1 Analysis for total phenolic content (TPC)

The total phenolic content of wheat is measured using the Folin Ciocalteu method, which is a colorimetric method first described by Singleton and Rossi (1965) as a replacement to the Folin-Denis method for determining total phenols. This method involves electron transfer from the phenolic antioxidants to sodium phosphomolybdate/sodium phosphotungstate complex in an alkaline medium (sodium carbonate). This results in a color change of the phosphomolybdate/phosphotungstate reagent from yellow to dark blue. The extent of color change is directly proportional to the amount of phenolic compounds donating electrons, and this can be measured spectrophotometrically at 765 nm to determine the total phenolic content (Singleton et al., 1999). While this assay conventionally used the cuvette-based UV-visible spectrophotometer, it has been adapted for use with a 96-well microtiter plate reader leading to high throughput analysis (Ainsworth and Gillespie, 2007).

The Folin Ciocalteu method is based on electron transfer method. Several analytes such as ascorbic acid, tyrosine-containing peptides or proteins and reducing sugars can similarly react with the Folin Ciocalteu reagent, leading to overestimation (Singleton et al., 1999; Waterhouse, 2002). Due to this specificity of the method for phenolic compounds, it may be considered as a measure of the total reducing capacity of the sample being analyzed (Ainsworth and Gillespie, 2007). The accuracy of phenolic content estimation can be significantly enhanced using an effective phenolic extraction procedure that eliminates approximately 85–95% of potentially interfering compounds. Folin Ciocalteu assay is a popular method for determining the total phenolic content of wheat and its products (Liu et al., 2010; Yu and Beta, 2015). It is important to note that in determining total phenolic content for wheat, ferulic acid is used as a standard but not gallic acid, since ferulic acid is the predominant phenolic compound in wheat.

4.2 Analysis of total anthocyanins content (TAC)

Anthocyanins are present mostly in pigmented wheat grains. The two methods generally used to estimate total anthocyanin content of pigmented wheat grains are: (1) the pH differential method and (2) the direct method. The pH differential method is based on anthocyanins undergoing reversible structural transformations under different pH conditions, leading to changes in color. The flavylium form of anthocyanins dominates at pH 1.0 with the color ranging from orange to purple, while the hemiketal form which is colorless dominates at pH 4.5 (Basílio and Pina, 2016; Giusti and Wrolstad, 2001). These changes in color can be monitored spectrophotometrically to estimate the total anthocyanin content of pigmented wheat grains.

For example, a known volume of pigmented wheat extract is dissolved in two buffer solutions maintained at pH 1.0 (0.03 M potassium chloride buffer) and pH 4.5 (0.4 M sodium acetate buffer). Their corresponding absorbances are then measured with a spectrophotometer at the wavelength for maximum absorption (λmax) of the predominant anthocyanin in the sample. The total anthocyanin content of the wheat extract can be estimated by difference using the Beer-Lambert's law, provided that the absorbances of the two solutions lie within the linear range. It is therefore important to use sufficiently diluted extracts for the assay. The pH-differential method can be used to rapidly and accurately estimate the total anthocyanin content of wheat extracts, even in the presence of polymerized degraded pigments and other interfering compounds. This is because absorbance is read both at the λmax of the predominant anthocyanin (520 nm for cyanidin 3-*O*-glucoside) and at 700 nm to correct for haze formation or interfering substances in the extracts (Yu and Beta, 2015).

The total anthocyanin content (mg/L) of pigmented wheat is calculated as:

$$\frac{A \times MW \times DF \times 10^3}{\varepsilon \times L}$$

Where A (Absorbance) = $(A_{520nm} - A_{700nm})_{pH\ 1.0} - (A_{520nm} - A_{700nm})_{pH\ 4.5}$; MW = molecular weight of cyanidin 3-*O*-glucoside (449.2 g/mol); DF = dilution factor; L = path length (1 cm); ε = molar extinction coefficient for cyanidin 3-*O*-glucoside = 26,900 $L.cm^{-1}.mol^{-1}$ (Yu and Beta, 2015).

For the direct method, however, absorbance of the extract is measured in acidified methanol (1 M HCl in methanol, 15:85, v/v, pH 1.0) at the λmax of the predominant anthocyanin, and the total anthocyanin content is calculated as follows:

$$\frac{A \times MW \times \text{volume} \times 10^2}{\varepsilon \times \text{sample weight}}.$$

Total anthocyanin contents of pigmented wheat measured by the two methods appear comparable (Yu and Beta, 2015).

4.3 Analysis of total flavonoids content (TFC)

Three colorimetric methods commonly used to determine total flavonoid content of samples are the: (1) aluminium chloride ($AlCl_3$) method, (2) 2,4-dinitrophenylhydrazine (DNPH) method and (3) ceric sulphate or cerium (IV) sulphate [$Ce(SO_4)_2$] method. Flavonoids form colored complexes with metal ions such as Al^{3+} (Malešev and Kuntić, 2007) as well as DNPH (Chang et al., 2002). The absorbance of the colored complexes can be measured spectrophotometrically and used to estimate the total flavonoid content of a sample. Among the different flavonoid classes, flavones and flavonols form stable complexes with $AlCl_3$, while flavanones and flavanonols form stable complexes with dinitrophenyl hydrazine (DNPH) (Chang et al., 2002).

Two versions of the $AlCl_3$ method are commonly used. In the first method, appropriately diluted flavonoid extracts are reacted with 10% (w/v) aqueous $AlCl_3$ and 1 M aqueous potassium acetate (CH_3COOK), and diluted with distilled deionized water (ddH_2O). The reaction mixture is incubated at ambient temperature for 30 min and the absorbance is measured at 415 nm against a blank solution in which distilled deionized water (ddH_2O) replaces the 10% $AlCl_3$. A calibration curve, plotted using the predominant flavonoid in the sample, is then used to estimate the total flavonoid content (TFC) (Chang et al., 2002).

In the second method, the appropriately diluted extracts are added to 5% (w/v) aqueous sodium nitrite ($NaNO_2$) and vortexed to mix. Freshly prepared 10% (w/v) aqueous $AlCl_3$ is added and the mixture is left to react for 6 min. Finally, 1 M aqueous sodium hydroxide (NaOH) is added, followed by dilution with ddH_2O and the mixture is vortexed again. The reaction mixture can be centrifuged to remove cloudy precipitate. Absorbance of the reaction mixture is measured at 510 nm against a reagent blank and the TFC is estimated from a calibration curve of the predominant flavonoid which is used as a standard (Herald et al., 2012). This second method has been adapted to a 96-well microplate format (Herald et al., 2012).

In the DNPH method, the diluted extracts are added to 1% DNPH reagent and methanol at 50°C for 50 min. The reaction mixture is cooled to ambient temperature, mixed with 1% (w/v) potassium hydroxide (KOH) in 70% methanol and incubated at ambient temperature for 2 min. An aliquot of the reaction mixture is diluted with methanol, vortexed and centrifuged to remove precipitates. The supernatant may be diluted further and the absorbance measured at 495 nm. The predominant flavanone or flavanonol is used as a standard for estimating the TFC (Chang et al., 2002).

The $Ce(SO_4)_2$ method involves reacting an aliquot of the extract with 25 mM $Ce(SO_4)_2$ solution and diluting with ddH_2O. The diluted mixture is vortexed, allowed to react for 15 min at ambient temperature, and the absorbance measured at 320 nm using ddH_2O as blank. The TFC is estimated from a calibration curve plotted using the predominant flavonoid (Hassan et al., 2013).

4.4 Analysis of total condensed tannins content (TCTC)

The vanillin-HCl method is the most widely used method to estimate total condensed tannin (proanthocyanidin) content of cereals including wheat. The method of Herald et al. (2014) is described here. In the assay, two aliquots of the extract representing sample test and sample control are incubated in a water bath at 30°C for 5 min. Freshly prepared working vanillin reagent (1% vanillin + 8% HCl, 1:1 v/v) is added at 1 min intervals to the sample test while 4% HCl is added to the sample control. The reaction mixtures are incubated for 20 min at 30°C and the absorbance measured at 500 nm against methanol as blank. Since color development is time dependent, the 1 min intervals between measurements must be strictly adhered to, for accuracy and precision to be obtained. Net absorbance (absorbance of sample test minus absorbance of sample control) is used to estimate total tannin content from a standard curve plotted using catechin (Herald et al., 2014). The method has been adapted for measurement using the 96-well microplate reader (Herald et al., 2014).

4.5 Analysis of total carotenoid content (TCC)

Total carotenoid content of wheat carotenoid extracts is estimated by measuring the absorbance of the diluted extract at 450 nm. The Beer-Lambert's law is applied to compute the concentration using lutein as a standard. Alternatively, the concentration can be estimated from a calibration curve of lutein since it is the predominant carotenoid in wheat (Ndolo and Beta, 2013).

4.6 Analysis of phenolic compounds in wheat grains

Liquid chromatography-mass spectrometry (LCMS) and gas chromatography-mass spectrometry (GCMS) are powerful analytical tools that can be used to identify and quantify constituent phenolic compounds in wheat grains. Since phenolic compounds are non-volatile, polar organic compounds, they are more easily analyzed using LCMS than GCMS. GCMS can be used for their analysis only when the phenolic compounds are derivatized into volatile compounds. To analyze by liquid chromatography, the grains are milled into flour and the compounds are extracted using organic solvents as described in the previous section. The extracts are filtered through micron-pore size filters (0.45 μm or 0.2 μm) and then injected onto a chromatographic column where the constituent compounds are separated by gradient elution using at least two mobile phase systems (A and B). Mobile phase A is usually acidified ultrapure water while mobile phase B is a polar organic solvent (usually methanol or acetonitrile) which may or may not be acidified.

The concentration of the acid (formic acid or acetic acid) added varies depending on the method used, but typically ranges between 0.1 and 4.5% v/v (Hosseinian et al., 2008; Liu et al., 2010; Yu and Beta, 2015). Higher acid concentrations are mostly used for anthocyanin analysis in order to maintain them in the flavylium form, while phenolic acids and flavonoids require lower acid concentrations (Hosseinian et al., 2008; Liu et al., 2010; Yu and Beta, 2015). Reversed-phase columns packed with non-polar material are used to separate phenolic compounds. Therefore, the highly polar compounds elute first, while the less polar compounds elute later. The elution gradient program, flow rate and column temperature usually depend on the type of column used.

In our laboratory, a Waters 2695 HPLC system equipped with a photodiode array detector (PDA) (Waters 2996), an autosampler (Waters 717 Plus), and a quaternary pump, and coupled to a quadrupole-time of flight mass spectrometer (Q-Tof-MS) (Waters *Micromass Micro*), has been used to determine phenolic composition of different types of wheat grains (Hosseinian et al., 2008; Liu et al., 2010; Ndolo and Beta, 2014; Yu and Beta, 2015). The PDA is used to monitor the UV-visible spectral characteristics of the compounds, while the Q-Tof-MS is used to determine their masses. Hydroxybenzoic acids are usually monitored at 265–280 nm while hydroxycinnamic acids and most flavonoids are monitored at 320–350 nm. Since anthocyanins are pigments, they are usually monitored within the visible spectral range (450–550 nm) (Hosseinian et al., 2008; Ndolo and Beta, 2014).

Gas chromatography, on the other hand, requires analytes to be in the gas phase since the eluent is a gas (helium or nitrogen gas). The phenolic compounds are therefore derivatized into volatile compounds before they are injected onto the column to be analyzed. This limits gas chromatographic applications for phenolic analysis, especially when the compounds of interest are heat labile. This is because exposure of the compounds to high temperatures may degrade them. One of the most preferred derivatization methods for phenolic acids and flavonoids such as (+)-catechin is silylation which produces thermostable trimethylsilyl phenolic derivatives for GCMS analysis (Proestos et al., 2006). It is worth noting that alkylresorcinols, which are phenolic lipids, are mostly analyzed by gas chromatography (Andersson et al., 2008; Landberg et al., 2008; Ross et al., 2001).

In LCMS and GCMS, identification is achieved by comparing the retention times, UV-visible spectra and mass spectra of the analyte compounds with those of authentic standards, as well as data reported in the literature for compounds for which standards are unavailable. GCMS data library may also be acquired to assist in identification. Quantification is usually done by comparing peak areas of chromatographic peaks of analytes with those of authentic standards that are used to plot calibration curves (Yu and Beta, 2015). In GCMS analysis, compounds may also be quantified as percent relative abundance.

4.7 Analysis of carotenoids in wheat grains

Carotenoids in wheat grains are analyzed mostly by liquid chromatography as previously explained for phenolic compounds. The grains are milled into flour and the carotenoids are extracted as previously described. The extract is filtered through a nylon filter (0.45 μm or 0.2 μm) and injected onto the chromatographic column where separation occurs when the mobile phase is applied. In our lab (Ndolo and Beta, 2013), HPLC was used to separate wheat carotenoids on a YMC carotenoid S-3 column (4.6 × 100 mm, 3 μm) (Waters, Milford, MA, USA). The mobile phase was a binary solvent system (A and B). Solvent A comprised methanol/methyl-*t*-butyl ether (MTBE)/water (81:15:4 v/v/v) and solvent B comprised methanol/MTBE (90:10 v/v). Gradient elution was used at a flow rate of 1.0 mL/min, and separation was completed in 15 min. The compounds were monitored at a wavelength of 450 nm (Abdel-Aal et al., 2007; Ndolo and Beta, 2013).

4.8 Analysis of tocochromanols in wheat grains

Tocols from wheat can be identified and quantified using normal-phase liquid chromatography. As an example, a method reported by Labuschagne et al. (2017) used a mobile phase comprising *n*-hexane/ethyl acetate/acetic acid (97.3:1.8:0.9 v/v/v) pumped isocratically at a flow rate of 1.6 mL/min, to separate wheat tocols on a Phenomenex Luna Silica column (250 mm × 4.6 mm, 5 μm). A fluorescence detector was used to monitor the separated compounds at excitation wavelength of 290 nm and emission wavelength of 330 nm. Separation of tocols is usually achieved within 20 min (Schwartz et al., 2008). Authentic standards were used to plot calibration curves from which the identified compounds were quantified.

5 Nutritional value of phytochemical antioxidants in wheat

5.1 Effect of phenolic compounds on nutritional value of wheat foods

Proanthocyanidins are among the most potent antioxidants known in wheat. This is because of the number of OH-groups on the molecules that can release hydrogen atoms or protons to react with and quench reactive-free radicals and other oxidants. Proanthocyanidins are known to affect the nutritional value of foods in the following two ways.

5.1.1 Chelation of metal ions

Phenolic compounds such as phenolic acids, flavonoids, anthocyanins and proanthocyanidins chelate or complex with nutritionally important metal ions such as iron (Fe^{3+}) (Chobot and Hadacek, 2010), copper (Cu^{2+}) and zinc (Zn^{2+}) (Hosseinian et al., 2008), that are present in foods, making them bio-unavailable for absorption when consumed together (Santos-Buelga and Scalbert, 2000). These cations are micronutrients of immense physiological importance. For example, Fe is an important micronutrient used by the body to produce hemoglobin, a protein found in red blood cells, which transports oxygen from the lungs to the tissues. Deficiency of Fe resulting from binding with phenolic compounds can lead to Fe-deficiency anemia (Abbaspour et al., 2014).

Cu^{2+} on the other hand is an important trace element that helps in maintaining the strength and integrity of the skin barrier, blood vessels, epithelial and connective tissue within the body. It also helps in the production of hemoglobin, myelin and melanin, and helps in the normal functioning of the thyroid gland (Osredkar and Sustar, 2011). Therefore binding of phenolic compounds to Cu^{2+} causing their deficiency can lead to consequent negative effects on human health. Zinc serves as an important component of many enzymes and protein. It is required for normal growth and development in humans. At the cellular level, zinc is essential for proliferation, differentiation and apoptosis. It also helps in immunity, intermediary metabolism, DNA metabolism and repair, reproduction, vision, taste, and cognition and behavior, as well as nerve function (Maret and Sandstead, 2006). Zinc deficiency due to its binding to phenolic compounds will be detrimental to health.

The inhibition of absorption of metal ions due to phenolic binding depends on factors such as the type (phenolic acid, flavonoid or proanthocyanidin) and source (e.g. from wheat or black tea) of the phenolic compound, concentration of the phenolic compound, nature of food (single meal vs. multi-meal) and presence of other inhibitors (e.g. phytate) or enhancers (e.g. ascorbic acid) of absorption of the micronutrient (Hurrell and Egli, 2010; Santos-Buelga and Scalbert, 2000). Phenolic acids with catechol moiety (ortho-dihydroxy group) such as caffeic acid or galloyl moiety (adjacent trihydroxy group) such as gallic acid have better cation-chelating ability than phenolic acids with no catechol or galloyl moiety such as ferulic, vanillic or syringic acids (Andjelković et al., 2006; Iwasaki et al., 2011). Hydroxycinnamic acids chelate with cations better than hydroxybenzoic acids (Andjelković et al., 2006). The presence of buffers and sequestrants such as EDTA reduces the metal-chelating ability of phenolic acids.

It is noteworthy that although phenolic compounds function as antioxidants by scavenging reactive oxygen species that could otherwise cause oxidative damage to physiological molecules, in the presence of Fe^{2+} or Cu^{2+}, the phenolic compounds can form phenolic-metal ion complexes that exert pro-oxidant effect (Andjelković et al., 2006; Iwasaki et al., 2011). This calls for caution when introducing exogenous sources of phenolic antioxidants into food systems that are high sources of iron, copper and other mineral elements or cations.

5.1.2 Binding of proteins

One of the most important physiological properties of phenolic compounds such as phenolic acids, flavonoids and proanthocyanins is their ability to interact with proteins and precipitate them (Rawel et al., 2002; Rawel and Rohn, 2010). The proteins may include nutritionally relevant proteins, enzymes, hormones and toxins. Phenolic acids, particularly the hydroxycinnamic acids (caffeic acid, ferulic acid and sinapic acid), have demonstrated the ability to interact both covalently and non-covalently with proteins (Rawel et al., 2002). The non-covalent interactions include hydrogen bonding, hydrophobic interactions and ionic interactions (Rawel et al., 2005; Rawel and Rohn, 2010). The nature of phenolic-protein interaction depends on the protein structure (i.e., amino acid sequence of the protein and its molecular conformation), pH, temperature and ionic strength of the medium. While the hydrogen bonding occurs between OH-groups on the phenol ring and α-NH_2-groups of the constituent amino acids in the protein, the ionic bond occurs between the carboxyl group of the phenolic acid and basic groups of the constituent amino acids, particularly for arginine and lysine (Rawel and Rohn, 2010). Flavonoids similarly interact with proteins either covalently or non-covalently, with non-covalent interactions the same as those described for phenolic acids above (Papadopoulou et al., 2005; Świeca et al., 2013).

Proanthocyanins are termed 'condensed tannins' due to their particularly strong ability to bind to proteins and precipitate them (Hagerman, 1992). They are believed to cross-link with proteins through hydrophobic interactions and hydrogen-bonding (Girard et al., 2018). Proteins with high levels of proline such as salivary protein, gelatin and collagen have high affinities for proanthocyanidins. As with phenolic acids and flavonoids, factors such as the structure and molecular weight of the proanthocyanidins, their number of OH-groups and galloylation affect the strength of the proanthocyanidin-protein interactions (Santos-Buelga and Scalbert, 2000).

Phenolic-protein interactions potentially decrease protein digestibility (Świeca et al., 2013) either by making nutritionally relevant proteins less susceptible to digestive proteases when they bind to phenolic compounds, or the digestive enzymes may bind to the phenolics and lose their hydrolytic activity (Papadopoulou et al., 2005). They may result in a reduction in net protein utilization and consequent reduction in animal growth and development, especially for proanthocyanidins (Mole et al., 1993; Santos-Buelga and Scalbert, 2000). Since wheat is not a rich source of proanthocyanidins, the phenolic acid-protein and flavonoid-protein interactions will be more significant than proanthocyanidin-protein interactions.

It is interesting to note that phenolic-protein interactions serve a protective function for phenolic compounds against enzymatic and pH-induced degradation in the small intestine (Buitimea-Cantúa et al., 2018). Most of the phenolic compounds in foods that are not absorbed in the small intestine are therefore transported to the colon where they are metabolized by the colonic microbiota and contribute to the antioxidant environment in the colon. These phenolic metabolites also exhibit beneficial health effects such as anti-proliferative activity in some tumor cell lines (Buitimea-Cantúa et al., 2018).

5.1.3 Binding to dietary fiber

A larger concentration of phenolic acids in wheat exists in the bound form compared to the free acids (Liyana-Pathirana and Shahidi, 2006). These bound phenolic acids, most of which are hydroxycinnamates such as ferulic acid, bind to dietary fiber components such as arabinoxylans in the wheat, which serve to protect them from degradation by digestive enzymes or pH in the small intestine. They are subsequently transported to the colon where they exhibit antioxidant protection for colonic cells (Apea-Bah and Beta, 2019). The health benefits of dietary fiber have been reported by many studies. However, it is worth noting that in cereals such as wheat, their constituent bound phenolic acids, especially hydroxycinnamates exist as co-passengers to dietary fiber components (arabinoxylans) (Apea-Bah and Beta, 2019) and jointly elicit beneficial health effects in the colon of consumers, as similarly explained above for phenolic-protein interactions.

From the foregoing, it is evident that health-promoting properties of dietary nutrients usually occur through the protective function of the nutrients when bound to phenolic compounds.

5.2 Effect of carotenoids on nutritional value of wheat foods

Carotenoids in wheat grains have demonstrable antioxidant properties in that they protect biological molecules such as proteins, fats and DNA from free radical-induced oxidative damage, and consequently prevent or inhibit oxidative stress-related diseases (Gammone et al., 2015). They also have anti-inflammatory properties. For example, lutein, which is the predominant carotenoid in wheat grains, has been reported to reduce blood lipid peroxidation and C-reactive protein in healthy non-smokers (Wang et al., 2013). C-reactive protein is a marker for inflammation. Lutein also suppresses cholesterol accumulation, reduces oxidation of low-density lipoprotein, which otherwise could initiate formation of foam cells in arteries leading to cardiovascular disease. It also inhibits formation of other inflammatory cytokines in the aorta of guinea pigs (Kim et al., 2011). Lutein, zeaxanthin and β-carotene, which are all carotenoids in wheat, have been demonstrated to protect against early atherosclerosis (Dwyer et al., 2004). This means that wheat carotenoids exhibit both antioxidant and anti-inflammatory properties in healthy people and can potentially protect against cardiovascular disease.

Wheat carotenoids (lutein and zeaxanthin) have also been reported to protect the eye against diseases such as age-related macular disease (AMD) and cataracts (Calvo, 2005). This protection is done through two mechanisms. First, they protect the retina directly by absorbing the blue wavelength of the ultra-violet light, which has high energy and can otherwise damage retinal cells. Secondly, wheat carotenoids protect the eye indirectly as antioxidants by scavenging reactive free radicals and other oxidants that can oxidatively damage the retina (Krinsky et al., 2003; Toomey and McGraw, 2016). Because of this protective role, lutein and zeaxanthin are concentrated in the macula which is the functional center of the retina (Bernstein et al., 2016). They are therefore termed macular pigments. Therefore wheat carotenoids (lutein and zeaxanthin) work similarly to vitamin A in protecting the eye and maintaining vision.

The European Food Safety Authority (EFSA) Panel on Food Additives and Nutrient Sources added to Food has established an acceptable daily intake of 1 mg/kg bodyweight/day for lutein and 0.75 mg/kg bodyweight/day for zeaxanthin. Since these two carotenoids are generally recognized as safe (GRAS) for human consumption, food manufacturers are permitted to use their synthetic forms as food additives in Europe (Bernstein et al., 2016; EFSA, 2011).

5.3 Effect of tocochromanols on nutritional value of wheat foods

Like the other phytochemicals discussed above, tocols are potent antioxidants with demonstrated beneficial health effects. For example, tocols protect against radical-induced oxidative damage that leads to cancer, aging, arthritis and cataracts (Rizvi et al., 2014). They inhibit platelet hyper-aggregation, which otherwise could lead to atherosclerosis and consequent cardiovascular disease. They also reduce the production of prostaglandins which are mediators of inflammation that can cause platelet clumping if not regulated (Rizvi et al., 2014). Vitamin E is usually located in cell membranes and the membranes of cellular organelles where they serve as the first line of defense against radical-induced peroxidation of the phospholipid bilayer and other constituent lipid molecules in the membranes. They maintain cell membrane structure and orderliness by ensuring tight packing of the constituent lipids and enhancing stability of the cell (Rizvi et al., 2014). The different tocopherol vitamers (α, β, γ, δ) work together more efficiently in providing antioxidant protection against lipid peroxidation and other oxidative stress-related conditions than α-tocopherol alone when compared weight by weight (Liu et al., 2002). While α-tocopherol prevents initiation of lipid peroxidation, γ-tocopherol prevents its propagation.

Vitamin E interacts with a number of micronutrients, including vitamin C, selenium, β-carotene, vitamin B3 (niacin) and glutathione (Niki et al., 1995). They also interact with thiols and other compounds to enhance their cellular antioxidant protection (Rizvi et al., 2014). For example, vitamin E works synergistically with the carotenoids β-carotene and lycopene to suppress growth of prostate cancer (Siler et al., 2004). It also works effectively with vitamin C to prevent prostate cancer and cardiovascular disease in men (Gaziano et al., 2009; Sesso et al., 2008). Vitamins C and E are the most potent hydrophilic and lipophilic antioxidants known and therefore combining them helps in treating many oxidative stress-related non-communicable diseases.

The recommended dietary allowance (RDA) for vitamin E for different age groups is as follows: 15 mg/day for both male and female adults including pregnant women (above 14 years), 11 mg/day for males and females aged 11-13, 7 mg/day for 4–8-year-old children, 6 mg/day for 1–3 year-old children, 5 mg/day for 7–12-month-old infants and 4 mg/day for infants (0–6 months old). Lactating mothers require a higher RDA of 19 mg/day (Rizvi et al., 2014). Individuals whose physiological systems are more exposed to oxidants such as smokers will require higher levels of vitamin E and other antioxidants to protect them from oxidative stress-related complications.

In comparison, the levels of vitamin E (α-tocopherol) reported in wheat ranged between 0.8 μg/g and 14.4 μg/g on dry weight basis (Table 5). Hidalgo and Brandolini (2010) reported α-tocopherol content of 1.7–4.3 μg/g on dry weight basis for bread crumbs, 1.7–3.5 μg/g on dry weight basis for water biscuit and 2.2 μg/g on dry weight basis for pasta, all three products prepared from Einkorn wheat variety *Monlis*. α-tocopherol was not detected in pasta prepared from durum wheat semolina (Hidalgo and Brandolini, 2010). This means that, an adult male or female will require at least about 3.5 kg of bread, 4.3 kg of water biscuit or 6.8 kg of pasta prepared from Einkorn wheat, to meet their RDA for vitamin E. This explains why other rich sources of vitamin E are required in addition to a wheat-based meal. While the other tocols do not show vitamin E activity, they contribute to the antioxidant capacity of wheat.

6 Conclusion and future trends

Whole grain wheat has a diverse supply of hydrophilic and lipophilic antioxidants which include phenolic compounds (phenolic acids, flavonoids, anthocyanins which are present only in pigmented wheat, alkylresorcinols, and to a lesser extent proanthocyanidins), carotenoids (mainly lutein and zeaxanthin) and tocochromanols (α, β, γ and δ - tocopherols and tocotrienols). This diversity of antioxidants function to protect consumers from radical-induced oxidative damage caused by various free radicals produced endogenously from metabolic processes and exogenously from sunlight and other chemicals and environmental pollutants. They also chelate metal ions which could catalyze oxidation reactions in the physiological system, thereby providing antioxidant protection. However, phenolic compounds might exert pro-oxidant effect when bound to heavy metal ions. This effect could be prevented by the diverse antioxidant system in wheat.

While some of these antioxidants are absorbed in the small intestine, others cross the proximal region of the gastrointestinal tract (stomach and small intestine) unchanged and reach the distal end (colon) where they are catabolized by colonic microbiota into metabolites which provide antioxidant protection for the distal end of the gastrointestinal tract.

Apart from vitamin E which is a nutrient, all the other antioxidants in wheat are non-nutrient phytochemicals and therefore do not have recommended dietary allowance for humans, although they play essential antioxidant protective roles that help to maintain good health. Various factors affect the concentrations of the phytochemical antioxidants reported for wheat and these include genotype of the wheat and environmental or geographic factors as well as postharvest processes. The methods of extraction and analysis of the antioxidants may also affect the concentrations reported, since most extraction and analytical methods are not standardized.

Future work on wheat phytochemical antioxidants could focus on standardizing methods of extraction and analysis so as to promote comparability of results across different laboratories. Research efforts in mathematical modeling that could establish a relationship or correlation between the hydrophilic antioxidants on one end and the lipophilic antioxidants on the other end will be most useful in defining and characterizing the overall antioxidant properties of whole grain wheat.

7 Where to look for further information

Further information on the antioxidants in wheat can be obtained from the book titled 'wheat antioxidants' edited by Yu and published in 2008. Several other books have chapters that discuss the antioxidants in wheat. For example, the book titled 'Flour and breads and their fortification in health and disease prevention', edited by Preedy and Watson, and published in 2019, has Chapter 6 that provides information on flours and breads from colored (black, purple and blue) wheat varieties. Similarly, Chapter 10 of the book titled 'Nuts and Seeds in Health and Disease Prevention (Second Edition)', edited by Preedy and Watson and published in 2020, gives an update on the phenolic composition and antioxidant properties of purple wheat.

8 References

Abbaspour, N., Hurrell, R. and Kelishadi, R. (2014). Review on iron and its importance for human health. *Journal of Research in Medical Sciences* 19(2), 164–174.

Abdel-Aal, E.-S. M. and Rabalski, I. (2008). Bioactive compounds and their antioxidant capacity in selected primitive and modern wheat species. *Open Agriculture Journal* 2, 7–14.

Abdel-Aal, E.-S. M., Young, J. C. and Rabalski, I. (2006). Anthocyanin composition in black, blue, pink, purple, and red cereal grains. *Journal of Agricultural and Food Chemistry* 54(13), 4696–4704.

Abdel-Aal, E.-S. M., Young, J. C., Rabalski, I., Hucl, P. and Fregeau-Reid, J. (2007). Identification and quantification of seed carotenoids in selected wheat species. *Journal of Agricultural and Food Chemistry* 55(3), 787–794.

Ainsworth, E. A. and Gillespie, K. M. (2007). Estimation of total phenolic content and other oxidation substrates in plant tissues using Folin-Ciocalteu reagent. *Nature Protocols* 2(4), 875–877.

Andersson, A. A. M., Kamal-Eldin, A., Fraś, A., Boros, D. and Åman, P. (2008). Alkylresorcinols in wheat varieties in the HEALTHGRAIN diversity screen. *Journal of Agricultural and Food Chemistry* 56(21), 9722–9725.

Andjelković, M., Van Camp, J., De Meulenaer, B., Depaemelaere, G., Socaciu, C., Verloo, M. and Verhe, R. (2006). Iron-chelation properties of phenolic acids bearing catechol and galloyl groups. *Food Chemistry* 98(1), 23–31.

Apea-Bah, F. B. and Beta, T. (2019). Bound phenolic constituents as co-passengers of dietary fibre. In: Beta, T. and Camire, M. E. (Eds), *Cereal Grain-Based Functional*

Foods: Carbohydrate and Phytochemical Components (pp. 278–304). Cambridge, UK: Royal Society of Chemistry.

Awika, J. M. and Rooney, L. W. (2004). Sorghum phytochemicals and their potential impact on human health. *Phytochemistry* 65(9), 1199–1221.

Basílio, N. and Pina, F. (2016). Chemistry and photochemistry of anthocyanins and related compounds: a thermodynamic and kinetic approach. *Molecules* 21(11), 1502.

Bernstein, P. S., Li, B., Vachali, P. P., Gorusupudi, A., Shyam, R., Henriksen, B. S. and Nolan, J. M. (2016). Lutein, zeaxanthin, and meso-zeaxanthin: the basic and clinical science underlying carotenoid-based nutritional interventions against ocular disease. *Progress in Retinal and Eye Research* 50, 34–66.

Beta, T., Li, W. and Apea-Bah, F. B. (2019). Flour and bread from black, purple and blue-colored wheats. In: Preedy, V. R. and Watson, R. R. (Eds), *Flour and Breads and Their Fortification in Health and Disease Prevention* (2nd ed., pp. 75–88). London, UK: Academic Press.

Beta, T., Qiu, Y., Liu, Q., Borgen, A. and Apea-Bah, F. B. (2020). Purple wheat (*Triticum* spp.) seeds: phenolic composition and antioxidant properties. In: Preedy, V. R. and Watson, R. R. (Eds), *Nuts and Seeds in Health and Disease Prevention* (2nd ed., pp. 103–25). London, UK: Academic Press.

Buitimea-Cantúa, N. E., Gutiérrez-Uribe, J. A. and Serna-Saldivar, S. O. (2018). Phenolic-protein interactions: effects on food properties and health benefits. *Journal of Medicinal Food* 21(2), 188–198.

Bunzel, M. (2010). Chemistry and occurrence of hydroxycinnamate oligomers. *Phytochemistry Reviews* 9(1), 47–64.

Burkhardt, S. and Böhm, V. (2007). Development of a new method for the complete extraction of carotenoids from cereals with special reference to durum wheat (*Triticum durum* Desf.). *Journal of Agricultural and Food Chemistry* 55(21), 8295–8301.

Calvo, M. M. (2005). Lutein: a valuable ingredient of fruit and vegetables. *Critical Reviews in Food Science and Nutrition* 45(7–8), 671–696.

Chang, C.-C., Yang, M. H., Wen, H. M. and Chern, J. C. (2002). Estimation of total flavonoid content in propolis by two complementary colorimetric methods. *Journal of Food and Drug Analysis* 10(3), 178–182.

Chobot, V. and Hadacek, F. (2010). Iron and its complexation by phenolic cellular metabolites: from oxidative stress to chemical weapons. *Plant Signaling and Behavior* 5(1), 4–8.

Colombo, M. L. (2010). An update on vitamin E, tocopherol and tocotrienol-perspectives. *Molecules* 15(4), 2103–2113.

Dwyer, J. H., Paul-Labrador, M. J., Fan, J., Shircore, A. M., Merz, C. N. B. and Dwyer, K. M. (2004). Progression of carotid intima-media thickness and plasma antioxidants: the Los Angeles Atherosclerosis Study. *Arteriosclerosis, Thrombosis, and Vascular Biology* 24(2), 313–319.

Dykes, L. and Rooney, L. W. (2007). Phenolic compounds in cereal grains and their health benefits. *Cereal Foods World* 52(3), 105–111.

EFSA. (2011). EFSA panel on food additives and nutrient sources added to food. Scientific opinion on the re-evaluation of lutein preparations other than lutein with high concentrations of total saponified carotenoids at levels of at least 80%. *EFSA Journal* 9(5), 2144.

FAO. (2019). Food Outlook-Biannual report on global food markets. In: Ben-Belhassen, B. and Abbassian, A. (Eds), *Food Outlook* (pp. 1–155). Rome, Italy: Food and Agriculture Organization of the United Nations.

Feillet, P., Autran, J.-C. and Icard-Verniere, C. (2000). Mini review pasta brownness: an assessment. *Journal of Cereal Science* 32(3), 215–233.

Feng, Y., McDonald, C. E. and Vick, B. A. (1988). C-glycosylflavones from hard red spring wheat bran. *Cereal Chemistry* 65(6), 452–456.

Fratianni, A., Caboni, M., Irano, M. and Panfili, G. (2002). A critical comparison between traditional methods and supercritical carbon dioxide extraction for the determination of tocochromanols in cereals. *European Food Research and Technology* 215(4), 353–358.

Fu, B. X. (2008). Asian noodles: history, classification, raw materials, and processing. *Food Research International* 41(9), 888–902.

Gammone, M. A., Riccioni, G. and D'Orazio, N. (2015). Carotenoids: potential allies of cardiovascular health? *Food and Nutrition Research* 59, (26762), http://doi.org/10.3402/fnr.v59.26762.

Gaziano, J. M., Glynn, R. J., Christen, W. G., Kurth, T., Belanger, C., MacFadyen, J., Bubes, V., Manson, J. E., Sesso, H. D. and Buring, J. E. (2009). Vitamins E and C in the prevention of prostate and total cancer in men: the Physicians' Health Study II randomized controlled trial. *JAMA* 301(1), 52–62.

Girard, A. L., Bean, S. R., Tilley, M., Adrianos, S. L. and Awika, J. M. (2018). Interaction mechanisms of condensed tannins (proanthocyanidins) with wheat gluten proteins. *Food Chemistry* 245, 1154–1162.

Giusti, M. M. and Wrolstad, R. E. (2001). Characterization and measurement of anthocyanins by UV-visible spectroscopy. *Current Protocols in Food Analytical Chemistry* F1(2), F1.2.1–F1.2.13.

Gocmen, D., Inkaya, A. N. and Aydin, E. (2009). Flat breads. *Bulgarian Journal of Agricultural Science* 15(4), 298–306.

Hagerman, A. E. (1992). Tannin-protein interactions. In: Ho, C.-T., Lee, C. Y. and Huang, M.-T. (Eds), *Phenolic Compounds in Food and Their Effects on Health I* (pp. 236–247). Washington DC: ACS Publications.

Hassan, S. M., Al Aqil, A. A. and Attimarad, M. (2013). Determination of crude saponin and total flavonoids content in guar meal. *Advancement in Medicinal Plant Research* 1(2), 24–28.

Hengtrakul, P., Lorenz, K. and Mathias, M. (1991). Alkylresorcinol homologs in cereal grains. *Journal of Food Composition and Analysis* 4(1), 52–57.

Hentschel, V., Kranl, K., Hollmann, J., Lindhauer, M. G., Böhm, V. and Bitsch, R. (2002). Spectrophotometric determination of yellow pigment content and evaluation of carotenoids by high-performance liquid chromatography in durum wheat grain. *Journal of Agricultural and Food Chemistry* 50(23), 6663–6668.

Herald, T. J., Gadgil, P., Perumal, R., Bean, S. R. and Wilson, J. D. (2014). High-throughput micro-plate HCl-vanillin assay for screening tannin content in sorghum grain. *Journal of the Science of Food and Agriculture* 94(10), 2133–2136.

Herald, T. J., Gadgil, P. and Tilley, M. (2012). High-throughput micro plate assays for screening flavonoid content and DPPH-scavenging activity in sorghum bran and flour. *Journal of the Science of Food and Agriculture* 92(11), 2326–2331.

Heredia-Sandoval, N. G., del Carmen Granados-Nevárez, M., de la Barca, A. M. C., Vásquez-Lara, F., Malunga, L. N., Apea-Bah, F. B., Beta, T. and Islas-Rubio, A. R. (2019).

Phenolic acids, antioxidant capacity, and estimated glycemic index of cookies added with Brewer's spent grain. *Plant Foods for Human Nutrition* 75, 41-47.

Hidalgo, A. and Brandolini, A. (2010). Tocols stability during bread, water biscuit and pasta processing from wheat flours. *Journal of Cereal Science* 52(2), 254–259.

Hidalgo, A., Brandolini, A., Pompei, C. and Piscozzi, R. (2006). Carotenoids and tocols of einkorn wheat (*Triticum monococcum* ssp. *monococcum* L.). *Journal of Cereal Science* 44(2), 182–193.

Hidalgo, A., Fongaro, L. and Brandolini, A. (2014). Wheat flour granulometry determines colour perception. *Food Research International* 64, 363-370.

Hosseinian, F. S., Li, W. and Beta, T. (2008). Measurement of anthocyanins and other phytochemicals in purple wheat. *Food Chemistry* 109(4), 916–924.

Hurrell, R. and Egli, I. (2010). Iron bioavailability and dietary reference values. *American Journal of Clinical Nutrition* 91(5), 1461S–1467S.

Ioset, J. R., Urbaniak, B., Ndjoko-Ioset, K., Wirth, J., Martin, F., Gruissem, W., Hostettmann, K. and Sautter, C. (2007). Flavonoid profiling among wild type and related GM wheat varieties. *Plant Molecular Biology* 65(5), 645–654.

Iwasaki, Y., Hirasawa, T., Maruyama, Y., Ishii, Y., Ito, R., Saito, K., Umemura, T., Nishikawa, A. and Nakazawa, H. (2011). Effect of interaction between phenolic compounds and copper ion on antioxidant and pro-oxidant activities. *Toxicology in Vitro* 25(7), 1320–1327.

Kamal-Eldin, A. and Appelqvist, L. A. (1996). The chemistry and antioxidant properties of tocopherols and tocotrienols. *Lipids* 31(7), 671–701.

Kim, J. E., Leite, J. O., DeOgburn, R., Smyth, J. A., Clark, R. M. and Fernandez, M. L. (2011). A lutein-enriched diet prevents cholesterol accumulation and decreases oxidized LDL and inflammatory cytokines in the aorta of guinea pigs. *Journal of Nutrition* 141(8), 1458–1463.

Kim, K.-H., Tsao, R., Yang, R. and Cui, S. W. (2006). Phenolic acid profiles and antioxidant activities of wheat bran extracts and the effect of hydrolysis conditions. *Food Chemistry* 95(3), 466–473.

Kozubek, A. and Tyman, J. H. P. (1999). Resorcinolic lipids, the natural non-isoprenoid phenolic amphiphiles and their biological activity. *Chemical Reviews* 99(1), 1–26.

Krinsky, N. I., Landrum, J. T. and Bone, R. A. (2003). Biologic mechanisms of the protective role of lutein and zeaxanthin in the eye. *Annual Review of Nutrition* 23, 171–201.

Krygier, K., Sosulski, F. and Hogge, L. (1982). Free, esterified, and insoluble-bound phenolic acids. 1. Extraction and purification procedure. *Journal of Agricultural and Food Chemistry* 30(2), 330–334.

Labuschagne, M., Mkhatywa, N., Johansson, E., Wentzel, B. and Van Biljon, A. (2017). The content of tocols in South African wheat; impact on nutritional benefits. *Foods* 6(11), http://doi.org/10.3390/foods6110095.

Lampi, A. M., Nurmi, T., Ollilainen, V. and Piironen, V. (2008). Tocopherols and tocotrienols in wheat genotypes in the HEALTHGRAIN diversity screen. *Journal of Agricultural and Food Chemistry* 56(21), 9716–9721.

Lampi, A. M., Nurmi, T. and Piironen, V. (2010). Effects of the environment and genotype on tocopherols and tocotrienols in wheat in the HEALTHGRAIN diversity screen. *Journal of Agricultural and Food Chemistry* 58(17), 9306–9313.

Landberg, R., Kamal-Eldin, A., Salmenkallio-Marttila, M., Rouau, X. and Åman, P. (2008). Localization of alkylresorcinols in wheat, rye and barley kernels. *Journal of Cereal Science* 48(2), 401–406.

Leenhardt, F., Lyan, B., Rock, E., Boussard, A., Potus, J., Chanliaud, E. and Remesy, C. (2006). Genetic variability of carotenoid concentration, and lipoxygenase and peroxidase activities among cultivated wheat species and bread wheat varieties. *European Journal of Agronomy* 25(2), 170–176.

Li, L., Shewry, P. R. and Ward, J. L. (2008). Phenolic acids in wheat varieties in the HEALTHGRAIN diversity screen. *Journal of Agricultural and Food Chemistry* 56(21), 9732–9739.

Liu, M., Wallin, R., Wallmon, A. and Saldeen, T. (2002). Mixed tocopherols have a stronger inhibitory effect on lipid peroxidation than α-tocopherol alone. *Journal of Cardiovascular Pharmacology* 39(5), 714–721.

Liu, Q., Qiu, Y. and Beta, T. (2010). Comparison of antioxidant activities of different colored wheat grains and analysis of phenolic compounds. *Journal of Agricultural and Food Chemistry* 58(16), 9235–9241.

Liyana-Pathirana, C. and Shahidi, F. (2005). Optimization of extraction of phenolic compounds from wheat using response surface methodology. *Food Chemistry* 93(1), 47–56.

Liyana-Pathirana, C. M. and Shahidi, F. (2006). Importance of insoluble-bound phenolics to antioxidant properties of wheat. *Journal of Agricultural and Food Chemistry* 54(4), 1256–1264.

Lu, Y., Luthria, D., Fuerst, E. P., Kiszonas, A. M., Yu, L. and Morris, C. F. (2014). Effect of processing on phenolic composition of dough and bread fractions made from refined and whole wheat flour of three wheat varieties. *Journal of Agricultural and Food Chemistry* 62(43), 10431–10436.

Ma, D., Sun, D., Li, Y., Wang, C., Xie, Y. and Guo, T. (2015). Effect of nitrogen fertilisation and irrigation on phenolic content, phenolic acid composition, and antioxidant activity of winter wheat grain. *Journal of the Science of Food and Agriculture* 95(5), 1039–1046.

Malešev, D. and Kuntić, V. (2007). Investigation of metal-flavonoid chelates and the determination of flavonoids via metal-flavonoid complexing reactions. *Journal of the Serbian Chemical Society* 72(10), 921–939.

Malunga, L. N. and Beta, T. (2016). Isolation and identification of feruloylated arabinoxylan mono-and oligosaccharides from undigested and digested maize and wheat. *Heliyon* 2(5), e00106, https://doi.org/10.1016/j.heliyon.2016.e00106.

Malunga, L. N., Izydorczyk, M. and Beta, T. (2017). Effect of water-extractable arabinoxylans from wheat aleurone and bran on lipid peroxidation and factors influencing their antioxidant capacity. *Bioactive Carbohydrates and Dietary Fibre* 10, 20–26.

Maret, W. and Sandstead, H. H. (2006). Zinc requirements and the risks and benefits of zinc supplementation. *Journal of Trace Elements in Medicine and Biology: Organ of the Society for Minerals and Trace Elements* 20(1), 3–18.

McCallum, J. A. (1989). *Biochemistry of Phenolic Compounds in Wheat Grain (Triticum aestivum L.)*. PhD thesis. Christchurch, New Zealand: University of Canterbury.

McCallum, J. A. and Walker, J. R. L. (1990). Proanthocyanidins in wheat bran. *Cereal Chemistry* 67(3), 282–285.

Mole, S., Rogler, J. C. and Butler, L. G. (1993). Growth reduction by dietary tannins: different effects due to different tannins. *Biochemical Systematics and Ecology* 21(6–7), 667–677.

Moore, J. and Hao, J. (2012). Antioxidant and health promoting properties of wheat (Triticum spp.). In: Yu, L. L., Tsao, R. and Shahidi, F. (Eds), *Cereals and Pulses: Nutraceutical Properties and Health Benefits* (pp. 113–130). Ames, IA: Wiley-Blackwell.

Moore, J., Hao, Z., Zhou, K., Luther, M., Costa, J. and Yu, L. L. (2005). Carotenoid, tocopherol, phenolic acid, and antioxidant properties of Maryland-grown soft wheat. *Journal of Agricultural and Food Chemistry* 53(17), 6649–6657.

Ndolo, V. U. and Beta, T. (2013). Distribution of carotenoids in endosperm, germ, and aleurone fractions of cereal grain kernels. *Food Chemistry* 139(1–4), 663–671.

Ndolo, V. U. and Beta, T. (2014). Comparative studies on composition and distribution of phenolic acids in cereal grain botanical fractions. *Cereal Chemistry Journal* 91(5), 522–530.

Nicoletti, I., Martini, D., De Rossi, A., Taddei, F., D'Egidio, M. G. and Corradini, D. (2013). Identification and quantification of soluble free, soluble conjugated, and insoluble bound phenolic acids in durum wheat (*Triticum turgidum* L. var. *durum*) and derived products by RP-HPLC on a semimicro separation scale. *Journal of Agricultural and Food Chemistry* 61(48), 11800–11807.

Niki, E., Noguchi, N., Tsuchihashi, H. and Gotoh, N. (1995). Interaction among vitamin C, vitamin E, and beta-carotene. *American Journal of Clinical Nutrition* 62(6) (Suppl.), 1322S–1326S.

Osredkar, J. and Sustar, N. (2011). Copper and zinc, biological role and significance of copper/zinc imbalance. *Journal of Clinical Toxicology, Suppl.* 3(001), https://doi.org/10.4172/2161-0495.S4173-4001.

Panfili, G., Fratianni, A. and Irano, M. (2003). Normal phase high-performance liquid chromatography method for the determination of tocopherols and tocotrienols in cereals. *Journal of Agricultural and Food Chemistry* 51(14), 3940–3944.

Panfili, G., Manzi, P. and Pizzoferrato, L. (1994). High-performance liquid chromatographic method for the simultaneous determination of tocopherols, carotenes, and retinol and its geometric isomers in Italian cheeses. *Analyst* 119(6), 1161–1165.

Papadopoulou, A., Green, R. J. and Frazier, R. A. (2005). Interaction of flavonoids with bovine serum albumin: a fluorescence quenching study. *Journal of Agricultural and Food Chemistry* 53(1), 158–163.

Peh, H. Y., Tan, W. S. D., Liao, W. and Wong, W. S. F. (2016). Vitamin E therapy beyond cancer: tocopherol versus tocotrienol. *Pharmacology and Therapeutics* 162, 152–169.

Pietta, P. G. (2000). Flavonoids as antioxidants. *Journal of Natural Products* 63(7), 1035–1042.

Piironen, V., Lampi, A.-M., Ekholm, P., Salmenkallio-Marttila, M. and Liukkonen, K.-H. (2009). Micronutrients and phytochemicals in wheat grain. In: Khan, K. and Shewry, P. R. (Eds), *Wheat – Chemistry and Technology* (Vol. 4). American Associate of Cereal Chemists International: Minneapolis, MN, USA.

Proestos, C., Sereli, D. and Komaitis, M. (2006). Determination of phenolic compounds in aromatic plants by RP-HPLC and GC-MS. *Food Chemistry* 95(1), 44–52.

Qiu, Y., Liu, Q. and Beta, T. (2009). Antioxidant activity of commercial wild rice and identification of flavonoid compounds in active fractions. *Journal of Agricultural and Food Chemistry* 57(16), 7543–7551.

Rawel, H. M., Czajka, D., Rohn, S. and Kroll, J. (2002). Interactions of different phenolic acids and flavonoids with soy proteins. *International Journal of Biological Macromolecules* 30(3–4), 137–150.

Rawel, H. M., Meidtner, K. and Kroll, J. (2005). Binding of selected phenolic compounds to proteins. *Journal of Agricultural and Food Chemistry* 53(10), 4228–4235.

Rawel, H. M. and Rohn, S. (2010). Nature of hydroxycinnamate-protein interactions. *Phytochemistry Reviews* 9(1), 93–109.

Rizvi, S., Raza, S. T., Ahmed, F., Ahmad, A., Abbas, S. and Mahdi, F. (2014). The role of vitamin E in human health and some diseases. *Sultan Qaboos University Medical Journal* 14(2), e157–e165.

Ross, A. B. (2012). Present status and perspectives on the use of alkylresorcinols as biomarkers of wholegrain wheat and rye intake. *Journal of Nutrition and Metabolism* 2012, 462967, https://doi.org/10.1155/2012/462967.

Ross, A. B., Kamal-Eldin, A. and Åman, P. (2004). Dietary alkylresorcinols: absorption, bioactivities, and possible use as biomarkers of whole-grain wheat-and rye-rich foods. *Nutrition Reviews* 62(3), 81–95.

Ross, A. B., Kamal-Eldin, A., Jung, C., Shepherd, M. J. and Åman, P. (2001). Gas chromatographic analysis of alkylresorcinols in rye (*Secale cereale* L) grains. *Journal of the Science of Food and Agriculture* 81(14), 1405–1411.

Rupasinghe, H. P. V., Wang, L., Huber, G. M. and Pitts, N. L. (2008). Effect of baking on dietary fibre and phenolics of muffins incorporated with apple skin powder. *Food Chemistry* 107(3), 1217–1224.

Santos-Bue.g. C. and Scalbert, A. (2000). Proanthocyanidins and tannin-like compounds - nature, occurrence, dietary intake and effects on nutrition and health. *Journal of the Science of Food and Agriculture* 80(7), 1094–1117.

Schwartz, H., Ollilainen, V., Piironen, V. and Lampi, A.-M. (2008). Tocopherol, tocotrienol and plant sterol contents of vegetable oils and industrial fats. *Journal of Food Composition and Analysis* 21(2), 152–161.

Sen, C. K., Khanna, S., Rink, C. and Roy, S. (2007). Tocotrienols: the emerging face of natural vitamin E. *Vitamins and Hormones* 76, 203–261.

Sesso, H. D., Buring, J. E., Christen, W. G., Kurth, T., Belanger, C., MacFadyen, J., Bubes, V., Manson, J. E., Glynn, R. J. and Gaziano, J. M. (2008). Vitamins E and C in the prevention of cardiovascular disease in men: the Physicians' Health Study II randomized controlled trial. *JAMA* 300(18), 2123–2133.

Shewry, P. R. and Hey, S. J. (2015). The contribution of wheat to human diet and health. *Food and Energy Security* 4(3), 178–202.

Shewry, P. R., Piironen, V., Lampi, A. M., Nyström, L., Li, L., Rakszegi, M., Fras, A., Boros, D., Gebruers, K., Courtin, C. M., Delcour, J. A., Andersson, A. A., Dimberg, L., Bedo, Z. and Ward, J. L. (2008). Phytochemical and fiber components in oat varieties in the HEALTHGRAIN diversity screen. *Journal of Agricultural and Food Chemistry* 56(21), 9777–9784.

Siler, U., Barella, L., Spitzer, V., Schnorr, J. R., Lein, M., Goralczyk, R. and Wertz, K. (2004). Lycopene and vitamin E interfere with autocrine/paracrine loops in the Dunning prostate cancer model. *FASEB Journal* 18(9), 1019–1021.

Singleton, V. L., Orthofer, R. and Lamuela-Raventós, R. M. (1999). [14] Analysis of total phenols and other oxidation substrates and antioxidants by means of Folin-Ciocalteu reagent. *Methods in Enzymology* 299, 152–178.

Singleton, V. L. and Rossi, J. A. (1965). Colorimetry of total phenolics with phosphomolybdic-phosphotungstic acid reagents. *American Journal of Enology and Viticulture* 16, 144–158.

Świeca, M., Gawlik-Dziki, U., Dziki, D., Baraniak, B. and Czyż, J. (2013). The influence of protein-flavonoid interactions on protein digestibility *in vitro* and the antioxidant quality of breads enriched with onion skin. *Food Chemistry* 141(1), 451–458.

Toomey, M. B. and McGraw, K. J. (2016). The effects of sun exposure on carotenoid accumulation and oxidative stress in the retina of the House finch (*Haemorhous mexicanus*). *Avian Research* 7(1), 5.

Trono, D., Pastore, D. and Di Fonzo, N. (1999). Carotenoid dependent inhibition of durum wheat lipoxygenase. *Journal of Cereal Science* 29(1), 99–102.

Wang, M. X., Jiao, J. H., Li, Z. Y., Liu, R. R., Shi, Q. and Ma, L. (2013). Lutein supplementation reduces plasma lipid peroxidation and C-reactive protein in healthy nonsmokers. *Atherosclerosis* 227(2), 380–385.

Waterhouse, A. L. (2002). Determination of total phenolics. *Current Protocols in Food Analytical Chemistry,* 1.1, I1.1.1–I1.1.8.

Werner, S. and Böhm, V. (2011). Bioaccessibility of carotenoids and vitamin E from pasta: evaluation of an in vitro digestion model. *Journal of Agricultural and Food Chemistry* 59(4), 1163–1170.

Yu, L. (2008). *Wheat Antioxidants* (288 pp). Hoboken, NJ: Wiley.

Yu, L. and Beta, T. (2015). Identification and antioxidant properties of phenolic compounds during production of bread from purple wheat grains. *Molecules* 20(9), 15525–15549.

Yu, L., Nanguet, A. L. and Beta, T. (2013). Comparison of antioxidant properties of refined and whole wheat flour and bread. *Antioxidants* 2(4), 370–383.

Zhang, Y., Wang, L., Yao, Y., Yan, J. and He, Z. (2012). Phenolic acid profiles of Chinese wheat cultivars. *Journal of Cereal Science* 56(3), 629–635.

Zhou, K., Yin, J. J. and Yu, L. L. (2005). Phenolic acid, tocopherol and carotenoid compositions, and antioxidant functions of hard red winter wheat bran. *Journal of Agricultural and Food Chemistry* 53(10), 3916–3922.

Chapter 3

Advances in understanding the nutritional value of lipids in wheat

Marina Carcea, Council for Agricultural Research and Economics (CREA), Italy

1 Introduction

Lipids are usually defined in terms of the physical operation used to isolate them. In fact, when plant or animal tissues are extracted with a nonpolar solvent (e.g. ether, chloroform, benzene or an alkane), a portion of the material usually dissolves, and the components of this soluble fraction are called lipids. Lipids include a variety of different structural types, namely carboxylic acids (or fatty acids); mono-, di- and tri-acylglycerols (triglycerides or neutral fats); phospholipids; glycolipids; waxes; terpenes; steroids and alkylresorcinols. Wheat kernels contain all of these in different amounts. The American Oil Chemists' Society (AOCS) online Lipid Library provides a narrower definition of lipids as 'fatty acids (FA) and their derivatives, and substances related biosynthetically or functionally to these compounds', which has been adopted by many researchers in the field.

Lipids are, in fact, the most structurally diverse class of biological molecules and can be classified as simple or complex, depending on the number of products obtained from hydrolysis. Simple lipids include fatty acids as well as mono-, di- and tri-acylglycerols (Pareyt et al., 2011).

Only a small portion of the total lipid fraction is made of free carboxylic acids. Most of them in the lipid fraction are found as esters of glycerol, that

http://dx.doi.org/10.19103/AS.2021.0087.06

is triacylglycerols, the most common being long-chain carboxylic acids. Carboxylic acids may also contain one or more carbon–carbon double bond. Those triacylglycerols that are liquid at room temperature are generally known as oils and those that are solid are usually called fats. The carboxylic acids that are obtained by hydrolysis of naturally occurring fats and oils usually have unbranched chains with an even number of carbon atoms. The most common carboxylic acids obtained from fats and oils are the C14-C16 and C18 acids.

Glycolipids contain one or more monosaccharide moiety bound by a glycosidic linkage to a hydrophobic moiety, whereas most phospholipids are structurally derived from a glycerol derivative known as a phosphatidic acid. In a phosphatidic acid, two hydroxyl groups of glycerol are ester linked to fatty acids and one terminal hydroxyl group is ester linked to phosphoric acid. Most waxes are esters of long-chain fatty acids and long-chain alcohols. Alkylresorcinols (AR) (1,3-dihydroxy-5-n-alkylbenzenes) are a group of phenolic lipids mainly found in the outer parts of rye and wheat kernels. More than 99% of total AR content is located in an intermediate layer of the caryopsis that includes the hyaline layer, the testa and the inner pericarp (particularly the outer cuticle of testa and inner cuticle of pericarp), whereas no AR are found in the endosperm or in the germ (Landberg et al., 2008).

The primary function of triacylglycerols is as a source of chemical energy. Dietary lipids provide an average energy intake of 9 kcal/g which is about twice that provided by carbohydrates or proteins. They have other important biological functions in the cells including signaling and as structural components of membranes.

Cereal lipids are nutritionally valuable thanks to their composition. They are rich in polyunsaturated fatty acids and other bioactive compounds such as tocols, sterols, carotenoids and alkylresorcinols (Caboni, 2012).

2 Lipids in wheat kernels

Lipids are a small component in the wheat grain composition. Durum wheat (*Triticum durum* Desf.) kernels contain about 2.4–3.8% d.b. of lipids (Caboni, 2012), whereas soft wheat kernels (*Triticum aestivum* L.) contain 2.03–2.50 % d.b. (USDA, 2020). Other *Triticum* species present similar average values (emmer 2.13% m.b., einkorn 2.08% m.b. (USDA)) even if Ruibal-Mendieta et al. (2005) showed that, on average, spelt (*Triticum aestivum ssp.spelta*) wholemeals and milling fractions were higher in lipids and unsaturated fatty acids compared to soft wheat. Majewska et al. (2018) reached the same conclusion. Triticale, a man-made cereal from wheat and rye hybridization, is reported as having a total lipid content of 2.09% m.b. (USDA, 2020).

Roughly, two thirds (66%) of lipids are contained in the germ, 15% are in the bran (particularly in the aleuronic layer) and about 20% are distributed in the endosperm, partly within the starch granules. The starchy endosperm of the mature wheat grain comprises three major cell types, namely sub-aleurone cells, prismatic cells and central cells which differ in their content of functional components including lipids (Shewry et al., 2020). Lipids are small molecules that are distributed in all cells and tissues of the wheat grains and their different amounts in the different tissues are closely related to the metabolic specificities of each tissue in the grain.

From a chemical point of view, the most abundant wheat grain fraction is composed of nonpolar lipids, mainly storage acylglycerols or triglycerides. The polar fraction is made up of glycolipids and phospholipids: digalactosyldiglycerides are a major component of glycolipids, whereas lysophosphatidylcholines and phosphatidylcholines are major components of phospholipids (Pomeranz, 1987). Three types of glycolipids, namely glycosylglyceride, glycosylsterol and glycosylceramide, occur in higher plants: they may be regarded as natural surfactants, important in food processes, such as breadmaking, where an appropriate balance between different types of lipids is crucial (Melis et al., 2019). Glycosylceramides in soft wheat grain were studied by Fujino and Ohnishi (1983). The phospholipid profile was determined on lipids extracted from the whole grains of 10 genotypes of tetraploid wheats (*Triticum dicoccon* Shrank and *T. durum* Desf.) and 10 genotypes of hexaploid wheats (*T. spelta* L. and *T. aestivum* L.) by Pelillo et al. (2010). They found that phosphatidylethanolamine, phosphatidylinositol and phosphatidylcholine were the phospholipids present in major amounts, accounting for more than 70% of the total phospholipids. In all species, phospholipids represented about 10% of total lipids.

According to differences in solubility, wheat lipids can be separated into free and bound lipids: free lipids can be extracted with nonpolar solvents such as ether or petroleum ether. For extraction of bound (to proteins, starch and other components) lipids, polar solvents, such as water-saturated-butanol or a mixture of chloroform-methanol-water, are required. Lipids extracted by nonpolar solvents are arbitrarily defined as free and those extracted by polar solvents are bound. The free and bound lipids can in turn be fractionated into nonpolar and polar. Most polar lipids of a wheat kernel are present in the starchy endosperm whereas the nonpolar lipids are predominantly present in the germ and aleurone tissues (Pareyt et al., 2011). The fatty acids of wheat lipids are mostly unsaturated (C18:2, C18:1, C18:3 and C16:1) and two of them are essential (linoleic 18:2 and linolenic 18:3), that is, they cannot be synthesized by the human organism.

Lipid content, lipid classes and fatty acid levels in wheat kernels depend on a set of factors, some of which are genetic, such as species and variety (Armanino et al., 2002, Moore et al., 2015, Majewska et al., 2018), whereas others depend on the environment and are related to pedoclimatic conditions, agronomic practices and maturity level (Armanino et al., 2002; Beleggia et al., 2013; Min et al., 2017; Marion and Saulnier, 2020). For example, durum wheat and hard red wheat generally have higher lipid content than soft white wheat and the levels of fatty acids are different in durum and soft wheat. The knowledge of lipid components of wheat is useful in differentiating between *Triticum durum* Desf. and *Triticum aestivum* L., namely the determination of the percentages of methyl esters of the different unsaturated fatty acids with 18-carbon atoms (C_{18}), of sterol fraction and of the other components (Armanino et al., 2002). Beleggia et al. (2013) identified the interaction genotype × year × treatment as the main contributor to the variability of the fatty acid levels observed in 24 durum wheat samples, especially for linoleic, oleic and stearic acids. Armanino et al. (2002) linked the fatty acid profile of 135 samples of durum wheat to the cultivar, the geographic origin and the harvest year. Regarding climatic conditions, it has been seen that cold weather favors an increase of lipid content in wheat and a higher degree of unsaturation in fatty acids, due to the need for membrane fluidification (Nejadsadeghi et al., 2015). A high night temperature also induces changes in grain starch metabolism and alters lipid accumulation in winter wheat (Impa et al., 2020). Other kinds of biotic and abiotic stresses, such as salt, drought, pathogens and others, can influence the level of saturated and unsaturated fatty acids in plants (Upchurch, 2008). Moreover, different extraction and analytical methods can also account for differences found in the literature (Caboni, 2012; Rosicka-Kaczmarek et al., 2015).

Hargin and Morrison (1980) determined the quantitative distribution of 23 classes of acyl lipids in the germ, aleurone, starch and endosperm non-starch fractions of soft wheat (soft English, hard English and US Hard Red Spring) and US Amber durum. All wheats had similar quantities of acyl lipids in the whole kernel (2.8–3.2%), germ (25.7–30.5%), starch (0.8–1.2%) and endosperm non-starch fractions (0.8–1.1%). Lipids in the germ and aleurone consisted of triglycerides (60.3–79.3%), other nonpolar lipids (5.6–12.0%) and phospholipids (13.6–17.9%). Starch lipids were almost exclusively lysophospholipids (89.4–94.4%). Greater variation was found in the endosperm of non-starch lipids which consisted of triglycerides (13.7–34.1%), other nonpolar lipids (33.2–48.5%), glycolipids (18.6–38.3%) and phospholipids (21.9–35.3%). Lipid-profiling techniques were utilized to quantitatively determine the polar lipid species present in soft whole-wheat meal, flour and starch belonging to two cultivars and 146 polar lipid species were identified (Finnie et al., 2009).

Lipidomic analyses of milling and pearling fractions from soft wheat grains were carried out to determine differences in composition related to the spatial distribution of lipids in grain. Free fatty acids and triacylglycerols were major components in all fractions, but the relative contents of polar lipids varied, particularly those of lysophosphatidylcholine and digalactosyldiglyceride, which were enriched in flour fractions. Minor phospholipids were instead enriched in bran and offal fractions. The most abundant fatty acids in the analyzed acyl lipids were C16:0 and C18:2 and their combinations, including C36:4 and C34:2 (Gonzalez-Thuillier et al., 2015).

Bioimaging combined with biochemical analysis of fractions produced by sequential pearling of whole wheat grains were used to determine the distributions of components within the endosperm tissue and gradients were found for lipids (Tosi et al., 2018): a higher content of triacylglycerols was found in the aleurone with a decrease in concentration occurring from the outside to the center of the starchy endosperm. Also, the ratio of unsaturated to saturated fatty acids decreased from the aleurone to the center of grain.

In a recent paper by Narducci et al. (2019), the level of variation in lipids and their fatty acids was determined in the grains of ten popular durum wheat cultivars commercially grown in Italy: samples were harvested for two consecutive years to account for differences due to changes in climatic conditions. Total lipid content ranged from 2.97% d.b. to 3.54% d.b. the first year and from 3.10% d.b. to 3.50% d.b. in the second year. Six main fatty acids were detected in all samples. In the order of decreasing amounts, these were: linoleic (C18:2) > palmitic (C16:0) ≈ oleic (C18:1) > linolenic (C18:3) > stearic (C18:0) > palmitoleic (C16:1). This distribution did not change during the two years the cultivars were grown. As it can be seen in Table 1, linoleic acid (C18:2) was present in amounts starting from 1.14 g/100 g d.b. to 0.50 g/100 g d.b. throughout all samples, with a mean of 0.68 g/100 g d.b. and a standard deviation (SD) of 0.16 g/100 g d.b. Palmitic acid (C16:0) and oleic (C18:1) acid were detected in equal amounts. Palmitic acid ranged from 0.36 g/100 g d.b.

Table 1 Fatty acids in 10 Italian durum wheat cultivars, grown in different locations for two consecutive years (g/100 g sample, d.b.)

Fatty acid	Min	Max	Mean	*SD*
Palmitic (C16:0)	0.17	0.36	0.24	*0.04*
Palmitoleic (C16:1)	0.004	0.007	0.005	*0.001*
Stearic (C18:0)	0.01	0.03	0.016	*0.005*
Oleic (C18:1)	0.17	0.43	0.24	*0.07*
Linoleic (C18:2)	0.50	1.14	0.68	*0.16*
Linolenic (C18:3)	0.14	0.06	0.08	*0.02*
Saturated	0.19	0.39	0.26	*0.04*
Monounsaturated	0.18	0.44	0.25	*0.07*
Polyunsaturated	0.57	1.28	0.76	*0.18*
Ratio unsaturated/saturated	3.15	4.44	3.83	*0.38*

d.b. = dry basis; SD = standard deviation. From: Narducci et al. (2019)

to 0.17 g/100 g d.b., with a mean 0.24 g/100 g d.b. (SD 0.04) and oleic from 0.43 g/100 g d.b. to 0.17 g/100 g d.b., with a mean 0.24 g/100 g d.b. (SD 0.07). Linolenic acid (C18:3) ranged from 0.14 g/100 g d.b. to 0.06 g/100 g d.b., with a mean 0.08 g/100 g d.b. (SD 0.02). Stearic acid (C18:0) ranged from 0.01 g/100 g d.b. to 0.03 g/100 g d.b., with a mean 0.016 (SD 0.005). Finally, palmitoleic acid (C16:1) was detected in very small amounts starting from 0.007 g/100 g d.b. to 0.004 g/100 g d.b., with a mean of 0.005 g/100 g d.b. (SD 0.001). Polyunsaturated acids were preponderant over saturated and monounsaturared in all samples ranging from 0.57 g/100 g d.b. to 1.28 g/100 g d.b. Total monounsaturated and total saturated, whose levels were roughly similar ($p<0.01$), covered from 0.18 g/100 g d.b. to 0.44 g/100 g d.b. and from 0.19 g/100 g d.b. to 0.39 g/100 g d.b., respectively. The unsaturated/saturated ratio ranged from 3.15 g/100 g d.b. to 4.44 g/100 g d.b. considering all samples, with a mean of 3.83 g/100 g d.b.

Concerning fatty acids composition, Bottari et al. in 1999 (Caboni, 2012) observed the presence of more than 60 peaks by gas chromatography and mass spectrometry (GC-MS) and identified fatty acids with even numbers of carbon atoms from C12 to C30 as well as from C15 and C17. However, the major fat components were saturated and unsaturated C16 and C18, and particularly C16:0, C18:1 and C18:2, which together represented around 90% of the total composition.

The USDA database and other works also reported small amounts of C14:0 in durum wheat kernels (USDA 0.003 g/100 g f.m., corresponding to 0.0035 g/100 g d.b.). There are publications reporting about other fatty acids as well (C17, C20, C22 and C24), some in kernels (Beleggia et al., 2013) and others in germ oil (Zarroug et al., 2015; Güven and Kara, 2016). However, C16:0, C18:0, C18:1, C18:2 and C18:3 are constantly reported by all published works and are regarded as the most important fatty acids in durum wheat, the others amounting to about 1–2% in total.

Lipids in wheat kernels of varying size were determined by Chien-Mei Chiu and Pomeranz (1967). They found that total lipid content per kernel depended primarily on kernel size and that concentrations of free and bound lipids were higher in small kernels than in large ones.

Alkylresorcinol composition allows the differentiation of *Triticum* spp. having different degrees of ploidy. Total alkylresorcinol content was assessed in wholegrain flours of 15 varieties each of bread wheat, durum, spelt, emmer and einkorn grown in four different environments. Bread wheat (761 ± 92 μg/g d.m.) and spelt (743 ± 57 μg/g d.m.) belonging to the hexaploid species showed higher alkylresorcinols concentrations than the tetraploid durum (654 ± 48 μg/g d.m.) while the concentrations found in the diploid einkorn (737 ± 91 μg/g d.m.) and the tetraploid emmer (697 ± 94 μg/g d.m.) did not significantly differ from the other species (Ziegler et al., 2015).

3 Significance of lipids in wheat technology and products

It is well known that wheat kernels are usually milled into flours which are then used to manufacture widely consumed foods worldwide such as bread, pasta, noodles and bakery products.

Kernel hardness is an important quality characteristic used in classifying wheat cultivars. Differences in endosperm texture among *Triticum aestivum* L. or between *Triticum aestivum* L. and *Triticum durum* Desf. affect their milling behavior and the properties of the obtained flours or semolina. It is accepted that the presence or absence of puroindolines is responsible for differences in endosperm texture (Greenwell and Schofield, 1986; Pauly et al., 2013) and several results confirm that starch-surface polar lipids are involved in the interaction of puroindolines with wheat starch and kernel hardness (Rosicka-Kaczmarek et al., 2015; Ma et al., 2016). Qin et al. (2019) studied the accumulation of glycolipids in wheat grain and their role in hardness during grain development and concluded that the content of polar lipids in the starch surface was closely related to grain hardness.

The two predominant techniques for grinding wheat grains are stone milling and roller milling. Stone mills are the oldest attrition mills used for making wholegrain flours, which simultaneously use compression, shear and abrasion techniques to grind wheat kernels between two stones and produce a theoretical extraction rate of 100%. In contrast to this, the process of roller milling involves separation of the endosperm from the bran and germ followed by gradual size reduction of the endosperm which then produces different fractions (refined flour/semolina and brans) with different compositions. Wheat (*Triticum aestivum* L.) flour contains approximately 2.0–2.5 lipids depending on the genetic and environmental characteristics of grains, milling conditions and lipid extraction method (Pareyt et al., 2011). In a recent paper by Carcea et al. (2020), eight commercial mixes of soft wheat grains were stone milled and roller milled and the milling products were analyzed, among other parameters, for their total lipids and alkylresorcinols content. The range of lipids in the grains was between 1.8% d.m. and 3.0% d.m. and the total lipid values in milled samples was 1.8–3.0% d.m. in stone-milled flour, 0.7–1.7% d.m. in the roller-milled refined flour, 6.1–7.5% d.m. in the coarse bran and 3.9–6.9% d.m. in the fine bran. These results are justified by the fact that the wheat germ during roller milling is ground and ends up mainly in the two bran fractions, slightly more in the coarse one than in the fine one. In terms of alkylresorcinols, expressed in mg/100 g d.m., there were 32.1–47.5 mg/100 g d.m. in grains and stone-milled wholegrain flour, 168.4–251.2 mg/100 g d.m. in the coarse bran and 68.1–106.6 mg/100g d.m. in fine bran. In the refined flour, they were below the limit of detection.

Shelf life of whole wheat flour is shorter compared to white flour due to the presence of lipids and lipid-degrading enzymes. Lypolitic degradation leads to reduction in functionality, palatability and nutritional properties (Doblado-Maldonado et al., 2012). Wholegrain wheat flour is facing the quality challenge of lipid rancidity which negatively affects its nutritional, sensory and technological properties. The main cause of lipid rancidity is endogenous esterases and lipases. Wei et al. (2020) evaluated 66 European wheat varieties for their esterase and lipase activity and discovered that there are varieties more suitable for lipid-stable wholegrain products and suggested a new prospect for breeding selection based on low lipase activity.

Similarly, as a consequence of its high lipid content and lipase activity level, wheat bran deteriorates rapidly upon storage. Shelf life studies of durum wheat bran as a by-product of debranning/milling of wheat showed an increase in free fatty acid content and a decrease of linoleic acid during 30-day storage (Cardenia et al., 2018). Pearling prior to milling has been experimented as a way to lower the free fatty acids content and lipase activity levels in wheat bran, therefore contributing to improving storage stability (De Brier and Delcour, 2017).

The lipid profile has been identified in soft wheat as a way to differentiate a whole grain flour from a refined one. In the work by Geng et al. (2015), galactolipids and phospholipids were proposed to be potential discriminatory compounds for refined flour, whereas γ-oryzanol, alkylresorcinols, triglycerides and diglycerides could distinguish whole wheat flour from a refined one based on principal component analysis.

Intrinsic lipids of refined flours even from single soft wheat varieties of different baking quality have been studied by several researchers (Fisher at al., 1964, 2006; Macmurray and Morrison, 1970; Bell et al., 1987). They are either starch (30–40%) or non-starch (60–70%) lipids and they play a different role depending on the stage of product manufacturing and preservation. For example, starch lipids are important in bread staling, but they do not affect bread loaf volume or crumb structure as they are only set free at the end of the baking process and prior to that they are not available because of their location inside the starch granules (Melis and Delcour, 2020).

When flour is used to prepare a dough for breadmaking with no external lipids added, it has been seen that during the first stages of mixing, about 70% of the total flour lipids become bound to or trapped within the gluten fraction (Pareyt et al., 2011). Protein–lipid interactions in wheat gluten were also studied by Carcea and Schofield (1996). Moreover, it has been observed that the coupled oxidation of cysteine-containing storage proteins (free SH groups of gluten proteins) and polyunsaturated fatty acid by lipoxygenases (LOX) induces changes in the rheological properties of doughs and especially improves dough-mixing tolerance (Marion et al., 2003). Lipids also play a role in the decrease in baking quality of flours with aging which has been connected to an increase in free fatty acids content (Mense and Faubion, 2017).

Lipids, and non-starch lipids in particular, are important for breadmaking because they adsorb to the surface of gas bubbles formed during the proving stage of breadmaking, stabilizing the gas cells and improving gas retention within the dough. This behavior contributes to the volume and texture of the loaf (Min et al., 2017; Janssen et al., 2020). The technological role of lipids is well documented for wheat flours. It was demonstrated by MacRitchie (2016) and by earlier works (Pomeranz, 1987) that nonpolar lipids of wheat flour negatively impact bread volume, whereas polar lipids, above a threshold content, have a positive effect. All experiments that involved defatting of flours or extraction of lipids, their fractionation and subsequent reintroduction in wheat flour did not account for the complexity of lipid distribution in the grain. They proved that the effects could be related to the foaming properties of the aqueous phase of bread dough and to a mechanism related to the formation and stabilization of gas bubble cells in doughs (Salt et al., 2018). Angioloni and Collar (2011) studied the significance of lipid binding in wheat flour, doughs and bread, and found that free lipids, starch lipids and bound lipids were, respectively, the major, intermediate and minor lipid fractions in flour, dough and bread. They also found that lipids bound to proteins during dough mixing are translocated and bound to starch during baking. Phospholipids and galactolipids have been reported to have beneficial properties for breadmaking, whereas free fatty acids and triacylglycerols are considered detrimental (Gonzalez-Thuillier et al., 2015; Min et al., 2020). The effect of bran fractions on dough and end-product quality was studied in whole wheat bread by Khalid et al. (2017). They found bran lipids to be detrimental to most breadmaking characteristics.

Oxidation of flour lipids contributes to the formation of volatile compounds in bread crumb and potentially to the bread aroma (Birch et al., 2014). Lipids also play a role in bread storage. Some wheat flour lipids can form complexes with amylose during baking and it is reasonable to suggest that this has a softening effect on bread crumb (Pareyt at al., 2011). Endogenous lipids interactions with amylopectin during bread storage were also hypothesized (Gerits et al., 2015). Janssen et al. (2018) studied the lipid species distribution in the different stages of soft wheat straight dough breadmaking and observed differences in extractability.

In durum wheat (*Triticum durum* Desf.), the distribution of lipid classes in semolina reported by Caboni (2012) showed a total lipids content of 1.95 g/100 g d.b., of which 1.15 g/100 g d.b. were free and the rest were bound lipids. The USDA (2020) database reports a total lipid value of 1.0 g m.b. for durum wheat semolina, whereas the Italian CREA database (CREA, 2020) indicates a total lipid value of 0.5 g/100 g m.b..

Lipids can affect the technological quality of durum wheat products such as couscous and pasta. In a study by Ounane et al. (2006) it was revealed that the cooking quality of couscous made from different durum wheat semolinas was partially dependent on the semolina-free lipid content and composition. The surface state of the couscous depended on the apolar lipid content, whereas polar lipids - and especially glycolipids - affected couscous texture in terms of firmness. Matsuo et al. (1986) established that durum semolina nonpolar lipids influence surface stickiness of microprocessed cooked spaghetti.

Untargeted lipidomics in conjunction with chemometric tools has been proven to have a significant potential for screening and detection of adulteration of durum wheat with soft wheat as it might happen in pasta making. Digalactosyl diglyceride 36:4 was shown to be an effective marker differentiating between authentic durum wheat and its adulterated admixture down to 3% adulteration level which is the maximum contamination level allowed by Italian legislation (Righetti et al., 2018).

Kaur et al. (2016) studied the physicochemical and rheological properties of starch and flour from different durum wheat varieties and their relationships with noodle quality; they concluded that noodle cohesiveness related positively to the proportion of amylose-lipid complexes.

4 Significance of wheat lipids and associated substances in human nutrition

Lipids are a minor nutritional component of wheat despite being important contributors to flour processing and, particularly, for breadmaking. In human nutrition, the lipid contribution of cereals is certainly a minor one especially for people consuming mostly refined cereal products. Wholegrain products consumption or consumption of products enriched with wheat bran, wheat germ or wheat oil may give a contribution considering that these products are much richer in lipids, besides fiber, vitamins, minerals and bioactive substances.

Epidemiological studies have shown that the consumption of whole grains can reduce the risk of metabolic disorders, in particular type 2 diabetes mellitus (Hu et al., 2020), cardiovascular diseases (Truswell, 2002; Hollænder et al., 2015) and some types of cancer (Aune et al., 2016). Evidence from recent intervention studies is also accumulating (Schutte et al., 2018; Eriksen et al., 2020). However, the underlying mechanisms remain a matter of debate. Prior to the application of metabolomics on wholegrains and health, explanations for why observational studies consistently found associations between eating more wholegrains and reduced risk of many diseases centered on the higher fiber, vitamin and mineral content as well as phenolic phytochemicals for their antioxidant capacity. To date, animal and human trials using metabolomics have led to new hypothesis around mechanistic effects of wholegrains; however, greater work is needed to establish if there is any causality between the phytochemicals found in grains and observed metabolic effects (Ross, 2015). Reddy et al. (2000) studied the preventive potential of wheat bran fractions against experimental colon carcinogenesis and concluded for the first time that the lipid fraction of wheat bran has strong colon tumor-inhibitor properties. However, the mechanisms by which the lipid fraction of wheat bran inhibited colon carcinogenesis remained to be elucidated as well as the identification of the biologically active constituents of the lipid fraction.

Besides their role as energy contributors, wheat lipids are important in human nutrition because they include essential and healthy (polyunsaturated) fatty acids, but also fat-soluble vitamins and other bioactive compounds such as sterols and alkylresorcinols. The nutritional role of fatty acids depends on the length of the carbon chain, the number of double bonds and the polyunsaturated fatty acids (PUFA) as well as the position of the unsaturation. PUFA of nutritional interest have a number of carbon atoms ≥ 18. Among these, linoleic acid (C18:2 n-6) and α-linolenic (C18:3 n-3) are defined as essential because they cannot be synthetized within our body and must be introduced into our diet. Wheat grains contain both of them. Fatty acids are known to play a part in many metabolic processes in humans. Esterified as phospholipids, they are fundamental components of cellular membranes, they regulate the activity of enzymes and receptors, they are precursors of molecules with important biological activities such as prostaglandins, and so on and they modulate the expression of genes. There is evidence to suggest that the lack of essential fatty acids or their wrong balance in the diet may lead to several illnesses including skin and cardiovascular diseases and osteoporosis (USDA, 2005).

Unsaturated fatty acids are found in membranes as esterified phosphoglycerides. The degree of unsaturation introduces 'kinks' in the hydrocarbon chain and affects the ability of the phosphoglycerides to be able to pack together. Chains containing multiple double bonds pack together better than chains containing few double bonds (Lunn and Theobald, 2006).

Some researchers studied the influence of ancient wheat species on human health and hypothesized that one of the reasons for the improvement of the lipid profile induced by a diet based on ancient or heritage cultivars in patients suffering from different pathologies is the better lipid profile of these cultivars, in particular their higher content in mono- and polyunsaturated fatty acids (Dinu et al., 2018). However, in a recent review by Spisni et al. (2019) on differential physiological responses elicited by ancient and heritage wheat cultivars compared to modern ones, analyzing the literature on lipid profiles they concluded that, to date, there is no scientific evidence of differences in lipid composition between heritage and modern cultivars, while extensive comparative studies of the differences in the lipid profile between ancient and modern wheats are still lacking. Simsek et al. in a recent study (2019) compared the chemical composition of historical versus modern hard red spring wheat cultivars (released between 1910 and 2013) and concluded that wheat-breeding strategies did not seem to affect the composition of modern cultivars as compared to historical ones.

4.1 Carotenoids

Carotenoids, besides being responsible for yellow, orange and red colors in plants and derived products, are also nutritionally important as bioactive substances, having provitamin and antioxidant activities. Lutein, zeaxanthin, β-cryptoxanthin, β-carotene and α-carotene are the carotenoids commonly found in wholegrain cereals (Adom et al., 2003). Lutein and zeaxanthin are the carotenoids found in the highest concentration in wheat and they are found at relatively high levels in durum wheat, its products and in einkorn wheat (Abdel-Aal et al., 2013). Lutein and zeaxanthin do not show provitamin A activity, but they have been identified as essential for eye health (Abdel-Aal et al., 2013). Carotenoids concentration differs according to kernel fraction, showing a decreasing amount in germ, bran and endosperm, respectively. Moreover, the kind of carotenoid varies, α-and β-carotene being mainly located in the germ, whereas lutein is equally distributed across the grain (Caboni, 2012). Processing into bread, pasta, and so on produces losses of original carotenoids.

4.2 Alkylresorcinols

Alkylresorcinols are phenolic lipids which are present in cereals and they have received research attention from several points of view: as markers of wholegrain flour, as markers of consumption of wholegrain foods and as single components responsible for the positive effects on metabolism in the consumption of wholegrain products.

Alkylresorcinols have been found in wheat (*Triticum* species, 489–1429 μg/g), rye (720–761 μg/g), triticale (439–647 μg/g) and barley (42–51 μg/g) but not in rice, oats, maize, sorghum or millet (Ross et al., 2003b; Ciccoritti et al., 2013). They are located in the outer cuticula of the testa/inner cuticula of the pericarp, and therefore they are found in detectable amounts only in whole grains and bran products of the previously mentioned cereals, whereas refined products only contain traces (Landberg et al., 2014). It was also determined that the alkylresorcinols of milling fractions closely follow the ash content and could be used as a marker of the presence of bran in flour. Moreau and Fang (2016), by using a sensitive HPLC method with fluorescence detection, reported for the first time that wheat germ oil contains moderate levels of alkylresorcinols, approximately 800–1500 μg/g of oil.

Alkylresorcinols are incorporated into human erythrocyte membranes in vivo (Linko and Adlercreutz, 2005). They are absorbed in humans in proportion to intake and have, therefore, been suggested and evaluated as biomarkers for wholegrain wheat and rye intake (Ross et al., 2003a, 2005). Moreover, results from different model studies have indicated that the alkylresorcinols may have some bioactivities, including enzyme inhibition, suppression of adipocyte lipolysis and inhibition of colon cancer tumor growth, but it is currently unknown whether they are bioactive in vivo or not (Landberg et al., 2014). The findings of Oishi et al. (2015) in mice suggested that wheat alkylresorcinols increase glucose tolerance and insulin sensitivity by suppressing hepatic lipid accumulation and intestinal cholesterol absorption, which subsequently suppresses diet-induced obesity. The same research group (Horikawa et al., 2017) showed that wheat alkylresorcinols reduce micellar solubility of cholesterol in vitro and increase cholesterol excretion in mice.

5 Wheat germ

The separation of the germ (embryonic axis and scutellum) during wheat milling is the starting point for the production of a lipid-rich product, the wheat germ, which can be used as such in food or for the extraction of wheat germ oil.

Wheat germ is, therefore, a by-product of wheat milling and it contains - besides proteins, carbohydrates, dietary fiber, vitamins and minerals - around 10% total lipids (USDA, 2020) including nonpolar lipids, glycolipids, phospholipids. It also contains alcohols, esters, alkene, aldehydes, tocopherols, n-alkanols, sterols, 4-methyl sterols, tryterpenols, hydrocarbons, pigments and volatile components (Jiang and Wang, 2005; Brandolini and Hidalgo, 2011; Ghafoor et al., 2017). The presence of high lipase activity and high content of unsaturated fatty acids induce rapid lipid rancidity in wheat germ during storage. Heat treatments are used to inactivate lipases and stabilize it (Boukid et al., 2018; Ling et al., 2019).

Given its richness in nutritionally valuable components, wheat germ has received the attention of research as a supplement on its own or as an ingredient to improve the nutritional quality of different cereal foods such as bread and pasta, and also other foods such as soft cheese and yoghurt (Khalid and Mosa, 2018; Gahruie et al., 2019).

The effects of the daily intake of wheat germ rich in lipids on gastrointestinal discomfort and gut microbiota was recently studied by Moreira-Rosário et al. (2020) in a randomized, double-blind, crossover, controlled trial where healthy adults consumed wheat bread enriched with 6 g of wheat germ compared to non-enriched wheat bread. The results showed that adding wheat germ to white wheat bread may promote a healthy gut bacterial microbiota and the gastrointestinal health. It can be hypothesized that wheat lipids also take part in the improvement. The effect of wheat germ on metabolic markers was studied in a systematic review and meta-analysis of randomized controlled trials by Liaqat et al. (2020), and this review revealed nonsignificant association between wheat germ interventions and metabolic markers. Furthermore, the protective effect of a fermented wheat germ extract in intestinal epithelial cells in the case of lipopolysaccharide-evoked inflammation was demonstrated by Karancsi et al. (2020). Wheat germ fermented with *Lactobacillus plantarum* dy-1 was used by Zhao et al. (2020) to improve the nutritional and sensorial qualities of bread. Yaseen et al. (2019) investigated the potential protection from Alzheimer's disease by wheat germ nano-form in a rat model and concluded that the wheat germ powder gave promising results. Recently, Azizova et al. (2020) studied the effects of supplementing wheat germ on the spectrum of lipoproteins and fatty acids in the blood of experimental hyperlipidemic animals and concluded that wheat germ can be prescribed to reduce the dose of statin.

6 Wheat oil

Wheat germ can in turn be used for the extraction of wheat oil by different techniques involving mechanical pressing or solvent extraction (El-Shami et al., 2011; Al-Obaidi et al., 2013). For a better-quality oil, techniques such as supercritical fluid fractionation, molecular distillation and other innovative methods have been used (Panfili et al., 2003; Durante et al., 2012; Ghafoor et al., 2017). Enzymes have also been experimented with to improve oil extraction (Xie et al., 2011). Oil can also be extracted directly from bran (Kumar and Krishna, 2015).

Wheat germ oil (WGO) finds utilization in foods, feeds, medicine, the cosmetic industry and biological insect control. Due to the presence of polyunsaturated fatty acids, minerals and bioactive substances such as vitamin E, carotenoids and phytosterols, wheat germ oil is nutritionally valuable (Hassanien et al., 2014). The cereal germ oils are in fact important sources of tocopherols (Górnaś et al., 2016) and this makes them more stable to oxidation compared to other oils with high amounts of unsaturated fatty acids. The main tocols of WGO are α-tocopherol and β-tocopherol. The antioxidant activity of tocopherols increases from α to δ whereas the vitamin activity is reduced: for this reason WGO is considered a good source of vitamin E (Trela and Szymańska, 2019). Niu et al. (2013) analyzed volatile compounds in wheat germ oil and they isolated and identified 44 volatile compounds, mainly alcohols, esters, alkenes and aldehydes, whereas the main component of the unsaponifiable matter was β-sitosterol.

Wheat germ oil has been used as an antimicrobial ingredient in bread by Debonne et al. (2018) and it was shown to actually extend the shelf life of bread for 3 days.

It has been reported that WGO has the ability to protect against hepatotoxicity induced by a variety of oxygen radical generators (Saleh et al., 2010; Saleh, 2016; Akool, 2019) and in general it is able to improve cellular defence against free radicals (Ghafoor et al., 2017). Field et al. (2010) found that feeding wheat germ meal and wheat germ oil reduced azoxymethane-induced colon tumors in rats. Hamdi (2019) studied the preventive role of wheat germ oil against testicular damage in male rats finding that wheat germ oil has a preventive role via its potential to counteract reactive oxygen species. An effect of wheat germ oil in enhancing kidney recovery processes in irradiated rats was also observed by Gazar et al. (2016).

Mirtalebi et al. (2020) studying the storage stability of wheat germ oil, indicated that encapsulation within nanostructured lipid carriers could be an effective delivery and protection system for wheat germ oil as a source of bioactive compounds. Nano-encapsulated wheat germ oil has been effectively used in the manufacture of functional cheese (Soliman et al., 2019).

7 Conclusion

From the previous sections, we can deduct that lipids and associated substances such as fat-soluble vitamins and other bioactive molecules are important components of wheat kernels, even if they are present in relatively small amounts. The germ is the richest part of the grain; however, lipids are also present in the endosperm and in the outer layers making up the bran. This implies that lipids have a technological and a nutritional role in all grain-milling products, that is, refined flours, wholegrain flours, bran fractions and germ-enriched flours. The nutritional value of wheat grain lipids is connected to their composition and in particular to their fatty acid composition. If the wheat germ is isolated from the bran fraction, wheat germ oil can be produced which not only finds utilization in foods but also in feeds, medicine, the cosmetic industry and biological insect control.

Considering recent scientific evidence of positive effects on human health in terms of the consumption of wholegrain products, those richer in lipids compared to refined products or germ-enriched wheat products and wheat germ oil, it can be said that lipids are nutritionally important components in wheat grains. Further research is needed to definitely assess their specific role in the observed phenomena.

8 Where to look for further information

General information on the chemical nature of molecules belonging to the lipids group can be found in any organic chemistry book. An example of a good one is Solomons, T. W. G., Fryhle, C. B. and Snyder, S. A., Organic Chemistry, 12th edition, John Wiley & Sons Inc., USA, 2016.

Detailed information on the levels of lipids in wheats as raw materials or on the lipid content of wheat-based foods can be found in food composition tables which are issued by different organisations in several countries and which are also available online. Examples are the Food Composition Tables of CREA in Italy (https://www.alimentinutrizione.it/tabelle-nutrizionali/ricerca-per-alimento, in Italian) and the USDA National Nutrient Database for Standard Reference in the USA (https://fdc.nal.usda.gov/). On the FAO website there is a section dedicated to the International Network of Food Data Systems (INFOODS) (http://www.fao.org/infoods/infoods/tables-and-databases/en/) where information on the food composition databases of several countries can be accessed.

There are several national societies worldwide that focus on the advancement of cereal science and technology which hold annual scientific meetings where results of research involving lipids are presented. The two key international ones are the International Association for Cereal Science and Technology (ICC) (https://icc.or.at/) based in Vienna, Austria and the Cereals and Grain Association (https://www.cerealsgrains.org/Pages/default.aspx) based in Eagan, Minnesota, USA. They organize a number of scientific initiatives such as periodical international conferences and symposia, workshops, working groups on specific topics, issuing of analytical standards, position documents, books and publications, so it is worth visiting their website to keep up with research trends in the cereals world which might also include wheat lipids. A society which is devoted to lipids in foods in general is the European Federation for the Science and Technology of Lipids (Euro Fed Lipid) (http://www.eurofedlipid.org/index.html). Its mission is the furthering of lipid science and technology and amongst its activities there is the organisation of international congresses and the publishing of the European Journal of Lipid Science and Technology (https://onlinelibrary.wiley.com/journal/14389312).

Other scientific journals specialised in cereal science and technology where publications on lipids in wheat can be found are the Journal of Cereal Science (https://www.sciencedirect.com/journal/journal-of-cereal-science), Cereal Chemistry (https://onlinelibrary.wiley.com/journal/19433638) and Cereal Foods World (https://www.cerealsgrains.org/publications/plexus/cfw/Pages/default.aspx).

9 References

Abdel-Aal, el-S. M., Akhtar, H., Zaheer, K. and Ali, R. Dietary sources of lutein and zeaxanthin carotenoids and their role in eye health. *Nutrients* 5(4), 1169–1185, 2013.

Adom, K. K., Sorrells, M. E. and Liu, R. H. Phytochemical profiles and antioxidant activity of wheat varieties. *Journal of Agricultural and Food Chemistry* 51(26), 7825–7834, 2003.

Akool, E.-S. Molecular mechanisms of the protective role of wheat germ oil against oxidative stress-induced liver disease. In: Watson, R. R. and Preedy, V. (Eds), *Dietary*

Interventions in Liver Disease: Foods, Nutrients and Dietary Supplements. Academic Press, Cambridge, MA, 233–238, 2019.

Al-Obaidi, L., Dunford, N. T. and Goad, C. Mechanical extraction of wheat germ oil. *Transactions of the ASABE* 56(5), 1871–1876, 2013.

Angioloni, A. and Collar, C. Significance of lipid binding on the functional and nutritional profiles of single and multigrain matrices. *European Food Research and Technology* 233(1), 141–150, 2011.

AOCS. http://lipidlibrary.aocs.org. Accessed on October 29th, 2020.

Armanino, C., De Acutis, R. and Rosa Festa, M. Wheat lipids to discriminate species, varieties, geographical origins and crop years. *Analytica Chimica Acta* 454(2), 315-326, 2002.

Aune, D., Keum, N., Giovannucci, E., Fadnes, L. T., Boffetta, P., Greenwood, D. C., Tonstad, S., Vatten, L. J., Riboli, E. and Norat, T. Whole grain consumption and risk of cardiovascular disease, cancer, and all cause and cause specific mortality: systematic review and dose-response meta-analysis of prospective studies. *BMJ* 353, i2716, 2016.

Azizova, D. M., Sabirova, R. A., Mavlyanov, I. R., Kulmanova, M. U., Soliev, A. and Abdullaev, B. S. Dietary supplement-wheat germ's effect on the spectrum of lipoproteins and fatty acids in the blood of experimental hyperlipidemic animals. *International Journal of Psychosocial Rehabilitation* 24(2), 2170–2186, 2020.

Beleggia, R., Platani, C., Nigro, F., De Vita, P., Cattivelli, L. and Papa, R. Effect of genotype, environment and genotype by environment interaction on metabolita profiling in durum wheat (*Triticum durum* Desf.) grain. *Journal of Cereal Science* 57(2), 183–192, 2013.

Bell, B. M., Daniels, D. G. H., Fearn, T. and Stewart, B. A. Lipid compositions, baking qualities and other characteristics of wheat varieties grown in the UK. *Journal of Cereal Science* 5(3), 277–286, 1987.

Birch, A. N., Petersen, M. A. and Hansen, Å. S. Review: aroma of wheat bread crumb. *Cereal Chemistry Journal* 91(2), 105–114, 2014.

Boukid, F., Folloni, S., Ranieri, R. and Vittadini, E. A compendium of wheat germ: separation, stabilization and food applications. *Trends in Food Science and Technology* 78, 120-133, 2018.

Brandolini, A. and Hidalgo, A. Wheat germ: not only a by-product. *International Journal of Food Sciences and Nutrition* 63(Suppl. 1), 71–74 2011.

Caboni, M. F. Durum wheat lipids. In: Sissons M., Abecassis J., Marchylo B. and Carcea M. (Eds), *Durum Wheat. Chemistry and Technology*. AACCInternational, St.Paul, MN, 106–113, 2012.

Carcea, M. and Schofield, J. D. Protein-lipid interactions in wheat gluten: reassessment of the occurence of lipid-mediated aggregation of protein in the gliadin fraction. *Journal of Cereal Science* 24, 101–113, 1996.

Carcea, M., Turfani, V., Narducci, V., Melloni, S., Galli, V. and Tullio, V. Stone milling versus roller milling in soft wheat: influence on products composition. *Foods* 9(1), 3, 2020.

Cardenia, V., Sgarzi, F., Mandrioli, M., Tribuzio, G., Rodriguez-Estrada, M. T. and Gallina Toschi, T. G. Durum wheat bran by-products: oil and phenolic acids to be valorized by industrial symbiosis. *European Journal of Lipid Science and Technology* 120(8), 1700209, 2018.

Chiu, C.-M. and Pomeranz, Y. Lipids in wheat kernels of varying size. *Journal of Food Science* 32(4), 422–425, 1967.

Ciccoritti, R., Carbone, K., Bellato, S., Pogna, N. and Sgrulletta, D. Content and relative composition of some phytochemicals in diploid, tetraploid and hexaploid *Triticum* species with potential nutraceutical properties. *Journal of Cereal Science* 57(2), 200-206, 2013.

CREA Italian Food Composition Tables. https://www.alimentinutrizione.it/tabelle-nutrizionali/000310. Accessed on October 29th, 2020.

Debonne, E., De Leyn, I., Verwaeren, J., Moens, S., Devlieghere, F., Eeckhout, M. and Van Bockstaele, F. The influence of natural oils of blackcurrant, black cumin seed, thyme and wheat germ on dough and bread technological and microbiological quality. *LWT - Food Science and Technology* 93, 212–219, 2018.

De Brier, N. and Delcour, J. A. Pearling affects the lipid content and composition and lipase activity levels of wheat (*Triticum aestivum* L.) roller milling fractions. *Cereal Chemistry Journal* 94(3), 588–593, 2017.

Dinu, M., Whittaker, A., Pagliai, G., Benedettelli, S. and Sofi, F. Ancient wheat species and human health: biochemical and clinical implications. *Journal of Nutritional Biochemistry* 52, 1–9, 2018.

Doblado-Maldonado, A. F., Pike, O. A., Sweley, J. C. and Rose, D. J. Key issues and challenges in whole wheat flour milling and storage. *Journal of Cereal Science* 56(2), 119–126, 2012.

Durante, M., Lenucci, M. S., Rescio, L., Mita, G. and Caretto, S. Durum wheat by-products as natural sources of valuable nutrients. *Phytochemistry Reviews* 11(2–3), 255–262, 2012.

El-Shami, S. M., El-Mallah, M. H., Hassanein, M. M. M. and Abdel-Razek, A. G. Elucidation of polynutrients and fatty acid patterns of wheat germ oil extracted by n-hexane and chloroform/methanol. *Journal of Applied Sciences Research* 7(12), 1840–1846, 2011.

Eriksen, A. K., Brunius, C., Mazidi, M., Hellström, P. M., Risérus, U., Iversen, K. N., Fristedt, R., Sun, L., Huang, Y., Nørskov, N. P., Knudsen, K. E. B., Kyrø, C., Olsen, A., Tjønneland, A., Dicksved, J. and Landberg, R. Effects of whole-grain wheat, rye, and lignan supplementation on cardiometabolic risk factors in men with metabolic syndrome: a randomized crossover trial. *American Journal of Clinical Nutrition* 111(4), 864–876, 2020.

Field, R., Verghese, M., Panala, V., Shackelford, L., Boateng, J. and Walker, L. T. Feeding wheat germ meal and wheat germ oil reduced azoxymethane-induced colon tumors in Fisher 344 male rats. *Research Journal of Phytochemistry* 4(1), 1–12, 2010.

Finnie, S. M., Jeannotte, R. and Faubion, J. M. Quantitative characterization of polar lipids from wheat whole meal, flour, and starch. *Cereal Chemistry Journal* 86(6), 637–645, 2009.

Fisher, N., Bell, B. M., Rawlings, C. E. B. and Bennett, R. The lipids of wheat.III-Further studies of the lipids of flours from single wheat varieties of widely varying baking quality. *Journal of the Science of Food and Agriculture* 17(8), 370–382, 2006.

Fisher, N., Broughton, M. E., Peel, D. J. and Bennett, R. The lipids of wheat.II-Lipids of flours from single wheat varieties of widely varying baking quality. *Journal of the Science of Food and Agriculture* 15(5), 325–341, 1964.

Fujino, Y. and Ohnishi, M. Sphingolipids in wheat grain. *Journal of Cereal Science* 1(2), 159–168, 1983.

Gahruie, H. H., Eskandari, M. H. and Mesbahi, G. Development of functional yogurt fortified with wheat germ and strawberry as functional ingredients. *Progress in Nutrition* 21, 388–398, 2019.

Gazar, A. A., Fahmy, H. A. and Mohamed, O. S. Possible effect of wheat germ oil or beta-carotene to enhance kidney recovery processes in irradiated rats. *International Journal of Research in Pharmaceutical Sciences* 7(4), 266–273, 2016.

Geng, P., Harnly, J. M. and Chen, P. Differentiation of whole grain from refined wheat (*T. aestivum*) flour using lipid profile of wheat bran, germ and endosperm with UHPLC-HRAM mass spectrometry. *Journal of Agricultural and Food Chemistry* 63(27), 6189–6211, 2015.

Gerits, L. R., Pareyt, B., Masure, H. G. and Delcour, J. A. A lipase based approach to understand the role of wheat endogenous lipids in bread crumb firmness evolution during storage. *LWT - Food Science and Technology* 64(2), 874–880, 2015.

Ghafoor, K., Özcan, M. M., Al-Juhaimi, F., Babiker, E. E., Sarker, Z. I., Ahmed, I. A. M. and Ahmed, M. A. Nutritional composition, extraction and utilization of wheat germ oil: a review. *European Journal of Lipid Science and Technology* 119(7), 1600160, 2017.

Greenwell, P. and Schofield, J. D. A starch granule protein associated with endosperm softness in wheat. *Cereal Chemistry Journal* 63, 379–380, 1986.

González-Thuillier, I., Salt, L., Chope, G., Penson, S., Skeggs, P., Tosi, P., Powers, S. J., Ward, J. L., Wilde, P., Shewry, P. R. and Haslam, R. P. Distribution of lipids in the grain of wheat (cv. Hereward) determined by lipidomic analysis of milling and pearling fractions. *Journal of Agricultural and Food Chemistry* 63(49), 10705–10716, 2015.

Górnaś, P., Rudzińska, M., Raczyk, M. and Soliven, A. Lipophilic bioactive compounds in the oils recovered from cereal by-products. *Journal of the Science of Food and Agriculture* 96(9), 3256–3265, 2016.

Güven, M. and Kara, H. H. Some chemical and physical properties, fatty acid composition and bioactive compounds of wheat germ oils extracted from different wheat cultivars. *Journal of Agricultural Sciences* 22, 433–443, 2016.

Hamdi, H. The preventive role of wheat germ oil against sertraline-induced testicular damage in male albino rats. *Andrologia* 51(10), e13369, 2019.

Hargin, K. D. and Morrison, W. R. The distribution of acyl lipids in the germ, aleurone, starch and non-starch endosperm of four wheat varieties. *Journal of the Science of Food and Agriculture* 31(9), 877–888, 1980.

Hassanien, M. M. M., Abdel-Razek, A. G., Rudzińska, M., Siger, A., Ratusz, K. and Przybylski, R. Phytochemical contents and oxidative stability of oils from non-traditional sources. *European Journal of Lipid Science and Technology* 116(11), 1563–1571, 2014.

Hollænder, P. L. B., Ross, A. B. and Kristensen, M. Wholegrain and blood lipid changes in apparently healthy adults: a systematic review and meta-analysis of randomized controlled studies. *American Journal of Clinical Nutrition* 102(3), 556–572, 2015.

Horikawa, K., Hashimoto, C., Kikuchi, Y., Makita, M., Fukudome, S. I., Okita, K., Wada, N. and Oishi, K. Wheat alkylresorcinols reduce micellar solubility of cholesterol in vitro and increase cholesterol excretion in mice. *Natural Product Research* 31(5), 578–582, 2017.

Hu, Y., Ding, M., Sampson, L., Willett, W. C., Manson, J. E., Wang, M., Rosner, B., Hu, F. B. and Sun, Q. Intake of whole grain foods and risk of type 2 diabetes: results from three prospective cohort studies. *BMJ* 370, m2206, 2020.

Impa, S. M., Vennapusa, A. R., Bheemanahalli, R., Sabela, D., Boyle, D., Walia, H. and Jagadish, S. V. K. High night temperature induced changes in grain starch metabolism alters starch, protein and lipid accumulation in winter wheat. *Plant, Cell and Environment* 43(2), 431–447, 2020.

Janssen, F., Wouters, A. G. B., Pareyt, B., Gerits, L. R., Delcour, J. A., Waelkens, E. and Derua, R. Wheat (*Triticum aestivum* L.) lipid species distribution in the different stages of straight dough bread making. *Food Research International* 112, 299–311, 2018.

Janssen, F., Wouters, A. G. B., Linclau, L., Waelkens, E., Derua, R., Dehairs, J., Moldenaers, P., Vermant, J. and Delcour, J. A. The role of lipids in determining the air-water interfacial properties of wheat, rye and oat dough liquor constituents. *Food Chemistry* 319, 126565, 2020.

Jiang, Y. and Wang, T. Phytosterols in cereal by-products. *Journal of the American Oil Chemists' Society* 82(6), 439–444, 2005.

Karancsi, Z. Móritz, A. V., Lewin, N., Veres, A. M., Jerzsele, A., Farkas, O., De Oliveira, M. R. Beneficial effect of a fermented wheat germ extract in intestinal epithelial cells in case of lipopolysaccharide-evoked inflammation. *Oxidative Medicine and Cellular Longevity*, 1482482, 2020.

Kaur, A., Shevkani, K., Katyal, M., Singh, N., Ahlawat, A. K. and Singh, A. M. Physicochemical and rheological properties of starch and flour from different durum wheat varieties and their relationships with noodle quality. *Journal of Food Science and Technology* 53(4), 2127–2138, 2016.

Khalid, K. H., Ohm, J.-B. and Simsek, S. Whole wheat bread: effect of bran fractions on dough and end-product quality. *Journal of Cereal Science* 78, 48–56, 2017.

Khalid, N. T. and Mosa, M. A. Effect of wheat germ on chemical, sensory and technological properties of soft cheese. *International Journal of Dairy Science* 13(1), 40–45, 2018.

Kumar, G. S. and Krishna, A. G. G. Studies on the nutraceuticals composition of wheat derived oils wheat bran oil and wheat germ oil. *Journal of Food Science and Technology* 52(2), 1145–1151, 2015.

Landberg, R., Kamal-Eldin, A., Salmenkallio-Marttila, M., Rouau, X. and Åman, P. Localization of alkylresorcinols in wheat, rye and barley kernels. *Journal of Cereal Science* 48(2), 401–406, 2008.

Landberg, R., Marklund, M., Kamal-Eldin, A. and Åman, P. An update on alkylresorcinols occurrence, bioavailability, bioactivity and utility as biomarkers. *Journal of Functional Foods* 7, 77–89, 2014.

Liaqat, H., Jeong, E., Kim, K. J. and Kim, J. Y. Effect of wheat germ on metabolic markers: a systematic review and meta-analysis of randomized controller trials. *Food Science and Biotechnology* 29(6), 739–749, 2020.

Ling, B., Ouyang, S. and Wang, S. Radio-frequency treatment for stabilization of wheat germ: storage stability and physicochemical properties. *Innovative Food Science and Emerging Technologies* 52, 158–165, 2019.

Linko, A. M. and Adlercreutz, H. Whole-grain rye and wheat alkylresorcinols are incorporated into human erythrocyte membranes. *British Journal of Nutrition* 93(1), 11–13, 2005.

Lunn, J. and Theobald, H. E. The health effects of dietary unsaturated fatty acids. *Nutrition Bulletin* 31(3), 178–224, 2006.

Ma, D., Qin, H., Ding, H., Zhang, J., Wang, C. and Guo, T. Surface lipids play a role in the interactions of puroindolines with wheat starch and kernel hardness. *Cereal Chemistry Journal* 93(5), 523–528, 2016.

Macmurray, T. A. and Morrison, W. R. Composition of wheat-flour lipids. *Journal of the Science of Food and Agriculture* 21(10), 520–528, 1970.

MacRitchie, F. Seventy years of research into breadmaking quality. *Journal of Cereal Science* 70, 123–131, 2016.

Majewska, K., Dabkowska, E., Grabowska, E., Tyburski, J. and Czaplicki, S. Composition of fatty acids in dark flour from spelt and common wheat grain grown organically in Poland. *Polish Journal of Natural Sciences* 33(1), 75–88, 2018.

Marion, D., Dubreil, L. and Douliez, J. P. Functionality of lipids and lipid-protein interactions in cereal-derived food products. *Oléagineux, Corps Gras, Lipides* 10(1), 47–56, 2003.

Marion, D. and Saulnier, L. Minor components and wheat quality: perspectives on climate changes. *Journal of Cereal Science* 94, 103001, 2020.

Matsuo, R. R., Dexter, J. E., Boudreau, A. and Daun, J. K. The role of lipids in determining spaghetti coking quality. *Cereal Chemistry Journal* 63(6), 484–489, 1986.

Melis, S. and Delcour, J. A. Impact of wheat endogenous lipids on the quality of fresh bread: key terms, concepts and underlying mechanisms. *Comprehensive Reviews in Food Science and Food Safety* 19(6), 3715–3754, 2020.

Melis, S., Reza Morales, W. R. and Delcour, J. A. Lipases in wheat flour bread making: importance of an appropriate balance between wheat endogenous lipids and their enzymatically released hydrolysis products. *Food Chemistry* 298(1250), 125002, 2019.

Mense, A. L. and Faubion, J. M. Effects of aging new crop wheat and flour on breadmaking quality and lipid composition. *Cereal Foods World* 62(1), 4–10, 2017.

Min, B., González-Thuillier, I., Powers, S. J., Wilde, P., Shewry, P. R. and Haslam, R. P. Effects of cultivar and nitrose nutrition on the lipid composition of wheat flour. *Journal of Agricultural and Food Chemistry* 65(26), 5427–5434, 2017.

Min, B., Salt, L., Wilde, P., Kosik, O., Hassall, K., Przewieslik-Allen, A., Burridge, A. J., Poole, M., Snape, J., Wingen, L., Haslam, R., Griffiths, S. and Shewry, P. R. Genetic variation in wheat grain quality is associated with differences in the galactolipid content of flour and the gas bubble properties of dough liquor. *Food Chemistry X* 6, 100093, 2020.

Mirtalebi, M., Rajaei, A., Bahmaei, M. and Khosroushahi, A. Y. Storage stability of wheat germ oil encapsulated within nanostructured lipid carriers. *Journal of Nanostructures* 10(2), 268–278, 2020.

Moore, C. M., Richards, R. A. and Rebetzke, G. J. Phenotipic variation and QTL analysis for oil content and protein concentration in bread wheat (*Triticum aestivum* L.). *Euphytica* 204(2), 371–382, 2015.

Moreau, R. A. and Fang, X. Analysis of alkylresorcinols in wheat germ oil via HPLC and fluorescence detection: cochromatography with tocols. *Cereal Chemistry Journal* 93(3), 293–298, 2016.

Moreira-Rosário, A., Marques, C., Pinheiro, H., Norberto, S., Sintra, D., Teixeira, J. A., Calhau, C. and Azevedo, L. F. Daily intake of wheat germ-enriched bread may promote a healthy gutbacterial microbiota: a randomised controller trial. *European Journal of Nutrition* 59(5), 1951–1961, 2020.

Narducci, V., Finotti, E., Galli, V. and Carcea, M. Lipids and fatty acids in Italian durum wheat (*Triticum durum* Desf.) cultivars. *Foods* 8(6), 223, 2019.

Nejadsadeghi, L., Maali-Amiri, R., Zeinali, H., Ramezanpour, S. and Sadeghzade, B. Membrane fatty acid compositions and cold-induced responses in tetraploid and hexaploid wheats. *Molecular Biology Reports* 42(2), 363–372, 2015.

Niu, L. Y., Jiang, S. T., Pan, L. J. and Pang, M. Characterization of wheat germ oil in terms of volatile compounds, lipid composition, thermal behavior and structure. *International Journal of Food Properties* 16(8), 1740–1749, 2013.

Oishi, K., Yamamoto, S., Itoh, N., Nakao, R., Yasumoto, Y., Tanaka, K., Kikuchi, Y., Fukudome, S., Okita, K. and Takano-Ishikawa, Y. Wheat alkylresorcinols suppress high-fat, high-sucrose diet-induced obesity and glucose intolerance by increasing insulin sensitivity and cholesterol excretion in male mice. *Journal of Nutrition* 145(2), 199-206, 2015.

Ounane, G., Cuq, B., Abecassis, J., Yesli, A. and Ounane, S. M. Effects of physicochemical characteristics and lipid distribution in Algerian durum wheat semolinas on the technological quality of couscous. *Cereal Chemistry Journal* 83(4), 377-384, 2006.

Panfili, G., Cinquanta, L., Fratianni, A. and Cubadda, R. Extraction of wheat germ oil by supercritical CO_2: oil and defatted cake characterization. *Journal of the American Oil Chemists' Society* 80(2), 157-161, 2003.

Pareyt, B., Finnie, S. M., Putseys, J. A. and Delcour, J. A. Lipids in breadmaking: sources, interactions and impact on bread quality. *Journal of Cereal Science* 54(3), 266-279, 2011.

Pauly, A., Pareyt, B., Fierens, E. and Delcour, J. A. Wheat (*Triticum aestivum* L. and *T. turgidum* L. ssp. *durum*) Kernel Hardness: II. implications for end-product quality and role of puroindolines therein. *Comprehensive Reviews in Food Science and Food Safety* 12(4), 427-438, 2013.

Pelillo, M., Ferioli, F., Iafelice, G., Marconi, E. and Caboni, M. F. Characterisation of the phospholipid fraction of hulled and naked tetraploid and hexaploid wheats. *Journal of Cereal Science* 51(1), 120-126, 2010.

Pomeranz, Y. Lipids. In: *Modern Cereal Science and Technology*. VCH Publishers, New York, 184-199, 1987.

Qin, H., Ma, D., Huang, X., Zhang, J., Sun, W., Hou, G., Wang, C. and Guo, T. Accumulation of glycolipids in wheat grain and their role in hardness during grain development. *Crop Journal* 7(1), 19-29, 2019.

Reddy, B. S., Hirose, Y., Cohen, L. A., Simi, B., Cooma, I. and Rao, C. V. Preventive potential of wheat bran fractions against experimental colon carcinogenesis: implications for human colon cancer prevention. *Cancer Research* 60(17), 4792-4797, 2000.

Righetti, L., Rubert, J., Galaverna, G., Hurkova, K., Dall'Asta, C., Hajslova, J. and Stranska-Zachariasova, M. A novel approach based on untargeted lipidomics reveals differences in the lipid pattern among durum and common wheat. *Food Chemistry* 240, 775-783, 2018.

Rosicka-Kaczmarek, J., Stasiuk, M., Nebesny, E. and Komisarczyk, A. Fluorimetric studies of the interactions of wheat puroindolines with polar lipids on the surface starch granules. *Journal of Cereal Science* 66, 53-58, 2015.

Ross, A. B. Whole grains beyond fibre: what can metabolomics tell us about mechanisms? *Proceedings of the Nutrition Society* 74(3), 320-327, 2015.

Ross, A. B., Becker, W., Chen, Y., Kamal-Eldin, A. and Åman, P. Intake of alkylresorcinols from wheat and rye in the United Kingdom and Sweden. *British Journal of Nutrition* 94(4), 496-499, 2005.

Ross, A. B., Kamal-Eldin, A., Lundin, E. A., Zhang, J. X., Hallmans, G. and Åman, P. Cereal alkylresorcinols are absorbed by humans. *Journal of Nutrition* 133(7), 2222-2224, 2003a.

Ross, A. B., Shepherd, M. J., Schüpphaus, M., Sinclair, V., Alfaro, B., Kamal-Eldin, A. and Aman, P. Alkylresorcinols in cereals and cereal products. *Journal of Agricultural and Food Chemistry* 51(14), 4111-4118, 2003b.

Ruibal-Mendieta, N. L., Delacroix, D. L., Mignolet, E., Pycke, J. M., Marques, C., Rozenberg, R., Petitjean, G., Habib-Jiwan, J. L., Meurens, M., Quetin-Leclercq, J., Delzenne, N. M. and Larondelle, Y. Spelt (*Triticum aestivum* ssp. *spelta*) as a source of breadmaking flours and bran naturally enriched in oleic acid and minerals but not phytic acid. *Journal of Agricultural and Food Chemistry* 53(7), 2751-2759, 2005.

Saleh, H. Modulatory effect of wheat germ oil on intestinal oxidative stress and DNA damage induced by carbon tetrachloride in mice. *Journal of Applied Pharmaceutical Science* 6(12), 067-074, 2016.

Saleh, Z. A., Ibrahim, K. S., Farrag, A.-R. H. and Shaban, E. E. Effect of carrot and wheat germ oil supplementation on antioxidant status of rats exposed to benzene. *Polish Journal of Food and Nutrition Sciences* 60(2), 175-181, 2010.

Salt, L. J., González-Thuillier, I.,Chope, G., Penson, S., Tosi, P., Haslam, R. P., Skeggs, P. K., Shewry, P. R. and Wilde, P. J. Instrinsic wheat lipid composition effects the interfacial and foaming properties of dough liquor. *Food Hydrocolloids* 75, 211-222, 2018.

Schutte, S., Esser, D., Hoevenaars, F. P. M., Hooiveld, G. J. E. J., Priebe, M. G., Vonk, R. J., Wopereis, S. and Afman, L. A. A. 12-wk whole-grain wheat intervention protects against hepatic fat: the Grandioos study, a randomized trial in overweight subjects. *American Journal of Clinical Nutrition* 108(6), 1264-1274, 2018.

Shewry, P. R., Wan, Y., Hawkesford, M. J. and Tosi, P. Spatial distribution of functional components in the starchy endosperm of wheat grains. *Journal of Cereal Science* 91, 102869, 2020.

Simsek, S., Budak, B., Schwebach, C. S. and Ovando-Martinez, M. Historical vs. modern hard red spring wheat: analysis of the chemical composition. *Cereal Chemistry Journal* 96(5), 937-949, 2019.

Soliman T.N., Farrag A.F., Zahran H.A.-H. and El-Salam M.E.-H.A. Preparation and properties nano-encapsulated wheat germ oil and its use in the manufacture of functional Labneh cheese. *Pakistan Journal of Biological Sciences* 22(7), 318-326, 2019.

Spisni, E., Imbesi, V., Giovanardi, E., Petrocelli, G., Alvisi, P. and Valerii, M. C. Differential physiological responses elicited by ancient and heritage wheat cultivars compared to modern ones. *Nutrients* 11(12), 2879, 2019.

Tosi, P., He, J., Lovegrove, A., González-Thuillier, I., Penson, S. and Shewry, P. R. Gradients in compositions in the starchy endosperm of wheat have implications for milling and processing. *Trends in Food Science and Technology* 82, 1-7, 2018.

Trela, A. and Szymańska, R. Less widespread plant oils as a good source of vitamin E. *Food Chemistry* 296, 160-166, 2019.

Truswell, A. S. Cereal grains and coronary heart disease. *European Journal of Clinical Nutrition* 56(1), 1-14, 2002.

Upchurch, R. G. Fatty acid unsaturation, mobilization and regulation in the response of plants to stress. *Biotechnology Letters* 30, 967-977, 2008.

USDA Dietary fats: total fat and fatty acids. In: *Dietary Reference Intakes for Energy, Carbohydrate, Fiber, Fat, Fatty Acids, Cholesterol, Protein and Aminoacids*. 2005. Available at: https://www.nal.usda.gov/sites/default/files/fnic_uploads/energy_full_report.pdf. Accessed on October 29th, 2020.

USDA Food Data Central. ,2020. https://fdc.nal.usda.gov/. Accessed on October 29th, 2020.

Wei, C. Y., Hund, A., Zhu, D. and Nyström, L. Exploring genetic dependance of lipase activity to improve the quality of whole-grain wheat. *Journal of the Science of Food and Agriculture* 100(7), 3120-3125, 2020.

Xie, M., Dunford, N. T. and Goad, C. Enzymatic extraction of wheat germ oil. *Journal of the American Oil Chemists' Society* 88(12), 2015-2021, 2011.

Yaseen, A. A., Al-Okbi, S. Y., Hussein, A. M. S., Mohamed, D. A., Mohammad, A. A., Fouda, K. A. and Mehaya, F. M. Potential protection from Alzheimer's disease by wheat germ and rice bran nano-form in rat model. *Journal of Applied Pharmaceutical Science* 9, 67–76, 2019.

Zarroug, Y., Mejri, J., Dhawefi, N., Ali, S. B. S., El Felah, M. and Hassouna, M. Comparison of chemical composition of two durum wheat (*Triticum durum* L.) and bread wheat (*Triticum aestivum* L.) germ oils. *Ekin Journal of Crop Breeding and Genetics* 1(1), 69–76, 2015.

Zhao, Y., Zhang, J., Wei, Y., Ai, L., Ying, D. and Xiao, X. Improvement of bread quality by adding wheat germ fermented with *Lactobacillus plantarum* dy-1. *Journal of Food Quality*, 9348951, 2020.

Ziegler, J. U., Steingass, C. B., Longin, C. F. H., Würschum, T., Carle, R. and Scweiggert, R. M. Alkylresorcinol composition allows the differentiation of *Triticum* spp. having different degrees of ploidy. *Journal of Cereal Science* 65, 244–251, 2015.

Chapter 4

Dietary fibers in the prevention of type 2 diabetes mellitus

Ellen E. Blaak, Maastricht University, The Netherlands

1 Introduction

Dietary fibers are a heterogenous group of food compounds. They include indigestible carbohydrates (i.e. cellulose, hemicelluose, ß-glucans, pectins, gums fructans and resistant starch) and lignins. Chemically, dietary fibers include non-starch polysaccharides such as cellulose, hemicellulose, ß-glucans, polyfructoses (i.e. inulin), natural gums and heteropolymers (i.e. pectin) as well as natural or synthetically produced oligosaccharides (i.e. fructo-oligosaccharides (FOS), galacto-oligosaccharides (GOS, Jones 2014)).

A series of systematic reviews and meta-analyses of relevant studies suggest a 15–30% decrease in all-cause and cardiovascular-related mortality, incidence of coronary heart disease, stroke incidence and mortality, type 2 diabetes (T2D), and colorectal cancer when comparing the highest dietary fiber-consuming consumers with the lowest dietary fiber-consuming consumers (Reynolds et al., 2019). A systematic review of studies or controlled trials showed that high-fiber diets are an important component in diabetes management, resulting in improvements in measures of glycaemic control, blood lipids, body weight and inflammation, as well as a reduction in premature mortality (Reynolds et al., 2020). In the PREDIMED-plus study, a multicenter primary prevention lifestyle intervention trial, improvement in the carbohydrate quality index (diet high in total dietary fiber) was associated with a slightly reduced body weight and improved cardiovascular risk factors after a 12-month follow-up (Martínez-Gonzalez et al., 2020).

http://dx.doi.org/10.19103/AS.2021.0087.09

Nevertheless, well-controlled long-term human intervention studies are not always consistent with respect to effects of specific fibers on body weight control, diabetes and cardiometabolic risk (Blaak et al., 2020, Canfora et al. 2019, Müller et al., 2020). This inconsistency emphasizes, among other factors, the heterogeneity of the food compounds within the broad group of dietary fibers. Dietary fibers are most commonly classified based on physicochemical composition and on properties like water solubility, viscosity, binding and bulking ability and fermentability (Müller et al., 2018). Generally, it can be stated that the main sources of soluble, viscous, more readily fermentable types of dietary fibers are fruits, certain vegetables and some products derived from barley and oats (the latter are rich in both insoluble dietary fibers and soluble β-glucans). The main sources of insoluble cereal fibers are bran products from corn and wheat, which contain cellulose, hemicelluloses, and lignin, a major compound of cell walls, and which are typically non gel-forming, non-viscous and with only moderate fermentability in the colon (Müller et al., 2018,

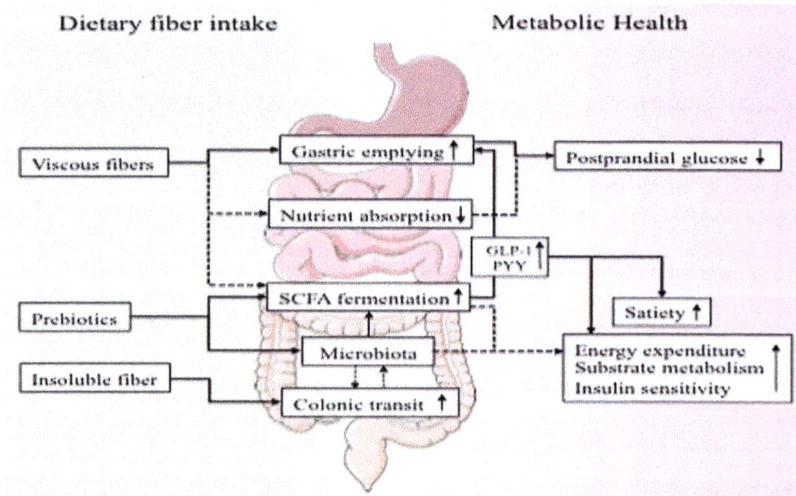

Figure 1 A global overview of the main mechanisms of action of insoluble, prebiotic and viscous fibers on gastro-intestinal and metabolic functioning. Viscous fibers may increase gastric emptying, may reduce nutrient absorption and increase SCFA fermentation. These effects may lead to a reduced postprandial glucose response and an increased release of incretin and satiety-stimulating hormones (GLP-1, PYY), which might influence energy intake and peripheral tissue metabolism. Prebiotics modulate microbiota composition and SCFA production thereby affecting energy expenditure, substrate metabolism and insulin sensitivity. Insoluble fibers may affect colonic transit thereby affecting microbial composition and in turn microbial metabolites may affect colonic transit. Solid lines indicate established effects, dashed lines indicate more controversial findings. Source: Müller et al., 2018. Abbreviations: SCFA, short-chain fatty acids; GLP-1, glucagon-like peptide 1; PYY, peptide YY.

Roager et al., 2019). It is important to note that most fiber-rich foods contain both soluble and insoluble (non-)fermentable fibers in varying ratios (McKee and Latner, 2000). Fiber properties strongly determine their effects on gastro-intestinal function and transit, their effects on gut microbial composition as well as their metabolic effects (for an overview, see Fig. 1). This chapter discusses the health effects of prebiotic, viscous as well as insoluble fibers (Table 1).

2 Underlying mechanisms of action of insoluble fibers

The underlying mechanisms of insoluble fibers in improving (GI) glucose homeostasis and reducing T2DM risk may (partly) be related to effects on gastro-intestinal transit (Müller et al., 2018). In this respect, it is widely accepted that laxation is one of the important health effects of insoluble and cereal fiber intake. Insoluble fibers increase fecal bulking as well as stool water content which in turn may affect peristalsis and mucus secretion (de Vries et al., 2015, Lawton et al., 2013, Vuksan et al., 2008). Previous research has estimated (by weighted regression analysis) that an increase in cereal fiber intake of 1 g/day may increase stool weight by 3.9 g/day (de Vries et al., 2015). Modulation of GI transit may affect nutrient absorption since fecal bulking affects the rheological structure of the food matrix which might impact the bioavailability and bio-accessibility of macronutrients within the matrix.

There is also a close relationship between gastro-intestinal and colonic transit and microbial composition (Roager et al., 2016, Müller et al., 2020) with consequences for host metabolism. Microbial change may also occur through effects on the colonic environment through, for instance, increased water availability as well as the trapping of nutrients within the stool matrix. Although microbial SCFA have been proposed as one of the mechanisms explaining effect on insulin sensitivity of insoluble dietary fibers (Lattimer and Haub, 2010), the degree of fermentability of insoluble fibers remains controversial (Weickert et al., 2011, Vetrani et al., 2016, Roager et al., 2019). It should be kept in mind that individual whole grain and cereal foods usually contain various amounts of (insoluble and soluble) dietary fiber, antioxidants, magnesium and bioactive phytochemicals, which might result in differential effects of different types of whole-grain foods on cardiometabolic health. Further studies are warranted to understand the underlying mechanisms linking gut motility, dietary fiber and host metabolism.

Habitual whole-grain consumption has also been suggested to cause lower body weight through multiple mechanisms such as lower energy density of whole-grain-based products, lower glycemic index, fermentation of non-digestible carbohydrates (satiety signals) and finally by modulating intestinal microflora. In that sense whole grain consumption can be recommended as one of the features of the diet that may help to control body weight (Giacco

Table 1 Overview of studies relating prebiotic, viscous and insoluble fibers to metablic outcomes

	Prebiotic dietary fiber	Viscous dietary fiber	Insoluble dietary fiber
Gastro-intestinal transit	Microbial metabolites may affect gastro-intestinal transit (Müller et al., 2018)	May affect gastric emptying (Dikeman and Fahey, 2006, Müller et al., 2018)	Consistent evidence for effect on GI-transit (de Vries et al., 2015, Lawton et al., 2013,Vuksan et al., 2008, Roager et al., 2016, Müller et al., 2020)
Effect on gut microbiota and fermentation	Yes, by definition	Controversial data, effects may be limited and indirect (as reviewed by Müller et al., 2018)	Limited and controversial (Lattimer and Haub, 2010, Weickert et al., 2011, Vetrani et al., 2016)
Glycemic control	improved postprandial glucose and insulin concentrations (Dewulf et al., 2013, Cani et al., 2009, Vulevic et al., 2013) or no changes (Pedersen et al., 2016, Canfora et al., 2017), may relate to different types of prebiotics, fermentation efficiency and site of fermentation, and heterogeneity in study populations (as reviewed in Canfora et al., 2019)	Yes, evidence from **RCT** (psyllium studies in T2D) (Gibb et al., 2015, Landin et al., 1992, Jovanovski et al., 2019) • studies with guar gum in T2D show no effect (Niemi et al., 1988, Uusitupa et al., 1984) • EFSA claim for ß-glcans in lowering postprandial glucose (EFSA, 2011)	Reduction in glycemia in **RCTs** (Musa-Veloso et al., 2018), not consistent (Weickert et al., 2005, Roager et al., 2019, Honsek et al., 2018), may depend on metabolic phenotype (Kabisch et al., 2019)
Body weight control	Inconsistent evidence (Kellow et al., 2014, Qu, 2019, John et al., 2018), • effects may depend on:Site of fermentation (Chambers et al., 2015, van der Beek et al., 2016) • Enterotype (Hjorth et al., 2018)	Modest improvement in body weight control (Jovanovski et al., 2019, 2020)	Probable in **observational studies** (Giacco, 2011, Della Pepa et al., 2018), • **RCTs**: may be different in overweight individuals (Pol et al., 2013), • depend on the type of fiber (Della Pepa et al., 2018) • probable but inconclusive due to heterogeneity in methodologies (Kissock et al., 2020)
Reduced diabetes risk	Inconsistent evidence effects may vary depend on different types of prebiotics, fermentation efficiency and site of fermentation, and heterogeneity in study populations (as reviewed in Canfora et al., 2019)	Psyllium and ß-glucan may reduce postprandial glucose (Gibb et al., 2015, Landin et al., 1992, Jovanovski et al., 2019, EFSA, 2011)	Consistent evidence from **observational studies** (Parker et al., 2013, Koh-Banerjee et al., 2004, Meyer et al., 2000, Schulze et al., 2007, Montonen et al., 2003, InterAct, 2015, Hu et al., 2020, Hardy et al., 2020).

et al., 2011). Indeed, observational evidence has consistently shown that a mean consumption of 2–3 servings/day (30–45 g/day) of whole grain is associated with a lower body mass index (BMI) and a decreased body weight gain over time (as reviewed Della Pepa et al., 2018).

Nevertheless, a meta-analysis of randomized control trials (RCTs) showed that habitual whole-grain intake does not induce any significant reduction in body weight or waist in overweight people in comparison to a refined cereal diet (Pol et al., 2013). A subsequent RCT confirmed these findings when comparing whole grain (90 g/day) obtained from different types of cereals (57% wheat, 21% rice and 16% oat) with a similar combination of refined grains during a 6-week intervention (Kirwan et al., 2016). A recent systematic review was performed on how different methods of reporting and calculating whole grain intake, including whole-grain food definitions, affect reported associations between whole-grain intake and body weight measures in adults. This review concluded that an association between whole grain and body weight measures remains likely, although precise associations are difficult to determine due to heterogeneity in methodologies to calculate whole-grain intake and an inability to formally compare studies (Kissock et al., 2020). Additionally, the effects of whole grain on body weight may be more important for specific fiber types like oat and rye, as recently reviewed (Della Pepa et al., 2018).

Foods rich in whole grains and cereals are consistently associated with a reduced risk of developing type 2 diabetes mellitus (T2DM) in observational studies (Parker et al., 2013, Koh-Banerjee et al., 2004, Meyer et al., 2000, Schulze et al., 2007, Montonen et al., 2003). A recent meta-analysis with more than 15 000 participants concluded that an increase in cereal fiber intake of 10 g/day decreased the risk of developing T2DM by 25% independent of BMI (InterAct, 2015). A meta-analysis of 17 prospective cohort studies also reported that an increase in cereal fiber intake of 2 g/day reduced the risk of developing T2DM by 6% (RR 0.94, 95% CI 0.93–0.96) (Yao et al., 2014). In three extensive cohort studies ((Nurses' Health Study (1984–2014), Nurses' Health Study II (1991–2017) and Health Professionals Follow-Up Study (1986 2016), United States)), it was shown that higher consumption of total whole grains and several commonly eaten whole-grain foods, including whole-grain breakfast cereal, oatmeal, dark bread, brown rice, added bran and wheat germ, was significantly associated with a lower risk of T2D (Hu et al., 2020). Finally, a recent meta-analysis across the United States, Europe and Asia showed protective dose-response risks for T2M for total dietary fiber which was most pronounced for cereal fiber (Hardy et al., 2020).

Only a few intervention studies investigated the impact of foods rich in insoluble fiber on GI transit and glucometabolic effects. In an acute crossover study, 14 healthy women consumed either bread enriched with 10.4 g wheat fibers or 10.6 g oat fibers or a white bread as a control (Weickert et al., 2005).

Intake of high-fiber breads increased early postprandial insulin peak and reduced postprandial glucose concentration, but stool consistency did not change. In another crossover study, 50 overweight or obese participants fulfilling one or more metabolic syndrome criteria consumed a whole grain diet (179 ±50 g/day) or a control diet with refined grain for 8 weeks. Circulating low-grade inflammatory markers and body weight (from 85.4 kg ±13.4 kg to 85.2 kg ±13.4 kg, $p < 0.001$) were reduced after the whole grain diet, which correlated with a reduced total energy intake. The whole grain intervention had no effect on fasting blood glucose, insulin, lipid or GLP-1 concentrations and no major changes in fecal gut microbiota or whole gut transit measured by radio-opaque markers were observed (Roager et al., 2019). A 12-week whole grain intervention prevented an increase in liver fat in overweight individuals (Schutte et al., 2018). A 2-year supplementation of insoluble fiber (15 g/day) in prediabetic individuals (impaired glucose tolerant) did not provide strong evidence of a beneficial effect on glycemic metabolism (Honsek et al., 2018). Nevertheless, a post-hoc analysis in this trial showed that individuals with impaired fasting glucose have the most pronounced effects on fasting glucose (Kabisch et al., 2019), suggesting that the effectiveness of intervention may depend on metabolic phenotype. In a meta-analysis of randomized controlled trials, whole-grain rice significantly attenuates the postprandial blood glucose response as compared to white rice (Musa-Veloso et al., 2018). Overall, a whole grain and a wheat fiber diet seems to be associated with slight positive health effects.

3 Dietary fiber intervention studies: prebiotic fibers

Specific dietary fibers termed prebiotics are fermented by the gut microbiota, thereby altering microbial composition and activity, thus conferring a beneficial physiologic effect on the host (Gibson et al., 2017). Well-studied prebiotics include fructans (inulin, fructo-oligosaccharides (FOS)), galacto-oligosaccharides (GOS) and other less known oligosaccharides (i.e., xylo-oligosaccharides, mannan-oligosaccharides). Intake of inulin, FOS and GOS results in an increased abundance of bacterial species such as *Bifidobacteria* and *Lactobacillus,* which have been proposed to be associated with beneficial health effects (Flamm et al., 2001, Ramirez-Farias et al., 2009, Davis et al., 2011). These genera have been reported to strengthen the gut barrier, improve host mucosal immunity, increase SCFA production and protect against opportunistic gut pathogens (Roberfroid, 2007). Recent advances in metagenomics have extended previous findings reporting that species such as *Faecalibacterium prausnitzii* and *Anaerostipes spp.* are increased upon prebiotic supplementation as well (Dewulf et al., 2013, Vandeputte et al., 2017).

Fructans (inulin, FOS) and GOS have been extensively studied in diet- or genetically-induced obese rodent models. Supplementation with GOS and

FOS in these rodent models improved glucose homeostasis, reduced serum lipids and weight gain (Delzenne et al., 2007, 2011, Slavin, 2013). A recent meta-analysis concluded that FOS intervention decreases hyperglycemia in obese rodent models (Le Bourgot et al., 2018). Proposed mechanisms are mainly based on an increase of colonic SCFA production, stimulation of enteroendocrine cell proliferation and increased gut hormone secretion and also on the modulation of the bile acid metabolism and a reduction of systemic inflammation (O'Connor et al., 2017, Florowska et al., 2016, Ejtahed et al., 2016). Human intervention studies (discussed below) are less conclusive.

In terms of body weight control, data on the effects of prebiotic supplementation on adiposity markers are controversial (Kellow et al., 2014). A meta-analysis of RCTs showed no effect on adiposity markers (Qu et al., 2021), or a slight significant decrease in BMI (John et al., 2018). A 10 g a day supplementation of an inulin-propionate ester, delivering additional 2.36 g of the SCFA propionate into the colon for 24 weeks, resulted in a significantly reduced weight gain compared to control (Chambers et al., 2015). The latter study indicates that the site of fermentation may be a determinant of effects on energy intake and body weight loss, as shown before (van der Beek et al., 2016) and reviewed more extensively (Blaak, 2020). Additionally, it has been shown that initial microbial composition may determine the response to dietary fiber intervention. An example of this is a study that showed that healthy, overweight adults with high Prevotella abundances lost more weight than subjects with low Prevotella abundances when consuming a diet rich in whole grain and fiber during a 6 month randomized controlled diet intervention (Hjorth et al., 2018). This data show that more personalised approaches in the field of prebiotics and body weight control are required.

Looking at diabetes and cardiometabolic risk, combined data in systematic reviews regarding prebiotic interventions to improve functional gastro-intestinal disorders or metabolic markers in T2D are potentially promising (Liu et al., 2017b), but data are not always consistent. Human intervention trials are inconclusive and either report improved postprandial glucose and insulin concentrations in healthy and obese volunteers (Dewulf et al., 2013, Cani et al., 2009, Vulevic et al., 2013) or no changes (Pedersen et al., 2016, Canfora et al., 2017). Different prebiotic types, fermentation efficiency and site of fermentation, and heterogeneity in study populations make it difficult to draw consistent conclusions (Blaak et al., 2020).

Previous human intervention studies have mainly focused on one specific type of fiber – fermentable oligosaccharides – to alter microbiota composition and elicit metabolic effects. The handful of longer-term human intervention studies using these fermentable fibers alone, however, showed only minor effects (Vulevic et al., 2013, Canfora and Blaak, 2015) or no effects (Canfora et al., 2017) on circulating SCFA concentrations and insulin sensitivity (indices). A 2017 study of healthy young individuals showed that fermentable oligosaccharides increased the abundance of bifidobacteria, but decreased the abundance of several other SCFA-producing bacteria, and even resulted in adverse effects on glucose homeostasis (Liu et al., 2017a). This study suggests that it is important to keep the diversity and abundance of SCFA-producing bacteria high, instead of stimulating only one bacteria genus (even if these bacteria are considered to be beneficial for gut and metabolic health). Furthermore, the response to intervention may depend on the microbial as well as the metabolic phenotype. A recent systematic review has summarized randomized controlled trials with supplementation of inulin-type fructans and GOS (from 7.4 g to 30 g/d during a period of 20 days up to 6 months) in healthy, obese and T2DM individuals. This meta-analysis showed that fasting insulin (mean difference −4.01 U/L; $P < 0.0001$) and fasting glucose (mean difference −0.42 mmol/L; $P = 0.09$) concentrations were only reduced in T2DM individuals but not in overweight/obese, normoglycemic participants (Liu et al., 2017b). On the other hand, other studies show a lack of response in the (pre) diabetic phenotype rather than a more pronounced response. A previous study showed that a 4-week oral butyrate administration altered metabolism and insulin sensitivity in lean but not obese insulin-resistant individuals (Bouter et al., 2018). The latter study suggests a putative resistance of the microbiome-host metabolism axis to dietary intervention in prediabetic individuals (Blaak, 2020); intervention with prebiotic fibers may thus take a longer time to become effective.

An intriguing study published in 2018 showed that a diet rich in complex fiber mixtures for 12 weeks decreased HbA1c levels, fasting blood levels of glucose and glucose tolerance by ~20% in individuals with T2DM (Zhao et al., 2018). Interestingly, the positive responders (bacterial strains that were significantly increased after the high fiber diet) showed an increased genetic microbial capacity to produce SCFAs through fiber fermentation. By contrast, the genome of negative responders (bacterial strains that were decreased after the intervention) harbors genes that are involved in the production of proteolytic metabolites such as indoles and hydrogen sulfide. These data indicate that a high carbohydrate fermentation may inhibit detrimental proteolytic fermentation. As recently reviewed this 'microbial substrate switch', through the whole colon, but most pronounced in the distal part of the colon where a large part of the proteolytic fermentation occurs, may promote cardiometabolic health (Canfora et al., 2019). Mechanistic support for a role of the site of fermentation as a determinant of metabolic effects comes from previous studies which showed that acute distal colonic acetate administration increased circulating acetate concentrations and pronouncedly increased fasting fat oxidation (25%), increased the concentration of satiety-stimulating hormone peptide YY and reduced concentration of the cytokine TNF-α in overweight males. Acetate administration in the proximal colon did not affect the metabolic profile (van der Beek et al., 2016).

In a recent paper (Liu et al., 2020), individuals with prediabetes who participated in a 12-week supervised exercise study were divided into responders and non-responders, based on the success of the exercise intervention at improving glucose homeostasis and insulin sensitivity indices. Shotgun metagenome sequencing of the feces (and the subsequent use of Kyoto Encyclopedia of Genes and Genomes orthology) demonstrated that both subgroups showed an increase in the activity of pathways involved in the fermentation of proteins, but that amino acid fermentation was shifted to an increased biosynthesis of SCFAs in the responders. In contrast, expression of genes with protein products involved in the production of metabolically detrimental compounds such as phenolic derivatives (indole and *p*-cresol) and sulfate was increased in non-responders. Besides an increased availability of saccharolytic substrates in the distal colon, the modification of proteolytic pathways by interventions might also be a strategy to improve host metabolism.

In summary, it is hard to draw a consistent conclusion on prebiotic interventions in relation to cardiometabolic health, which may relate to different types of prebiotics, fermentation efficiency and site of fermentation, as well as heterogeneity in study populations make it difficult to draw consistent conclusions.

4 Dietary fiber intervention studies: soluble viscous fibers

Soluble, viscous fibers include polysaccharides such as plant-derived pectins, ß-glucans, psyllium/ispaghula husks, natural gums, galactomannans and alginates. Once dissolved in water, viscous fibers form gels and/or thicken, a physico-chemical characteristic that may impact gastric emptying, intestinal motility and absorption rates of glucose, triglycerides and cholesterol (Dikeman and Fahey, 2006).

In terms of body weight control, randomized, controlled studies have reported that viscous and/or gel-forming fibers may improve body weight control, glycemic and insulinemic responses (Jenkins et al., 2000). A meta-analysis of RCTs on the role of dietary fiber in particular body weight management showed that dietary viscous fiber (within the context of an energy-restricted diet (Jovanovski et al., 2020) or when part of a non-restrictive ad libitum diet (Jovanovski et al., 2019)) modestly yet significantly improved body weight and other parameters of adiposity independently of caloric restriction.

Looking at glycemic control, diabetes and cardiometabolic risk, the viscous fiber psyllium has been reported to lower postprandial glucose concentrations and improve insulin sensitivity in healthy, obese and T2DM individuals. This positive effect on postprandial glycemia was reported in intervention studies ranging from 6 weeks to 6 months with doses of 10–14 g psyllium per day (Gibb et al., 2015). A meta-analysis of four studies (duration from 2 to 24 weeks) in T2DM individuals showed that psyllium intake decreased fasting glucose concentration by 37 mg/dL and reduced HbA1c (10.6 mmol/mol) compared to placebo (Gibb et al., 2015). Six-week supplementation of 10 g/d natural partly hydrolyzed guar gum also reduced fasting glucose and insulin concentration in healthy men (Landin et al., 1992) and HbA1c concentration in T2DM patients (Dall'Alba et al., 2013). A meta-analysis on 28 trials also showed that viscous fiber supplements improve conventional markers of glycemic control beyond usual care and should be considered in the management of T2D (Jovanovski et al., 2019).

However, other studies with hydrolyzed guar gum supplementation in T2DM patients reported no effect on glycemia (Niemi et al., 1988, Uusitupa et al., 1984). Acute reduction of subjective appetite rating and acute energy intake was reported after intake of pectins (mean dose 14.2 g/day and 4.8 g/day, respectively) and ß-glucans (mean dose 6.2 g/day and 5.8 g/day, respectively) compared to a control food (Wanders et al., 2011). The European Food Safety Agency (EFSA) has authorized the health claim that consumption of 4 g of ß-glucans (derived from barley and oats) per 30 g of available carbohydrates is sufficient to reduce postprandial glucose concentration in a range of clinical significance (EFSA, 2011). Effects on glucose metabolism viscosity rather than quantity of dietary fiber have been shown to predict the cholesterol-lowering effects in healthy individuals (Vuksan et al., 2011).

5 Conclusion

Overall, high intakes of dietary fiber and whole grains are associated with positive effects on metabolic health. Results from large prospective cohort studies consistently show that high intake of insoluble cereal fibers or whole-grain products rich in cereal fibers in particular may reduce insulin resistance (IR) and the risk of developing T2D by some 20–30%. Data on RCT with supplementation of insoluble fibers may be less consistent, which may relate to heterogeneity in methodologies to calculate whole-grain intake, specific type of fibers or impact of metabolic phenotype.

Data in systematic reviews regarding RCTs of prebiotic dietary fibers to improve metabolic markers in T2DM and obesity may be promising but provide (in general) a low level of evidence. This may be related to the heterogeneity of food compounds within the classification of prebiotic fibers, variation in fermentability effects on host metabolism and cardiometabolic health. It is increasingly clear that one size does not fit all and that response to dietary fiber intervention may depend on initial microbial and/or metabolic phenotype.

Dietary viscous fiber within the context of an energy-restricted diet or when part of an ad libitum diet modestly yet significantly improved body weight and other parameters of adiposity independently of caloric restriction and may improve glycemic regulation although data are not entirely consistent.

More research is required on the effectiveness of fiber intervention, taking into account individual characteristics, lifestyle factors, gastro-intestinal transit and individual fiber characteristic in order to obtain more targeted nutritional interventions. Given that most people across the world currently consume less than 20 g of dietary fiber per day, current evidence emphasizes the need for additional measures to increase the amount and the diversity of fiber intake.

7 References

Blaak, E. E. (2020). Current metabolic perspective on malnutrition in obesity: towards more subgroup-based nutritional approaches? *Proc. Nutr. Soc.* 79(3):331–7. doi: 10.1017/S0029665120000117. PMID: 32122428; PMCID: PMC7663313.

Blaak, E. E., Canfora, E. E., Theis, S., Frost, G., Groen, A. K., Mithieux, G., Nauta, A., Scott, K., Stahl, B., van Harsselaar, J., van Tol, R., Vaughan, E. E. and Verbeke, K. (2020). Short chain fatty acids in human gut and metabolic health. *Benef. Microbes* 11(5):411–55. doi: 10.3920/BM2020.0057. Epub 2020 Aug 31. PMID: 32865024.

Bouter, K., Bakker, G. J., Levin, E., Hartstra, A. V., Kootte, R. S., Udayappan, S. D., Katiraei, S., Bahler, L., Gilijamse, P. W., Tremaroli, V., Stahlman, M., Holleman, F., van Riel, N. A. W., Verberne, H. J., Romijn, J. A., Dallinga-Thie, G. M., Serlie, M. J., Ackermans, M. T., Kemper, E. M., Willems van Dijk, K., Backhed, F., Groen, A. K. and Nieuwdorp, M. (2018). Differential metabolic effects of oral butyrate treatment in lean versus metabolic syndrome subjects. *Clin. Transl. Gastroenterol.* 9(5):155.

Canfora, E. E. and Blaak, E. E. (2015). The role of polydextrose in body weight control and glucose regulation.*Curr. Opin. Clin. Nutr. Metab. Care* 18(4):395–400. doi: 10.1097/MCO.0000000000000184. PMID: 26049637.

Canfora, E. E., Meex, R. C. R., Venema, K. and Blaak, E. E. (2019). Gut microbial metabolites in obesity, NAFLD and T2DM. *Nat. Rev. Endocrinol.* 15(5):261–73. doi: 10.1038/s41574-019-0156-z. PMID: 30670819.

Canfora, E. E., van der Beek, C. M., Hermes, G. D. A., Goossens, G. H., Jocken, J. W. E., Holst, J. J., van Eijk, H. M., Venema, K., Smidt, H., Zoetendal, E. G., Dejong, C. H. C., Lenaerts, K. and Blaak, E. E. (2017). Supplementation of diet with galacto-oligosaccharides increases bifidobacteria, but not insulin sensitivity, in obese prediabetic individuals. *Gastroenterology* 153(1):87–97.e3. doi: 10.1053/j.gastro.2017.03.051. Epub 2017 Apr 8. PMID: 28396144.

Cani, P. D., Lecourt, E., Dewulf, E. M., Sohet, F. M., Pachikian, B. D., Naslain, D., De Backer, F., Neyrinck, A. M. and Delzenne, N. M. (2009). Gut microbiota fermentation of prebiotics increases satietogenic and incretin gut peptide production with consequences for appetite sensation and glucose response after a meal. *Am. J. Clin. Nutr.* 90(5):1236–43. doi: 10.3945/ajcn.2009.28095. Epub 2009 Sep 23. PMID: 19776140.

Chambers, E. S., Viardot, A., Psichas, A., Morrison, D. J., Murphy, K. G., Zac-Varghese, S. E., MacDougall, K., Preston, T., Tedford, C., Finlayson, G. S., Blundell, J. E., Bell, J. D., Thomas, E. L., Mt-Isa, S., Ashby, D., Gibson, G. R., Kolida, S., Dhillo, W. S., Bloom, S. R.,

Morley, W., Clegg, S. and Frost, G. (2015). Effects of targeted delivery of propionate to the human colon on appetite regulation, body weight maintenance and adiposity in overweight adults. *Gut* 64(11):1744–54. doi: 10.1136/gutjnl-2014-307913. Epub 2014 Dec 10. PMID: 25500202; PMCID: PMC4680171.

Dall'Alba, V., Silva, F. M., Antonio, J. P., Steemburgo, T., Royer, C. P., Almeida, J. C., Gross, J. L. and Azevedo, M. J. (2013). Improvement of the metabolic syndrome profile by soluble fibre–guar gum–in patients with type 2 diabetes: a randomised clinical trial. *Br. J. Nutr.* 110(9):1601–10. doi: 10.1017/S0007114513001025. Epub 2013 Apr 3. PMID: 23551992.

Davis, L. M., Martínez, I., Walter, J., Goin, C. and Hutkins, R. W. (2011). Barcoded pyrosequencing reveals that consumption of galactooligosaccharides results in a highly specific bifidogenic response in humans. *PLoS ONE* 6(9):e25200. doi: 10.1371/journal.pone.0025200. Epub 2011 Sep 26. PMID: 21966454; PMCID: PMC3180383.

de Vries, J., Miller, P. E. and Verbeke, K. (2015). Effects of cereal fiber on bowel function: a systematic review of intervention trials. *World J. Gastroenterol.* 21(29):8952–63. doi: 10.3748/wjg.v21.i29.8952. PMID: 26269686; PMCID: PMC4528039.

Della Pepa, G., Vetrani, C., Vitale, M. and Riccardi, G. (2018). Wholegrain intake and risk of Type 2 diabetes: evidence from epidemiological and intervention studies. *Nutrients* 10(9):1288. doi: 10.3390/nu10091288. PMID: 30213062; PMCID: PMC6163785.

Delzenne, N. M., Cani, P. D. and Neyrinck, A. M. (2007). Modulation of glucagon-like peptide 1 and energy metabolism by inulin and oligofructose: experimental data. *J. Nutr.* 137(11):2547S–51S. doi: 10.1093/jn/137.11.2547S. PMID: 17951500.

Delzenne, N. M., Neyrinck, A. M. and Cani, P. D. (2011). Modulation of the gut microbiota by nutrients with prebiotic properties: consequences for host health in the context of obesity and metabolic syndrome. *Microb. Cell Fact.* 10 (Suppl. 1):S10.

Dewulf, E. M., Cani, P. D., Claus, S. P., Fuentes, S., Puylaert, P. G., Neyrinck, A. M., Bindels, L. B., de Vos, W. M., Gibson, G. R., Thissen, J. P. and Delzenne, N. M. (2013). Insight into the prebiotic concept: lessons from an exploratory, double blind intervention study with inulin-type fructans in obese women(8). *Gut* 62(8):1112–21. doi: 10.1136/gutjnl-2012-303304. Epub 2012 Nov 7. PMID: 23135760; PMCID: PMC3711491.

Dikeman, C. L. and Fahey, G. C. (2006). Viscosity as related to dietary fiber: a review. *Crit. Rev. Food Sci. Nutr.* 46(8):649–63. doi: 10.1080/10408390500511862. PMID: 17092830.

EFSA Panel on Dietetic Products, Nutrition and Allergies (NDA) (2011). Scientific opinion on the substantiation of health claims related to beta-glucans from oats and barley and maintenance of normal blood ldl-cholesterol concentrations (ID 1236, 1299), increase in satiety leading to a reduction in energy intake (ID 851, 852), reduction of post-prandial glycaemic responses (ID 821, 824), and "digestive function" (ID 850) pursuant to article 13(1) of regulation (EC) no 1924/2006. EFSA J. 9:2207.

Ejtahed, H. S., Soroush, A. R., Angoorani, P., Larijani, B. and Hasani-Ranjbar, S. (2016). Gut microbiota as a target in the pathogenesis of metabolic disorders: a new approach to novel therapeutic agents. *Horm. Metab. Res.* 48(6):349–58. doi: 10.1055/s-0042-107792. Epub 2016 May 20. PMID: 27203411.

Flamm, G., Glinsmann, W., Kritchevsky, D., Prosky, L. and Roberfroid, M. (2001). Inulin and oligofructose as dietary fiber: a review of the evidence. *Crit. Rev. Food Sci. Nutr.* 41(5):353–62. doi: 10.1080/20014091091841. PMID: 11497328.

Florowska, A., Krygier, K., Florowski, T. and Dłużewska, E. (2016). Prebiotics as functional food ingredients preventing diet-related diseases. *Food Funct.* 7(5):2147–55. doi: 10.1039/c5fo01459j. Epub 2016 Mar 10. PMID: 26961814.

Giacco, R., Della Pepa, G., Luongo, D. and Riccardi, G. (2011). Whole grain intake in relation to body weight: from epidemiological evidence to clinical trials. *Nutr. Metab. Cardiovasc. Dis.* 21(12):901–8. doi: 10.1016/j.numecd.2011.07.003. PMID: 22036468.

Gibb, R. D., McRorie, J. W. Jr., Russell, D. A., Hasselblad, V. and D'Alessio, D. A. (2015). Psyllium fiber improves glycemic control proportional to loss of glycemic control: a meta-analysis of data in euglycemic subjects, patients at risk of type 2 diabetes mellitus, and patients being treated for type 2 diabetes mellitus. *Am. J. Clin. Nutr.* 102(6):1604–14. doi: 10.3945/ajcn.115.106989. PMID: 26561625.

Gibson, G. R., Hutkins, R., Sanders, M. E., Prescott, S. L., Reimer, R. A., Salminen, S. J., Scott, K., Stanton, C., Swanson, K. S., Cani, P. D., Verbeke, K. and Reid, G. (2017). Expert consensus document: the International Scientific Association for Probiotics and Prebiotics (ISAPP) consensus statement on the definition and scope of prebiotics. *Nat. Rev. Gastroenterol. Hepatol.* 14(8):491–502. doi: 10.1038/nrgastro.2017.75. PMID: 28611480.

Hardy, D. S., Garvin, J. T. and Xu, H. (2020). Carbohydrate quality, glycemic index, glycemic load and cardiometabolic risks in the US, Europe and Asia: a dose-response meta-analysis. *Nutr. Metab. Cardiovasc. Dis.* 30(6):853–71. doi: 10.1016/j.numecd.2019.12.050. Epub 2020 Jan 13. PMID: 32278608.

Hjorth, M. F., Roager, H. M., Larsen, T. M., Poulsen, S. K., Licht, T. R., Bahl, M. I., Zohar, Y. and Astrup, A. (2018). Pre-treatment microbial Prevotella-to-Bacteroides ratio, determines body fat loss success during a 6-month randomized controlled diet intervention. *Int. J. Obes. (Lond)* 42(2):284. doi: 10.1038/ijo.2018.1. Epub 2018 Feb 6. Erratum for (2017 Sep 08). *Int. J. Obes. (Lond)*; PMID: 29406520. PMCID: PMC7609312.

Honsek, C., Kabisch, S., Kemper, M., Gerbracht, C., Arafat, A. M., Birkenfeld, A. L., Dambeck, U., Osterhoff, M. A., Weickert, M. O. and Pfeiffer, A. F. H. (2018). Fibre supplementation for the prevention of type 2 diabetes and improvement of glucose metabolism: the randomised controlled Optimal Fibre Trial (OptiFiT). *Diabetologia* 61(6):1295–305 doi: 10.1007/s00125-018-4582-6. PMID: 29492637.

Hu, Y., Ding, M., Sampson, L., Willett, W. C., Manson, J. E., Wang, M., Rosner, B., Hu, F. B. and Sun, Q. (2020). Intake of whole grain foods and risk of type 2 diabetes: results from three prospective cohort studies. *BMJ* 370:m2206. doi: 10.1136/bmj.m2206. PMID: 32641435; PMCID: PMC7341349.

InterAct Consortium (2015). Dietary fibre and incidence of type 2 diabetes in eight European countries: the EPIC-InterAct Study and a meta-analysis of prospective studies. *Diabetologia* 58(7):1394–408. doi: 10.1007/s00125-015-3585-9. Epub 2015 May 29. PMID: 26021487; PMCID: PMC4472947.

Jenkins, D. J., Kendall, C. W., Axelsen, M., Augustin, L. S. and Vuksan, V. (2000). Viscous and nonviscous fibres, nonabsorbable and low glycaemic index carbohydrates, blood lipids and coronary heart disease. *Curr. Opin. Lipidol.* 11(1):49–56. doi: 10.1097/00041433-200002000-00008. PMID: 10750694.

John, G. K., Wang, L., Nanavati, J., Twose, C., Singh, R. and Mullin, G. (2018). Dietary alteration of the gut microbiome and its impact on weight and fat mass: a systematic review and meta-analysis. *Genes (Basel)* 9(3):167. doi: 10.3390/genes9030167. PMID: 29547587; PMCID: PMC5867888.

Jones, J. M. (2014). CODEX-aligned dietary fiber definitions help to bridge the 'fiber gap'. *Nutr. J.* 13:34. doi: 10.1186/1475-2891-13-34. PMID: 24725724; PMCID: PMC4007020.

Jovanovski, E., Khayyat, R., Zurbau, A., Komishon, A., Mazhar, N., Sievenpiper, J. L., Blanco Mejia, S., Ho, H. V. T., Li, D., Jenkins, A. L., Duvnjak, L. and Vuksan, V. (2019). Should viscous fiber supplements be considered in diabetes control? Results From a systematic review and meta-analysis of randomized controlled trials. *Diabetes Care* 42(5):755–66. doi: 10.2337/dc18-1126. PMID: 30617143.

Jovanovski, E., Mazhar, N., Komishon, A., Khayyat, R., Li, D., Blanco Mejia, S., Khan, T., L Jenkins, A., Smircic-Duvnjak, L., L Sievenpiper, J. and Vuksan, V. (2020) Can dietary viscous fiber affect body weight independently of an energy-restrictive diet? A systematic review and meta-analysis of randomized controlled trials. *Am. J. Clin. Nutr.* 111(2):471–85. doi: 10.1093/ajcn/nqz292. PMID: 31897475.

Kabisch, S., Meyer, N. M. T., Honsek, C., Gerbracht, C., Dambeck, U., Kemper, M., Osterhoff, M. A., Birkenfeld, A. L., Arafat, A. M., Hjorth, M. F., Weickert, M. O. and Pfeiffer, A. F. H. (2019). Fasting glucose state determines metabolic response to supplementation with insoluble cereal fibre: a secondary analysis of the optimal fibre trial (OptiFiT). *Nutrients* 11(10):2385. doi: 10.3390/nu11102385. PMID: 31590438; PMCID: PMC6835423.

Kellow, N. J., Coughlan, M. T. and Reid, C. M. (2014). Metabolic benefits of dietary prebiotics in human subjects: a systematic review of randomised controlled trials. *Br. J. Nutr.* 111(7):1147–61. doi: 10.1017/S0007114513003607. Epub 2013 Nov 13. PMID: 24230488.

Kirwan, J. P., Malin, S. K., Scelsi, A. R., Kullman, E. L., Navaneethan, S. D., Pagadala, M. R., Haus, J. M., Filion, J., Godin, J. P., Kochhar, S. and Ross, A. B. (2016). A whole-grain diet reduces cardiovascular risk factors in overweight and obese adults: a randomized controlled trial. *J. Nutr.* 146(11):2244–51. doi: 10.3945/jn.116.230508. Epub 2016 Oct 19. PMID: 27798329; PMCID: PMC5086786.

Kissock, K. R., Neale, E. P. and Beck, E. J. (2020). Whole grain food definition effects on determining associations of whole grain intake and body weight changes: a systematic review. *Adv. Nutr.*:122. doi: 10.1093/advances/nmaa122. Epub ahead of print. PMID: 33070194.

Koh-Banerjee, P., Franz, M., Sampson, L., Liu, S., Jacobs, D. R. Jr., Spiegelman, D., Willett, W. and Rimm, E. (2004). Changes in whole-grain, bran, and cereal fiber consumption in relation to 8-y weight gain among men. *Am. J. Clin. Nutr.* 80(5):1237–45. doi: 10.1093/ajcn/80.5.1237. PMID: 15531671.

Landin, K., Holm, G., Tengborn, L. and Smith, U. (1992). Guar gum improves insulin sensitivity, blood lipids, blood pressure, and fibrinolysis in healthy men. *Am. J. Clin. Nutr.* 56(6):1061–5. doi: 10.1093/ajcn/56.6.1061. PMID: 1442658.

Lattimer, J. M. and Haub, M. D. (2010). Effects of dietary fiber and its components on metabolic health. *Nutrients* 2(12):1266–89. doi: 10.3390/nu2121266.

Lawton, C. L., Walton, J., Hoyland, A., Howarth, E., Allan, P., Chesters, D. and Dye, L. (2013). Short term (14 days) consumption of insoluble wheat bran fibre-containing breakfast cereals improves subjective digestive feelings, general wellbeing and bowel function in a dose dependent manner. *Nutrients* 5(4):1436–55. doi: 10.3390/nu5041436. PMID: 23609776; PMCID: PMC3705356.

Le Bourgot, C., Apper, E., Blat, S. and Respondek, F. (2018). Fructo-oligosaccharides and glucose homeostasis: a systematic review and meta-analysis in animal models. *Nutr. Metab. (Lond.).* 15:9. doi: 10.1186/s12986-018-0245-3. PMID: 29416552; PMCID: PMC5785862.

Liu, F., Li, P., Chen, M., Luo, Y., Prabhakar, M., Zheng, H., He, Y., Qi, Q., Long, H., Zhang, Y., Sheng, H. and Zhou, H. (2017a). Fructooligosaccharide (FOS) and galactooligosaccharide (GOS) increase bifidobacterium but reduce butyrate producing bacteria with adverse glycemic metabolism in healthy young population. *Sci. Rep.* 7(1):11789. doi: 10.1038/s41598-017-10722-2. PMID: 28924143; PMCID: PMC5603605.

Liu, F., Prabhakar, M., Ju, J., Long, H. and Zhou, H. W. (2017b). Effect of inulin-type fructans on blood lipid profile and glucose level: a systematic review and meta-analysis of randomized controlled trials. *Eur. J. Clin. Nutr. Jan* 71(1):9–20. doi: 10.1038/ejcn.2016.156. Epub 2016 Sep 14. PMID: 27623982.

Liu, Y., Wang, Y., Ni, Y., Cheung, C. K. Y., Lam, K. S. L., Wang, Y., Xia, Z., Ye, D., Guo, J., Tse, M. A., Panagiotou, G. and Xu, A. (2020). Gut microbiome fermentation determines the efficacy of exercise for diabetes prevention. *Cell Metab. Jan* 31(1):77–91.e5. doi: 10.1016/j.cmet.2019.11.001. Epub 2019 Nov 27. PMID: 31786155.

Martínez-González, M. A., Fernandez-Lazaro, C. I., Toledo, E., Díaz-López, A., Corella, D., Goday, A., Romaguera, D., Vioque, J., Alonso-Gómez, Á. M., Wärnberg, J., Martínez, J. A., Serra-Majem, L., Estruch, R., Tinahones, F. J., Lapetra, J., Pintó, X., Tur, J. A., López-Miranda, J., Cano-Ibáñez, N., Delgado-Rodríguez, M., Matía-Martín, P., Daimiel, L., Sánchez, V. M., Vidal, J., Vázquez, C., Ros, E., Buil-Cosiales, P., Portoles, O., Soria-Florido, M., Konieczna, J., Navarrete-Muñoz, E. M., Tojal-Sierra, L., Fernández-García, J. C., Abete, I., Henríquez-Sánchez, P., Muñoz-Garach, A., Santos-Lozano, J. M., Corbella, E., Bibiloni, M. D. M., Becerra-Tomás, N., Barragán, R., Castañer, O., Fiol, M., García de la Hera, M., Belló-Mora, M. C., Gea, A., Babio, N., Fitó, M., Ruiz-Canela, M., Zazpe, I. and Salas-Salvadó, J. (2020). Carbohydrate quality changes and concurrent changes in cardiovascular risk factors: a longitudinal analysis in the PREDIMED-Plus randomized trial. *Am. J. Clin. Nutr.* 111(2):291–306. doi: 10.1093/ajcn/nqz298. PMID: 31868210.

McKee, L. H. and Latner, T. A. (2000). Underutilized sources of dietary fiber: a review. *Plant Foods Hum. Nutr.* 55(4):285–304. doi: 10.1023/a:1008144310986. PMID: 11086873.

Meyer, K. A., Kushi, L. H., Jacobs, D. R. Jr., Slavin, J., Sellers, T. A. and Folsom, A. R. (2000). Carbohydrates, dietary fiber, and incident type 2 diabetes in older women. *Am. J. Clin. Nutr.* 71(4):921–30. doi: 10.1093/ajcn/71.4.921. PMID: 10731498.

Montonen, J., Knekt, P., Järvinen, R., Aromaa, A. and Reunanen, A. (2003). Whole-grain and fiber intake and the incidence of type 2 diabetes. *Am. J. Clin. Nutr.* 77(3):622–9. doi: 10.1093/ajcn/77.3.622. PMID: 12600852.

Müller, M., Canfora, E. E. and Blaak, E. E. (2018). Gastrointestinal transit time, glucose homeostasis and metabolic health: modulation by dietary fibers(3). *Nutrients* 10(3):275. doi: 10.3390/nu10030275. PMID: 29495569; PMCID: PMC5872693.

Müller, M., Hermes, G. D. A., Emanuel, E. C., Holst, J. J., Zoetendal, E. G., Smidt, H., Troost, F., Schaap, F. G., Damink, S. O., Jocken, J. W. E., Lenaerts, K., Masclee, A. A. M. and Blaak, E. E. (2020). Effect of wheat bran derived prebiotic supplementation on gastrointestinal transit, gut microbiota, and metabolic health: a randomized controlled trial in healthy adults with a slow gut transit. *Gut Microbes* 12(1):1704141.

Musa-Veloso, K., Poon, T., Harkness, L. S., O'Shea, M. and Chu, Y. (2018). The effects of whole-grain compared with refined wheat, rice, and rye on the postprandial blood glucose response: a systematic review and meta-analysis of randomized controlled trials. *Am. J. Clin. Nutr.* 108(4):759–74. doi: 10.1093/ajcn/nqy112. PMID: 30321274.

Niemi, M. K., Keinänen-Kiukaanniemi, S. M. and Salmela, P. I. (1988). Long-term effects of guar gum and microcrystalline cellulose on glycaemic control and serum lipids in type 2 diabetes. *Eur. J. Clin. Pharmacol.* 34(4):427–9. doi: 10.1007/BF00542449. PMID: 3042430.

O'Connor, S., Chouinard-Castonguay, S., Gagnon, C. and Rudkowska, I. (2017). Prebiotics in the management of components of the metabolic syndrome. *Maturitas*. 104:11–8. doi: 10.1016/j.maturitas.2017.07.005. Epub 2017 Jul 14. PMID: 28923170.

Parker, E. D., Liu, S., Van Horn, L., Tinker, L. F., Shikany, J. M., Eaton, C. B. and Margolis, K. L. (2013). The association of whole grain consumption with incident type 2 diabetes: the women's health initiative observational study. *Ann. Epidemiol. Jun* 23(6):321–7. doi: 10.1016/j.annepidem.2013.03.010. Epub 2013 Apr 19. PMID: 23608304; PMCID: PMC3662533.

Pedersen, C., Gallagher, E., Horton, F., Ellis, R. J., Ijaz, U. Z., Wu, H., Jaiyeola, E., Diribe, O., Duparc, T., Cani, P. D., Gibson, G. R., Hinton, P., Wright, J., La Ragione, R. and Robertson, M. D. (2016). Host-microbiome interactions in human type 2 diabetes following prebiotic fibre (galacto-oligosaccharide) intake. *Br. J. Nutr.* 116(11):1869–77. doi: 10.1017/S0007114516004086. PMID: 27974055.

Pol, K., Christensen, R., Bartels, E. M., Raben, A., Tetens, I. and Kristensen, M. (2013). Whole grain and body weight changes in apparently healthy adults: a systematic review and meta-analysis of randomized controlled studies. *Am. J. Clin. Nutr.* 98(4):872–84. doi: 10.3945/ajcn.113.064659. Epub 2013 Aug 14. PMID: 23945718.

Qu, H., Song, L., Zhang, Y., Gao, Z. Y. and Shi, D. Z. (2021). The effect of prebiotic products on decreasing adiposity parameters in overweight and obese individuals: a systematic review and meta-analysis. *Curr. Med. Chem.* 28(2):419–31.

Ramirez-Farias, C., Slezak, K., Fuller, Z., Duncan, A., Holtrop, G. and Louis, P. (2009). Effect of inulin on the human gut microbiota: stimulation of Bifidobacterium adolescentis and Faecalibacterium prausnitzii. *Br. J. Nutr.* 101(4):541–50. doi: 10.1017/S0007114508019880. Epub 2008 Jul 1. PMID: 18590586.

Reynolds, A., Mann, J., Cummings, J., Winter, N., Mete, E. and Te Morenga, L. (2019). Carbohydrate quality and human health: a series of systematic reviews and meta-analyses. *Lancet*. 393(10170):434–45. doi: 10.1016/S0140-6736(18)31809-9. Erratum in: *Lancet* (2019 Feb 2) 393(10170):406. PMID: 30638909.

Reynolds, A. N., Akerman, A. P. and Mann, J. (2020). Dietary fibre and whole grains in diabetes management: systematic review and meta-analyses. *PLoS Med.* 17(3):e1003053. doi: 10.1371/journal.pmed.1003053. PMID: 32142510; PMCID: PMC7059907.

Roager, H. M., Hansen, L. B., Bahl, M. I., Frandsen, H. L., Carvalho, V., Gøbel, R. J., Dalgaard, M. D., Plichta, D. R., Sparholt, M. H., Vestergaard, H., Hansen, T., Sicheritz-Pontén, T., Nielsen, H. B., Pedersen, O., Lauritzen, L., Kristensen, M., Gupta, R. and Licht, T. R. (2016). Colonic transit time is related to bacterial metabolism and mucosal turnover in the gut. *Nat. Microbiol.* 1(9):16093.

Roager, H. M., Vogt, J. K., Kristensen, M., Hansen, L. B. S., Ibrügger, S., Mærkedahl, R. B., Bahl, M. I., Lind, M. V., Nielsen, R. L., Frøkiær, H., Gøbel, R. J., Landberg, R., Ross, A. B., Brix, S., Holck, J., Meyer, A. S., Sparholt, M. H., Christensen, A. F., Carvalho, V., Hartmann, B., Holst, J. J., Rumessen, J. J., Linneberg, A., Sicheritz-Pontén, T., Dalgaard,

M. D., Blennow, A., Frandsen, H. L., Villas-Bôas, S., Kristiansen, K., Vestergaard, H., Hansen, T., Ekstrøm, C. T., Ritz, C., Nielsen, H. B., Pedersen, O. B., Gupta, R., Lauritzen, L. and Licht, T. R. (2019). Whole grain-rich diet reduces body weight and systemic low-grade inflammation without inducing major changes of the gut microbiome: a randomised cross-over trial. *Gut.* 68(1):83–93. doi: 10.1136/gutjnl-2017-314786. Epub 2017 Nov 1. PMID: 29097438; PMCID: PMC6839833.

Roberfroid, M. (2007). Prebiotics: the concept revisited. *J. Nutr.* 137(3)(Suppl. 2):830S–7S. doi: 10.1093/jn/137.3.830S. PMID: 17311983.

Schulze, M. B., Schulz, M., Heidemann, C., Schienkiewitz, A., Hoffmann, K. and Boeing, H. (2007). Fiber and magnesium intake and incidence of type 2 diabetes: a prospective study and meta-analysis. *Arch. Intern. Med.* 167(9):956–65. doi: 10.1001/archinte.167.9.956. PMID: 17502538.

Schutte, S., Esser, D., Hoevenaars, F. P. M., Hooiveld, G. J. E. J., Priebe, M. G., Vonk, R. J., Wopereis, S. and Afman, L. A. (2018). A 12-wk whole-grain wheat intervention protects against hepatic fat: the Graandioos study, a randomized trial in overweight subjects. *Am. J. Clin. Nutr.* 108(6):1264–74. doi: 10.1093/ajcn/nqy204. PMID: 30541093.

Slavin, J. (2013). Fiber and prebiotics: mechanisms and health benefits. *Nutrients* 5(4):1417–35. doi: 10.3390/nu5041417. PMID: 23609775; PMCID: PMC3705355.

Uusitupa, M., Tuomilehto, J., Karttunen, P. and Wolf, E. (1984). Long term effects of guar gum on metabolic control, serum cholesterol and blood pressure levels in type 2 (non-insulin-dependent) diabetic patients with high blood pressure. *Ann. Clin. Res.* 16(Suppl. 43):126–31. PMID: 6398982.

van der Beek, C. M., Canfora, E. E., Lenaerts, K., Troost, F. J., Damink, S. W. M. O., Holst, J. J., Masclee, A. A. M., Dejong, C. H. C. and Blaak, E. E. (2016). Distal, not proximal, colonic acetate infusions promote fat oxidation and improve metabolic markers in overweight/obese men. *Clin. Sci. (Lond.)* 130(22):2073–82. doi: 10.1042/CS20160263. Epub 2016 Jul 20. PMID: 27439969.

Vandeputte, D., Falony, G., Vieira-Silva, S., Wang, J., Sailer, M., Theis, S., Verbeke, K. and Raes, J. (2017). Prebiotic inulin-type fructans induce specific changes in the human gut microbiota. *Gut* 66(11):1968–74. doi: 10.1136/gutjnl-2016-313271. Epub 2017 Feb 17. PMID: 28213610; PMCID: PMC5739857.

Vetrani, C., Costabile, G., Luongo, D., Naviglio, D., Rivellese, A. A., Riccardi, G. and Giacco, R. (2016). Effects of whole-grain cereal foods on plasma short chain fatty acid concentrations in individuals with the metabolic syndrome. *Nutrition* 32(2):217–21. doi: 10.1016/j.nut.2015.08.006. Epub 2015 Sep 2. PMID: 26706023.

Vuksan, V., Jenkins, A. L., Jenkins, D. J., Rogovik, A. L., Sievenpiper, J. L. and Jovanovski, E. (2008). Using cereal to increase dietary fiber intake to the recommended level and the effect of fiber on bowel function in healthy persons consuming North American diets. *Am. J. Clin. Nutr.* 88(5):1256–62. doi: 10.3945/ajcn.2008.25956. PMID: 18996860.

Vuksan, V., Jenkins, A. L., Rogovik, A. L., Fairgrieve, C. D., Jovanovski, E. and Leiter, L. A. (2011). Viscosity rather than quantity of dietary fibre predicts cholesterol-lowering effect in healthy individuals. *Br. J. Nutr.* 106(9):1349–52. doi: 10.1017/S0007114511001711. PMID: 21736815.

Vulevic, J., Juric, A., Tzortzis, G. and Gibson, G. R. (2013). A mixture of trans-galactooligosaccharides reduces markers of metabolic syndrome and modulates the fecal microbiota and immune function of overweight adults. *J. Nutr.* 143(3):324–31. doi: 10.3945/jn.112.166132. PMID: 23303873.

Wanders, A. J., van den Borne, J. J., de Graaf, C., Hulshof, T., Jonathan, M. C., Kristensen, M., Mars, M., Schols, H. A. and Feskens, E. J. (2011). Effects of dietary fibre on subjective appetite, energy intake and body weight: a systematic review of randomized controlled trials. *Obes. Rev.* 12(9):724–39. doi: 10.1111/j.1467-789X.2011.00895.x. PMID: 21676152.

Weickert, M. O., Arafat, A. M., Blaut, M., Alpert, C., Becker, N., Leupelt, V., Rudovich, N., Möhlig, M. and Pfeiffer, A. F. (2011). Changes in dominant groups of the gut microbiota do not explain cereal-fiber induced improvement of whole-body insulin sensitivity. *Nutr. Metab. (Lond.).* 8:90. doi: 10.1186/1743-7075-8-90. PMID: 22177085; PMCID: PMC3264513.

Weickert, M. O., Mohlig, M., Koebnick, C., Holst, J. J., Namsolleck, P., Ristow, M., Osterhoff, M., Rochlitz, H., Rudovich, N., Spranger, J. and Pfeiffer, A. F. (2005). Impact of cereal fibre on glucose-regulating factors. *Diabetologia* 48(11):2343–53. doi: 10.1007/s00125-005-1941-x. Epub 2005 Sep 20. PMID: 16172868.

Yao, B., Fang, H., Xu, W., Yan, Y., Xu, H., Liu, Y., Mo, M., Zhang, H. and Zhao, Y. (2014). Dietary fiber intake and risk of type 2 diabetes: a dose-response analysis of prospective studies. *Eur. J. Epidemiol.* 29(2):79–88. doi: 10.1007/s10654-013-9876-x. Epub 2014 Jan 5. PMID: 24389767.

Zhao, L., Zhang, F., Ding, X., Wu, G., Lam, Y. Y., Wang, X., Fu, H., Xue, X., Lu, C., Ma, J., Yu, L., Xu, C., Ren, Z., Xu, Y., Xu, S., Shen, H., Zhu, X., Shi, Y., Shen, Q., Dong, W., Liu, R., Ling, Y., Zeng, Y., Wang, X., Zhang, Q., Wang, J., Wang, L., Wu, Y., Zeng, B., Wei, H., Zhang, M., Peng, Y. and Zhang, C. (2018). Gut bacteria selectively promoted by dietary fibers alleviate type 2 diabetes. *Science* 359(6380):1151–6. doi: 10.1126/science.aao5774. PMID: 29590046.

Chapter 5

Fiber-associated wheat lignans and colorectal cancer prevention

Jingwen Xu, Shanghai Ocean University, China and Kansas State University, USA; and Weiqun Wang, Kansas State University, USA

1 Introduction

Cereal grains and cereal-processed products are staple foods for the most populations all over the world. Wherein, wheat is traditionally and largely consumed worldwide due to the large annual production and various well-developed bakery goods such as bread, tortilla, noodle and so on. Whole wheat is a health-promoting grain, containing dietary fibers (e.g. lignin) and secondary metabolites (e.g. lignans, phenolics, vitamins, microelements, sterols and tocopherols, etc.) that are beneficial for human health. Epidemiological studies suggest that ingestion of whole grains containing dietary fiber and phytochemicals is associated with reducing risk of many chronic diseases such as obesity, diabetes, cardiovascular disease and some types of cancer (Johnsen et al., 2009; Okarter and Liu, 2010; Lillioja et al., 2013; Jiang and Wang 2016; Xu et al., 2019; Davis et al., 2019).

Dietary lignans are fiber-associated polyphenols, and thus they are naturally present in fiber-rich cereals, oilseeds, nut, vegetables, fruits etc. It is well known that flaxseed and sesame seeds contain a high content of lignans up to 335 mg/100 g and 373 mg/100 g, respectively (Peñalvo et al., 2005a; Peterson et al., 2010). Among the top-consumed cereal grains, Smeds et al.

http://dx.doi.org/10.19103/AS.2021.0087.10

(2007) reported rye contained the highest level of lignans and then followed by wheat, triticale, oat, spelt wheat, Japanese rice, wild rice, buckwheat, barley, amaranth, corn, millet, quinoa, red rice and brown rice. The effect of fiber on preventing colorectal cancer is generally well documented, but its anticancer activity seems highly dependent upon the fiber source (Reddy et al., 1989, 1992). Wheat bran, for example, has been shown to prevent colorectal cancer consistently, while oat fiber did not (Reddy et al., 1989, 1992). Some vegetables such as brassica vegetables (cabbage, brussel sprouts, kale, asparagus and carrots, etc.) and fruits (apricot, citrus fruit and kiwi, etc.) also contain high lignan contents (Milder et al., 2005; Peñalvo et al., 2008). Furthermore, alcoholic beverages such as red wine and non-alcoholic beverages such as coffee also contain lignans (Thompson et al., 2006).

It is interesting that Tetens et al. (2013) investigated the main sources of lignans in five European countries and showed subjects ingested 1-2 mg/day of dietary lignans and were mainly from cereal grains, vegetables, fruits and berries. Although lignan content in the plant food is generally low, people consume plant foods quite often in the daily diet, resulting in significantly increased dietary exposure to lignans. Therefore, people can still benefit from dietary lignans for improving their health. Lignans have attracted extensive interest due to its health-promoting benefits such as antioxidant activity, phytoestrogen activity, hormone metabolism, cancer prevention and cardiovascular disease prevention (Adlercreutz, 2007; Aehle et al., 2011; Landete, 2012; Tresserra-Rimbau et al., 2014; Rodríguez-García et al., 2019).

Evidence shows dietary intake of lignans is associated with reducing risks of cancer, such as colorectal cancer (Nelson et al., 2002) and breast cancer in postmenopausal women (Buck et al., 2010; Ezzat et al., 2018). In addition, dietary intake of lignans is also associated with preventing osteoporosis and menopausal syndrome in postmenopausal women (Lemay et al., 2002; Wu et al., 2006). A significant reduction of both diastolic and systolic blood pressures in mildly hypertensive middle-aged men and women who had dietary intake of 60 mg of sesamin per day for four weeks has been reported by Miyawaki et al. (2009). Wang and Goodman (1999) presented that mammalian lignan enterolactone (ENL) functioned as an antioxidant against human low-density lipoprotein (LDL) oxidation, indicating reducing risk of cardiovascular disease. Similarly, a significant reduction of LDL was observed in 55 hypercholesterolemic adults who had taken 206 mg of lignan secoisolariciresionol diglycoside (SDG) over 8 weeks (Zhang et al., 2008).

In this chapter, we focus on wheat lignans given that wheat is a traditional and staple food for most populations worldwide. This chapter is needed with regards to advancing our understanding about wheat lignan biosynthesis and metabolism in the gut as well as the bioactive properties of antioxidant activity, anti-inflammatory effect and anti-cancer effect. In addition, the potential limitation and research gap of current studies with future research topics will be further discussed.

2 Wheat lignans

2.1 Biosynthesis of lignans in wheat

Lignans are diphenolic compounds formed by the coupling of two coniferyl alcohol residues in the presence of plant cell wall. They are plant secondary metabolites and are stereospecific dimers of cinnamic alcohols (monolignols) through the dimerization of two cinnamic acid residues and bonded at carbon 8 (C8-C8) (Davin and Lewis, 2003; Andersson et al., 2014; Peterson et al., 2010). Lignans are naturally present as free form or bound to polysaccharides in plants, such as secoisolariciresional (Sec), pinoresinol (Pin), syringaresinol (Syr), matairesinol (Mat), medioresinol (Med), lariciresinol (Lar), sesamin (Ses), 7′-hydroxymatairesinol, isolariciresinol and so on. Wherein, Sec, Mat, Lar, Pin and Syr are common plant lignans in human diet. Chemical structure of common lignans such as Mat, Sec and Pin are shown in Fig. 1. Secoisolariciresionol glucoside (SDG) (Fig. 2) is a dominant lignan in wheat bran (Wang et al., 2008). After ingestion, enterolignans can be further converted from SDG to enterodiol (EDL) and enterlactone (ENL) in the gut (Fig. 2). Therefore, both EDL and ENL are also named as mammalian lignans.

Biosynthesis of lignans in flaxseed has been well discovered by Touré and Xu (2010) and the biosynthesis of major lignans in plant food was reported by Rodríguez-García et al. (2019) (Fig. 3). However, so far, the biosynthesis of major lignans in wheat has not been well studied. Lignans, as the secondary metabolites, are generally biosynthesized via shikimate pathway. The shikimate pathway initially catalyzes phenylalanine with the help of enzyme phenylalanine ammonia lyase (PAL) to cinnamic acid. Then a series of phenolic acids are converted, then to coniferaldehyde, and to pinoresinol (Fig. 3). The enzyme pinoresinol/lariciresinol reductase (PLR) plays an important role in catalyzing the last two steps of lignan production through conversion of Pin to Lar and to Sec (Ayella et al., 2007; Ayella and Wang, 2010). We previously investigated the over-expression of PLR in regulating lignan biosynthesis in wheat and successfully transformed novel wheat lines with over-expressed PLR and accumulated health-promoting lignan SDG content (Ayella et al., 2007).

2.2 Analysis of wheat lignans

It is well known that lignans are mainly present in wheat bran. For example, the content of Lar in wheat grains varies from 45 μg/100 g to 95 μg/100 g (Smeds et al., 2009), while Lar content in wheat brans ranges from 220 μg/100 g to 459 μg/100 g (Durazzo et al., 2009). Lignan contents in a cereal grains such

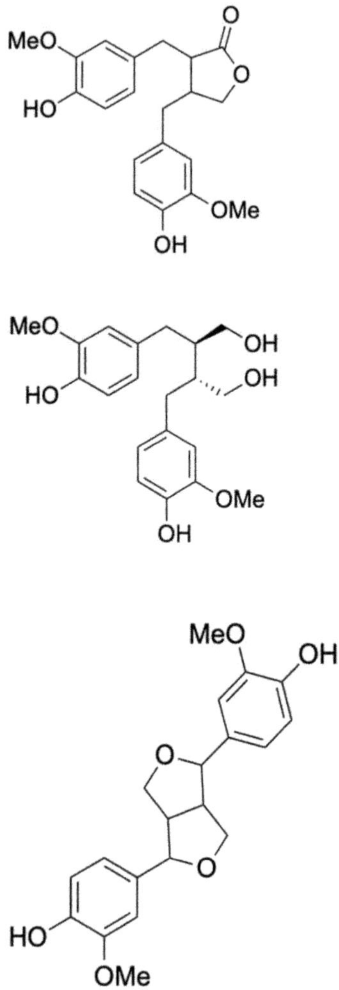

matairesinol

secoisolariciresional

pinoresinol

Figure 1 Chemical structures of major lignans including matairesinol (Mat), secoisolariciresional (Sec) and pinoresinol (Pin).

as rye, wheat and oat have been reported by Smeds et al. (2009). Lignans in plants are bound to cell wall molecules. In foods, lignans are mostly conjugated with glycosides, such as SDG. Lignan extraction mainly relies on various solvent reagents and different extraction methods, such as alkaline extraction, mild acid extraction, combination of alkaline and mild acid extraction, and so on (Smeds et al., 2007). However, strong alkaline or acid extraction has dual effects. On one side, for example, alkaline extraction releases lignans from cereal matrix through hydrolyzing, while on the other side, alkaline extraction destroys lignan structures and results in reducing lignan content (Smeds et al., 2007). Besides

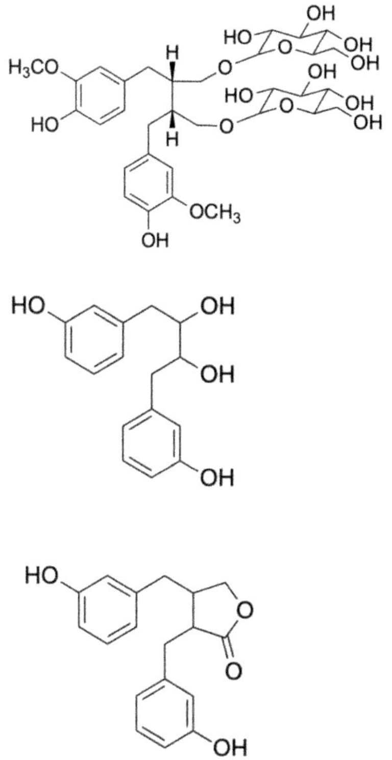

secoisolariciresionol diglycosid

enterodiol

enterolactone

Figure 2 Chemical structures of secoisolariciresionol diglycoside (SDG) and its metabolites, enterodiol (EDL) and enterolactone (ENL).

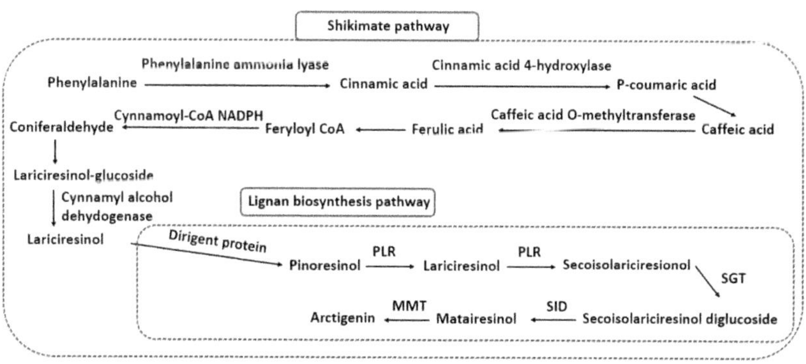

Figure 3 Lignan biosynthesis pathway. PLR: pinoresinol/lariciresinol reductase. SGT: secoisolariciresinol glycosyltransferase. SID: matairesinol O-methyltransferase. MMT: matairesinol O-methyltransferase (adapted from Rodríguez-García et al., 2019).

solvent extraction, enzymatic hydrolysis has also been considered as an effective way to extract lignans from bound glycosides (Nørskov and Knudsen, 2016).

Identification, quantification and qualification of lignan from foods are mainly dependent on the chromatographic techniques such as high performance liquid chromatography (HPLC) and gas chromatography (GC). Quantification and qualification of lignans are usually determined by chromatographic techniques combined with mass spectrometry, such as LC-MS (Dinelli et al., 2009; Smeds et al., 2009; Hanhineva et al., 2011; Su et al., 2019) and GC-MS (Peñalvo et al., 2005b). For example, we previously extracted lignans from four wheat cultivars including 'Madison', 'Erine', 'Betty' and 'Arapahoe' using methanol and detected SDG through HPLC (Qu et al., 2005). Lignans were also extracted from wheat cultivars by using 80% ethanol combined with alkaline (sodium hydroxide) and acid (hydrochloric acid) hydrolysis, and then identified by LC/MS (Dinelli et al., 2009). A C18 reversed phase column is mainly applied in separating target lignans compounds in HPLC system, and MS is usually equipped for further identification. Čukelj et al. (2010) developed a novel sensitive gas chromatography-electron capture detection (GC-ECD) method to analyze lignans in wheat bran in parts per billion (ppb). By using GC detection, capillary column is usually used in separating lignans.

The major lignan content varies among different studies due to the differences of wheat varieties and/or extraction and detection methods. For example, Andersson et al. (2014) reported that Syr was the dominant lignan at 372 μg/100 g in wheat bran, then followed by Lar at 62 μg/100 g, Pin at 37 μg/100 g, Sec at 35 μg/100 g, and Med at 30 μg/100 g. Major lignan contents in wheat and wheat-related foods are shown in Table 1. So far, only a few studies report the lignans in wheat-containing foods, such as breakfast cereal (Milder et al., 2005) and bread (Durazzo et al., 2009). Muir and Westcott (2000) demonstrated that SDG was stable in bread-baking process. However, the impact of food process (e.g., cooking method, thermal profile, food additives, etc.,) on wheat lignan has not been well studied. Thus, future studies regarding the effect of food processing on wheat lignan contents seem warranted.

3 Metabolism of wheat lignans in human

3.1 Conversion of plant lignans to mammalian lignans

After ingestion, lignan SDG can be digested and metabolized by intestinal bacteria and converted to lignan aglycones, then to mammalian lignans such as EDL and ENL (McCann et al., 2005; Wang et al., 2008; Peterson et al., 2010). In GI tract, gastrointestinal bacteria is mainly responsible for hydrolyzing sugar moiety of SDG for releasing Sec, then converted to END through dihydroxylation and demethylation reactions with the help of colonic microflora (Touré and Xu, 2010). For example, microorganism *Bifidobacteria*

Table 1 Major lignan contents in wheat and wheat-containing food (µg/100 g)

Wheat cultivars and wheat-containing food	SDG	Lar	Mat	Med	Sec	Syr	Pin	7′-hydroxymatairesinol	References
'Madison', soft red winter wheat	0.8	-	-	-	-	-	-	-	Qu et al. (2005)
'Ernie', soft red winter wheat	0.5	-	-	-	-	-	-	-	
'Betty', hard white winter wheat	0.4	-	-	-	-	-	-	-	
'Arapahoe', hard red winter wheat	UD	-	-	-	-	-	-	-	
Wheat (*Triticum aestivum* L.)	-	62	3	40	35	372	37	-	Peñalvo et al. (2005b)
'Aapeli' wheat	-	67.3	UD	55.9	28.5	-	75.0	10.8	Smeds et al. (2009)
'Aino' wheat	-	54.0	UD	58.2	21.7	-	67.3	10.7	
'Amaretto' wheat	-	59.2	UD	63.1	30.4	-	53.3	10.4	
'Bjame' wheat	-	79.9	UD	61.5	41.9	-	63.2	10.6	
'Bor 00703' wheat	-	54.6	UD	46.3	22.0	-	60.0	10.0	
'Kruunu' wheat	-	45.0	UD	56.4	20.0	-	56.9	8.47	
'Picolo' wheat	-	57.1	UD	81.6	28.8	-	57.8	9.99	
'Quama' wheat	-	76.8	UD	77.6	42.5	-	70.9	10.4	
'Zebra' wheat	-	95.1	UD	72.5	36.9	-	82.6	12.4	
Common wheat, germ	-	UD	9	UD	20	UD	-	-	Rodríguez-García et al. (2019)
Common wheat, refined flour	-	180	0.214	UD	20	0	-	-	
Common wheat, whole grain flour	-	100	0.9	30	20	370	-	-	
Hard wheat, semolin	-	UD	UD	UD	2	UD	-	-	
Bread (whole grain flour)	-	50	0.31	UD	8.68	UD	-	-	
Bread (refined flour)	-	10	0.123	UD	7.19	10	-	-	
Pasta, whole grain	-	UD	1.5	UD	5	UD	-	-	

UD: undetectable' -: unanalyzed.

(*Bifidobacterium pseudocatenulacum WC 401*) is responsible for transforming SDG to Sec in the gut (Roncaglia et al., 2011). Woting et al. (2010) demonstrated that microorganisms *Clostridium saccharogumia, Eggerthella lenta, Blautia producta* and *Lactonifactor longoviformis* are responsible for converting SDG to EDL and ENL. In addition, the conversion of SDG to EDL and ENL is highly dependent on the β-glucuronidase activity in the gut (Possemiers et al., 2007). In the gut, Pin can be converted to Lar and then to Sec and Mat, further to EDL and ENL (Andersson et al., 2014). Conversion of Sec, Mat, Lar, Pin and Syr to ENL and EDL in fecal incubation is effective and mostly is higher than 55%, except for Syr which is only 4% (Landete, 2012). However, little is known about the conversion of Med to ENL in the gut.

Some of lignan metabolites will be absorbed in the colon and conjugated with glucuronic acid or sulfate in the liver for enterohepatic circulation (Durazzo et al., 2013). Kuijsten et al. (2005) first investigated the pharmacokinetics of enterolignans by observing the enterolignan concentration in the plasma of 12 healthy men and women during 8–10 h after taking SDG at 1.31 μmol/kg body weight. They found EDL presented in the plasma for 14.8 ± 5.1 h with the maximum concentration up to 73 ± 40 nmol/L, and ENL in the plasma for 19.7 ± 6.2 h with the maximum concentration up to 56 ± 30 nmol/L. In addition, at least 40% of ingested SDG becomes bioavailable, indicating plasma enterolignans could be used as intermediate biomarkers of lignan exposure for evaluating the effect of lignan ingestions and biological functions (Kuijsten et al., 2005). Afterwards, lignans are excreted in urine as conjugated glucuronides and in feces as unconjugated form, wherein, ENL is the major compound found in feces (Heinonen et al., 2001; Kuijsten et al., 2005). EDL and ENL are excreted primarily as monoglucuronides in urine, which accounted for 95% and 85% respectively (Yoder et al., 2015).

4 Antioxidant and anti-inflammatory effects of wheat lignans

4.1 Antioxidant activity of wheat lignans

Wheat lignan SDG, EDL and ENL have been reported to contain antioxidant activities (Touré and Xu, 2010). The antioxidant property is dependent on the hydroxylation status of the aromatic rings, including reaction directly with lipid radicals and stabilization of cations through a hydrogen anion donation and scavenging of free radicals for lowering the rate of oxidation (Seifried et al., 2007). It has been reported that antioxidant activity of Sec, EDL and ENL was approximately 5-fold higher than vitamin E and SDG showed similar antioxidant activity to vitamin E *in vitro* (Prasad, 2000; Touré and Xu, 2010). *In vitro* studies have shown the antioxidant effect of EDL and ENL on lipid peroxidation (Hu et al., 2007), protection of SDG, Sec, EDL and ENL against AAPH-induced liposome lipid peroxidation (Hu et al., 2007), and the inhibition of Sec on lipid autoxidation in a homogeneous lipid media and β-carotene-linoleate emulsion system (Slavova-Kazakova et al., 2015).

However, to date, most of the studies have been focused on the antioxidant effect of wheat lignans on lipid oxidation, which appears to be associated with cardiovascular disease. The anti-cancer activity of wheat lignans via antioxidant mechanisms has not been well addressed, in terms of effective doses and mechanistic target, and so on.

4.2 Anti-inflammatory effects of wheat lignans

Chronic inflammation such as inflammatory bowel disease is an important causal factor for the development of colorectal cancer due to high oxidative stress present in the inflamed mucosa, which induces colorectal cancer development. Thus, preventing inflammation is a critically early target in colorectal cancer prevention.

Lignans have been demonstrated to have anti-inflammatory effect and regulate intestinal inflammatory response. During et al. (2012) tested the anti-inflammatory effect of lignans including SDG, Sec, Pin, Lar, Mat and hydroxymatairesinol on colorectal adenoma Caco-2 cells. Sec, Pin and ENL, but not SDG were taken up but partially conjugated in cells, wherein, Pin showed the most efficient conjugation in cells and strongest anti-inflammatory effect by blocking NF-κB pathway (During et al., 2012). The mechanism of anti-inflammatory effect of lignans is suggested to be related to Pin's molecular structure and/or intestinal metabolites.

Apart from wheat lignans, the effect of lignan SDG isolated from flaxseed on inflammatory bowel disease has also been investigated by Xu et al. (2016). In the research, isolated SDG from flaxseed was orally fed to male mice between 100 and 200 mg/kg/day for three days, resulting in a significant decrease in the disease activity index by suppressing dextran sulphate sodium (DSS)-induced erosion, ulceration, crypt abscesses and oedema, as well as decreasing colonic myeloperoxidase activity and TNF-α expression (Xu et al., 2016). In addition, Pietrofesa et al. (2016) showed the effect of flaxseed lignans SDG on asbestos-induced inflammation in mice, resulting in a downregulation of proinflammatory cytokines (IL-1β, IL-6, TNFα, HMGB1, and active TGFβ1) and cytokine receptors including TNFαR1 and TGFβR1.

The lignans involved in the studies aforementioned are naturally present in wheat bran. Thus, these studies might also represent the anti-inflammatory effect of wheat lignans to some extent. Meanwhile, anti-inflammatory effect of wheat lignans needs further investigation in terms of their effective doses and underlying mechanisms, as well as correlation to *in vivo* animal model and clinical trials if wheat lignans can be eventually used for chemopreventive agents for preventing colorectal cancer.

5 Wheat lignans and colon cancer prevention

Colon cancer has been reported to be the second most fatal cancer and the third leading cause of cancer death each year worldwide (Oduwole et al., 2003; Araújo et al., 2011; Siegel et al., 2020). About 70–90% colorectal cancer has been estimated in relation to dietary factors (Ahmed, 2004). Optimum diet such as high intake of vegetable, fruit and fiber, low calories, low or no alcohol drinking and tobacco smoking shows the potential to prevent colorectal cancer (Ahmed, 2004). Previously, wheat lignans such as SDG, EDL and ENL have been associated with reducing risk of colorectal cancer (Touré and Xu, 2010; Szewczyk et al., 2014). Herein, in this section, the association of dietary wheat lignans and colorectal cancer will be mainly focused on epidemiological and experimental studies.

5.1 Epidemiological studies

The study of association between lignan intake and colorectal cancer is primarily based on the following case-control studies. Cotterchio et al. (2006) conducted a case-control study in Canada to investigate the association between dietary lignan intake and risk of colorectal cancer based on 1095 cases and 1890 control subjects. High dietary lignan intake has been found to be associated with a significant decrease in the risk of colorectal cancer (Cotterchio et al., 2006). Another case-control study involving 532 cases with histological-confirmed colorectal adenoma and 503 control subjects conducted by Kuijsten et al. (2006) in the Netherlands showed a reducing risk of colorectal adenoma associated with high plasma enterolignans, in particular, EDL. Johnsen et al. (2009) reported a case-cohort study to investigate the association between plasma ENL and risk of colon and rectal cancer in Danish women and men (n = 57 053, including 21 178 men and 29 875 women) aged from 50 to 64. Double levels of ENL were associated with a reducing risk of colon cancer and a tendency of lower risk of rectal cancer for women, while double levels of ENL were associated with higher risk of colon and rectal cancer for men (Johnsen et al., 2009). The relationship between dietary lignans (Sec, Mat, and enterolignans) and colorectal cancer risk was also examined by Ward et al. (2010) based on 221 colorectal cases (125 male and 96 female) and 886 controls. Colorectal cancer risk was inversely associated with enterolignans and Sec intake among European women, but not in the European men (Ward et al., 2010). In addition, a case-control study conducted by Zamora-Ros et al. (2012) reported the association between lignan intake and risk of colorectal cancer in Spanish (cases = 424, control = 401), demonstrating an inverse association of intake of lignans with colorectal cancer risk in Spanish population. Furthermore, a case-control study was conducted by Kyrø et al. (2018) in Danish population with 416 women and 537 men who had colorectal cancer. During the follow-up study, 210 women and 325 men died (170 women and 215 men due to colorectal cancer) (Kyrø et al., 2018), indicating that the intake of ENL was associated with lower risk of mortality among women, but increasing risk of mortality among men. Taken together, the lignan intake is associated with reducing risk of colorectal cancer among women, but not in men, suggesting the lower risk of colorectal cancer associated with phytoestrogenic lignans. Although the difference of association of high levels of ENL with colon and rectal cancer risk is not fully understood, it has been postulated that such difference is caused by hormonal production, enterohepatic circulation, excretion of ENL from human body as well as estrogenic and anti-estrogenic effect of ENL (Žilić, 2016).

Apart from the inverse relationship between dietary lignan intake and risk of colorectal cancer, there are a few studies that did not find any association between lignan intake and colorectal cancer risk. For example, Zamora-Ros et al. (2015) demonstrated there was no association between the lignan intake and risk of colorectal cancer based on 409 histologically confirmed colorectal cancer cases in Barcelona (Spain). A cohort study involving 48 268 Swedish women aged from 30 to 49 found no association between dietary lignans and colorectal cancer risk (Hedelin et al., 2015). Ko et al. (2017) reported lignan ENL intake was not associated with colorectal cancer risk by observing two case-control studies involving Koreans with 101 cases and 391 controls and Vietnamese with 222 cases and 206 controls. Combining all studies, ethnic background and the amount of lignans ingested may have an impact on the relationship between lignan intake and colorectal cancer risk. In addition, the association between lignan intake and colorectal adenoma recurrence was also studied. Bobe et al. (2012) presented that lignan intake was not associated with colorectal adenoma recurrence, while a high lignan intake might increase the risk of adenoma recurrence in women. A summary of associations between dietary lignan intake and colorectal cancer risk or cancer recurrence is shown in Table 2.

It should be noted that lignans are generally fed as dietary supplements in the epidemiological studies. Little is known about the equivalent serving size of wheat lignans to the supplemental amount in those studies. Thus, further conversion of required wheat lignans in a case-control study for potential association of wheat lignans with colorectal cancer is highly demanded. Besides case-control studies, little is known about the meta-analysis of dietary lignan intake in colorectal cancer risk. Grosso et al. (2017) reported a meta-analysis regarding the association of dietary lignans intake and colorectal cancer risk. However, that meta-analysis cannot draw the conclusions of the association between dietary lignan intake and colorectal cancer mortality due to lack of

Table 2 Summary of epidemiological studies of association between dietary lignan intake and colorectal cancer risk or recurrence

Lignans	Population, age and country	No. of subjects	Association between lignans intake and colorectal cancer	References
Lignans	Canadian	Cases: 1095 Control: 1890	Reducing risk of colorectal cancer	Cotterchio et al. (2006)
EDL	People in the Netherlands, age: 18–75	Cases: 532 Control: 503	Reducing risk of colorectal cancer	Kuijsten et al. (2006)
ENL	Danish women (n = 21 178) and men (n = 29 875). age: 50–64	Cases: 57 053	Reducing risk of colorectal cancer for women, but higher risk of colorectal cancer for men	Johnsen et al. (2009)
Sec, Mat, Enterolignans	Europeans, women (n = 96), men (n = 125), age: 40–79	Cases: 221	Reducing risk of colorectal cancer for women, no association for men	Ward et al. (2010)
Lignans	Spanish	Cases: 424 Control: 401	Reducing risk of colorectal cancer	Zamora-Ros et al. (2012)
ENL	Danish, women (n = 416), men (n = 537)	Cases: 953	Lower risk of colorectal cancer for women, but increasing risk of colorectal cancer for men	Kyrø et al. (2018)
Lignans	Spanish	Cases: 409	No association	Zamora-Ros et al. (2015)
Lignans	Swedish, women, Age: 30–49	N = 48,268	No association	Hedelin et al. (2015)
ENL	Koreans, women (cases = 53, control = 206), men (cases = 48, control = 185)	Cases: 101 Control: 391	No association	Ko et al. (2017)
ENL	Vietnamese, women (cases = 117, control = 114), men (cases=105, control=92)	Cases: 222 Control: 206	No association	
Lignans	American, women	Cases: 929 Control: 930	No association between dietary lignans intake and colorectal adenoma recurrence, but high lignans intake associated with increasing risk of adenoma recurrence for women	Bobe et al. (2012)

prospective cohort studies. Another meta-analysis was reported by Jiang et al. (2016), indicating a non-linear inverse association between lignan intake and colorectal cancer risk according to sixteen identified studies. Thus, researchers in future are still warranted to address the association of wheat lignan intake and risk of colorectal cancer, cancer mortality and colorectal cancer recurrence.

5.2 Experimental studies

5.2.1 In vitro cell culture studies

We previously studied the effect of lignan ENL on human colonic Colo 205 cancer cells and found that ENL at 0.01–10 μM decreased Colo205 cancer cell growth number by inducing the detoxifying phase II enzymes such as NADPH: quinone reductase (Wang et al., 1998). We also studied an inhibitory effect of lignan metabolites EDL and ENL at 0–40 μmol/L for 72 h on human colorectal adenoma SW480 cells and found either EDL or ENL alone or a combination of EDL and ENL effectively decreased the SW480 cell number in a dose-dependent and time-dependent manner by inducing cell cycle arrest at S-phase through nontoxic mechanism (Qu et al., 2005). We further investigated the influence of lignan SDG and ENL 0–40 μmol/L for 48 h on human colonic SW480 cancer cells and found that SDG and ENL inhibited SW480 cell growth in a dose-dependent and time-dependent manner by cytostatic mechanism (Ayella et al., 2010). By using different colon cancer cell lines, Shin et al. (2019) investigated the effect of EDL on inhibiting human colorectal adenoma HT-29 and mouse colorectal adenoma CT26 cells, reporting that EDL induced both CT26 and HT-29 cell death in a dose-dependent and time-dependent manner through non-toxic mechanism by controlling the mitogen-activated protein kinase (MAPK) signaling pathway (Shin et al., 2019). In addition, EDL could down-regulate the phosphorylation of ERK, JNK and p38, which are involved in the well-known mitogen-activated protein kinase (MAPK) signaling pathway for cellular proliferation, apoptosis and migration (Shin et al., 2019). Effect of wheat lignans on inhibiting colorectal cancer *in vitro* is summarized in Table 3.

Apart from wheat lignans, lignans isolated from flaxseed or other plants are also reported to inhibit colorectal cancer *in vitro*. For example, flaxseed ENL inhibited Colo 201 cell growth by down-regulating a few proteins including Bcl-2 and PCNA protein (Danbara et al., 2005). Flaxseed EDL and ENL induced apoptosis and reduced cell proliferation in human colorectal Caco2 cells (Bommareddy et al., 2010). Furthermore, ethanolic lignan extracts from flaxseed flour also showed cytotoxic effect on colon carcinoma HCT cells (Khalaf et al., 2019). Lignan hydnocarpin isolated from *Lonicera japonica* also showed an inhibitory effect on SW480 cell growth by activating Wnt/β-catenin signaling (Lee et al., 2013). Lignan (-)-(2R, 3R)-1,4-O-diferuloylsecolariciresinol isolated

Table 3 Effect of wheat lignans on inhibiting colorectal cancer *in vitro*

Effect of wheat lignans on colorectal cancer *in vitro*				
Lignans	Cell line	Effect of lignans on cell growth	Mechanism	Reference
ENL	Colo 205	Inhibited	Induced the detoxifying phase II enzymes such as NADPH: quinone reductase (QR)	Wang et al. (1998)
EDL, ENL	SW480	Either EDL or ENL alone, or combination of EDL and ENL all inhibited	Induced cell cycle arrest at S-phase	Qu et al. (2005)
EDL, ENL	SW480	Both inhibited	Induced S-phase cell cycle arrest and apoptosis	Ayella et al. (2010)
EDL	HT-29	Inhibited	Down-regulated mitogen-activated protein kinase (MAPK) signaling pathway, and the phosphorylation of ERK, JNK and p38	Shin et al. (2019)

from *Alnus japonica* inhibited human colorectal cancer SW480 and HCT116 cells by inducing cell cycle G_1 phase arrest and suppressing β-catenin nuclear translocation and target gene expression through a reduction in FoxM1 protein (Dong et al., 2017).

Taken together, wheat lignans have confirmed an inhibitory effect on colorectal cancer *in vitro* through a few mechanisms, including inducing the detoxifying phase II enzymes, inducing cell cycle arrest and apoptosis, and controlling the MAPK signaling pathway. The health-promoting benefits of wheat lignans to humans such as antioxidant activity, anti-inflammation effect and colorectal cancer inhibition are shown in Fig. 4.

5.2.2 In vivo animal studies

Compared with *in vitro* studies, limited studies about the impact of wheat lignans on inhibiting colorectal cancer have been conducted *in vivo*. Buescher and Gallaher (2013) investigated the influence of wheat varieties (red vs. white) and state of refinement (whole vs. refined) on carcinogen-induced colonic precancerous lesions (aberrant crypt foci, ACF) in rats. They found that wheat varieties rather than state of refinement has more effect on colorectal cancer risk in rats, in particular, red hard wheat showed more protection in contrast to soft white wheat (Buescher and Gallaher, 2013). Furthermore, the effect of flaxseed lignans on inhibiting colorectal cancer *in vivo* has also been studied. For example, flaxseed SDG showed an inhibitory effect on formation of colonic ACF (Serraino and Thompson, 1992). SDG isolated from flaxseed also showed

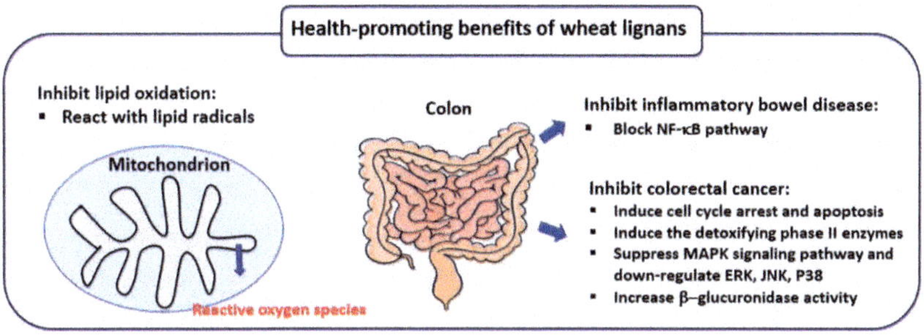

Figure 4 Schematic representation of health-promoting benefits by wheat lignans.

inhibitory effect on colorectal cancer in rats through different mechanisms, such as increased β-glucuronidase activity (Jenab and Thompson, 1996) and reduced angiogenesis (Nagel et al., 2012). Enterolignans were reported to inhibit the human colon cancer 201 cell growth in athymic mice compared to placebo-treated controls (Danbara et al., 2005).

Taken together, dietary lignans have shown an inhibitory effect on colorectal cancer in animal models, but the reported animal studies are limited. More studies regarding the effect of wheat lignans on colorectal cancer in animal models or even clinical intervention trials are needed.

6 Conclusion

Lignans are present in certain plant foods in nature. Among the cereals, wheat is a good source of lignans. Lignans are considered an important type of phytoestrogen in diet. After ingestion, lignans are metabolized into enterolignans, further to EDL and ENL, which have been reported to provide antioxidants, anti-inflammatory and anti-cancer activities. Both epidemiological and experimental (*in vitro* and *in vivo*) evidences show that wheat lignans are associated with preventing bowel inflammation and colorectal cancer. A few well-documented mechanisms of wheat lignans on inhibiting colorectal cancer have been reported including cell cycle arrest, apoptosis induction, and suppressed MAPK signaling pathway, and so on. However, animal and clinical intervention studies of lignans on colorectal cancer prevention are limited. More studies regarding wheat lignans and colorectal cancer as well as the underlying mechanisms appear warranted.

7 Where to look for further information

So far, most studies of wheat lignans on colorectal cancer prevention are conducted *in vitro* cell culture, *in vivo* animal experiments and epidemiological studies. Therefore, more clinical intervention trials regarding wheat lignans on colorectal cancer prevention seem warranted. Furthermore, limited contents of wheat lignans may be needed for new lignan-accumulated wheat variety development via genetically modified technology and/or traditional plant breeding. The fundamental knowledge of wheat lignans on anti-inflammatiory, anti-proliferative, and anti-cancer mechanisms with the potential safety and toxicity should also be conclusively elucidated.

The authors started the conception of the work and data collection at the Department of Food, Nutrition, Dietetics and Health, Kansas State University, Manhattan, KS 66506, USA. Drafting of this article and critical reversion occurred in the College of Food Science and Technology, Shanghai Ocean University, Shanghai 201306, China.

8 References

Aehle, E., Müller, U., Eklund, P. C., Willför, S. M., Sippl, W. and Dräger, B. (2011). Lignans as food constituents with estrogen and antiestrogen activity, *PhytoChemistry* 72(18), 2396–2405.

Adlercreutz, H. (2007). Lignans and human health, *Crit. Rev. Clin. Lab. Sci.* 44(5-6), 483–525.

Ahmed, F. E. (2004). Effect of diet, life style, and other environmental/chemopreventive factors on colorectal cancer development, and assessment of the risks, *J. Environ. Sci. Health C. Environ. Carcinog. Excotoxicol. Rev.* 22(2), 91–148.

Andersson, A. A. M., Dimberg, L., Åman, P. and Landberg, R. (2014). Recent findings on certain certain bioactive components in whole grain wheat and rye, *J. Cereal. Sci.* 59(3), 294–311.

Araújo, J. R., Gonçalves, P. and Martel, F. (2011). Chemopreventive effect of dietary polyphenols in colorectal cancer cell lines, *Nutr. Res.* 31(2), 77–87.

Ayella, A. K., Trick, H. N. and Wang, W. (2007). Enhancing lignan biosynthesis by over-expressing pinoresinol lariciresinol reductase in transgenic wheat, *Mol. Nutr. Food Res.* 51(12), 1518–1526.

Ayella, A. K. and Wang, W. (2010). Lignan biosynthesis enhancement in transgenic wheat. In: Bundgaards, K. and Isaksen, L. (Eds), *Agriculture Research and Technology*, Nova Science Publishers, Inc, Hauppauge, NY, pp. 441–450.

Ayella, A., Lim, S., Jiang, Y., Iwamoto, T., Lin, D., Tomich, J. and Wang, W. (2010). Cytostatic inhibition of cancer cell growth by lignan secoisolariciresinol diglucoside, *Nutr. Res.* 30(11), 762–769.

Bobe, G., Murphy, G., Albert, P. S., Sansbury, L. B., Lanza, E., Schatzkin, A. and Cross, A. J. (2012). Dietary lignan and proanthocyanidin consumption and colorectal adenoma recurrence in the polyp prevention trial, *Int. J. Cancer* 130(7), 1649–1659.

Bommareddy, A., Zhang, X. Y., Kaushik, R. S. and Dwivedi, C. (2010). Effects of components present in flaxseed on human colon adenocarcinoma Caco-2 cells: possible mechanisms of flaxseed on colon cancer development in animals, *Drug Discov. Ther.* 4(3), 184–189.

Buck, K., Zaineddin, A. K., Vrieling, A., Linseisen, J. and Chang-Claude, J. (2010). Meta-analyses of lignans and enterolignans in relation to breast cancer risk, *Am. J. Clin. Nutr.* 92(1), 141–153.

Buescher, M. I. and Gallaher, D. D. (2013). Wheat color (class), not refining, influences colon cancer risk in rats, *Nutr. Cancer* 66(5), 849–856.

Cotterchio, M., Boucher, B. A., Manno, M., Gallinger, S., Okey, A. and Harper, P. (2006). Dietary phytoestrogen intake is associated with reduced colorectal cancer risk, *J. Nutr.* 136(12), 3046–3053.

Čukelj, N., Jakasa, I., Sarajlija, H., Novotni, D. and Ćurić, D. (2010). Identification and quantification of lignans in wheat bran by gas chromatography-electron capture detection, *Talanta* 84(1), 127–132.

Danbara, N., Yuri, T., Tsujita-Kyutoku, M., Tsukamoto, R., Uehara, N. and Tsubura, A. (2005). Enterolactone induces apoptosis and inhibits growth of Colo 201 human colon cancer cells both in vitro and in vivo, *Anticancer. Res.* 25(3B), 2269–2276.

Davin, L. B. and Lewis, N. G. (2003). An historical perspective on lignan biosynthesis: monolignol, allylphenol and hydroxycinnamic acid coupling and downstream metabolism, *Phytochem. Rev.* 2(3), 257–288.

Davis, H., Su, X., Shen, Y., Xu, J., Wang, D., Smith, J. S., Aramouni, F., Wang, W. (2019). Phenotypic diversity of colored phytohemicals in sorghum accesions with various pericarp pigments. In: Watson, R. R. (Ed.), *Polyphenols in Plants* (2nd Edn). Academic Press, Cambridge, MA. pp. 123–131.

Dinelli, G., Carretero, A. S., Di Silvestro, R., Marotti, I., Fu, S., Benedettelli, S., Ghiselli, L. and Gutiérrez, A. F. (2009). Determination of phenolic compounds in modern and old varieties of durum wheat using liquid chromatography coupled with time-of-flight mass spectrometry, *J. Chrom. A* 1216(43), 7229–7240.

Dong, G. Z., Jeong, J. H., Lee, Y. I., Han, Y. E., Shin, J. S., Kim, Y. J., Jeon, R., Kim, Y. H., Park, T. J., Kim, K. I. and Ryu, J. H. (2017). A lignan induces lysosomal dependent degradation of FoxM1 protein to suppress β-catenin nuclear translocation, *Sci. Rep.* 7, 45951.

Durazzo, A., Azzini, E., Raguzzini, A., Maiani, G., Finocchiaro, F., Ferrari, B., Gianinetti, A. and Carcea, M. (2009). Influence of processing on the lignans content of cereal based foods, *Tec. Molitoria Int.* 60, 163–173.

Durazzo, A., Zaccaria, M., Polito, A., Maiani, G. and Carcea, M. (2013). Lignan content in cereals, buckwheat and derived foods, *Foods* 2(1), 53–63.

During, A., Debouche, C., Raas, T. and Larondelle, Y. (2012). Among plant lignans, pinoresinol has the strongest antiinflammatory properties in human intestinal Caco-2 cells, *J. Nutr.* 142(10), 1798–1805.

Ezzat, S. M., Shouman, S. A., Elkhoely, A., Attia, Y. M., Elsesy, M. S., El Senousy, A. S., Choucry, M. A., El Gayed, S. H., El Sayed, A. A., Sattar, E. A. and El Tanbouly, N. (2018). Anticancer potentiality of lignan rich fraction of six Flaxseed cultivars, *Sci. Rep.* 8(1), 544.

Grosso, G., Micek, A., Godos, J., Pajak, A., Sciacca, S., Galvano, F. and Giovannucci, E. L. (2017). Dietary flavonoid and lignan intake and mortality in prospective cohort studies: systematic review and dose-response meta-analysis, *Am. J. Epidem.* 185(12), 1304–1316.

Hanhineva, K., Rogachev, I., Aura, A. M., Aharoni, A., Poutanen, K. and Mykkänen, H. (2011).Identification of novel lignans in the whole grain rye bran by non-targeted LC-MS metabolite profiling, *Metabolomics* 8(3), 399–409.

Hedelin, M., Löf, M., Sandin, S., Adami, H. O. and Weiderpass, E. (2015). Prospective study of dietary phytoestrogen intake and the risk of colorectal cancer, *Nutr. Cancer* 68(3), 388–395.

Heinonen, S., Nurmi, T., Liukkonen, K., Poutanen, K., Wähälä, K., Deyama, T., Nishibe, S. and Adlercreutz, H. (2001). In vitro metabolism of plant lignans: new precursors of mammalian lignans enterolactone and enterodiol, *J. Agric. Food Chem.* 49(7), 3178–3186.

Hu, C., Yuan, Y. V. and Kitts, D. D. (2007). Antioxidant activities of the flaxseed lignan secoisolariciresinol diglucoside, its aglycone secoisolariciresinol and the mammalian lignans enterodiol and enterolactone in vitro, *Food Chem. Toxicol.* 45(11), 2219-2227. doi:10.1016/j.fct.2007.05.017.

Jenab, M. and Thompson, L. U. (1996). The influence of flaxseed and lignans on colon carcinogenesis and β-glucuronidase activity, *Carcinogenesis* 17(6), 1343 -1348.

Jiang, R., Botma, A., Rudolph, A., Hüsing, A. and Chang-Claude, J. (2016). Phyto-oestrogens and colorectal cancer risk: a systematic review and dose-response meta-analysis of observational studies, *Br. J. Nutr.* 116(12), 2115–2128.

Jiang, Y. and Wang, W. (2016). Potential mechanisms of cancer prevention by weight control. In: Gerstman, B. S. (Ed.), *Research on the Physics of Cancer: A Global Perspective*. World Scientific Publishing, Singapore, pp. 191–207.

Johnsen, N. F., Olsen, A., Thomsen, B. L. R., Christensen, J., Egeberg, R., Bach Knudsen, K. E., Loft, S., Overvad, K. and Tjønneland, A. (2009). Plasma enterolactone and risk of colon and rectal cancer in a case-cohort study of Danish men and women, *Cancer Causes Control* 21(1), 153–162.

Khalaf, H. H., El-Saadani, R. M., El-Amry, H. G. and El-Salam, R. S. (2019). Bioactive compounds of flaxseed as natural antioxidants and anticancer, *Ann. Agric. Sci. Mosht* 57(2), 395–404.

Ko, K. P., Yeo, Y., Yoon, J. H., Kim, C. S., Tokudome, S., Ngoan, L. T., Koriyama, C., Lim, Y. K., Chang, S. H., Shin, H. R., Kang, D., Park, S. K., Kang, C. H. and Yoo, K. Y. (2017). Plasma phytoestrogens concentration and risk of colorectal cancer in two different Asian populations, *Clin. Nutr.* 37(5), 1675–1682.

Kuijsten, A., Arts, I. C., Hollman, P. C., van't Veer, P. and Kampman, E. (2006). Plasma enterolignans are associated with lower colorectal adenoma risk, *Cancer Epidemiol. Biomarkers Prev.* 15(6), 1132–1136.

Kuijsten, A., Arts, I. C., Vree, T. B. and Hollman, P. C. (2005). Pharmacokinetics of enterolignans in healthy men and women consuming a single dose of secoisolariciresinol diglucoside, *J. Nutr.* 135(4), 795–801.

Kyrø, C., Frederiksen, K., Holm, M., Nørskov, N. P., Knudsen, K. E.,. Overvad, K., Tjønneland, A., and Olsen, A. (2018). Prediagnosis plasma concentrations of enterolactone and survival after colorectal cancer: the danish diet, cancer and health cohort, *Bri. J. Nutr.* 122(5), 552–563.

Landete, J. M. (2012). Plant and mammalian lignans: a review of source, intake, metabolism, intestinal bacteria and health, *Food Res. Int.* 46(1), 410–424.

Lee, M. A., Kim, W. K., Park, H. J., Kang, S. S. and Lee, S. K. (2013). Anti-proliferative activity of hydnocarpin, a natural lignan, is associated with the suppression of Wnt/β-catenin signaling pathway in colon cancer cells, *Bioorg. Med. Chem. Let.* 23(20), 5511–5514.

Lemay, A., Dodin, S., Kadri, N., Jacques, H. and Forest, J. C. (2002). Flaxseed dietary supplement versus hormone replacement therapy in hypercholesterolemic menopausal women, *Obstet. Gynecol.* 100(3), 495–504.

Lillioja, S., Neal, A. L., Tapsell, L. and Jacobs Jr., D. R. (2013). Whole grains, type 2 diabetes, diabetes, coronary heart disease, and hypertension: links to the aleurone preferred over indigestible fiber, *BioFactors* 39(3), 242–258.

McCann, M. J., Gill, C. I., McGlynn, H. and Rowland, I. R. (2005). Role of mammalian lignans lignans in the prevention and treatment of prostate cancer, *Nutr. Cancer* 52(1), 1–14.

Milder, I. E., Arts, I. C., van de Putte, B., Venema, D. P. and Hollman, P. C. (2005). Lignans contents of Dutch plant foods: a database including lariciresinol, pinoresinol, secoisolariciresinol and matairesinol, *Bri. J. Nutr.* 93(3), 393–402.

Miyawaki, T., Aono, H., Toyoda-Ono, Y., Maeda, H., Kiso, Y. and Moriyama, K. (2009). Antihypertensive effects of sesamin in humans, *J. Nutr. Sci. Vit.* 55(1), 87–91.

Muir, A. D. and Westcott, N. D. (2000). Quantitation of the lignan secoisolariciresinol diglucoside in baked goods containing flax seed or flax meal, *J. Agric. Food Chem.* 48(9), 4048–4052.

Nagel, J. M., Brinkoetter, M., Magkos, F., Liu, X., Chamberland, J. P., Shah, S., Zhou, J., Blackburn, G. and Mantzoros, C. S. (2012). Dietary walnuts inhibit colorectal cancer growth in mice by suppressing angiogenesis, *Nutrition* 28(1), 67–75.

Nelson, H. D., Humphrey, L. L., Nygren, P., Teutsch, S. M. and Allan, J. D. (2002). Postmenopausal hormone replacement therapy: scientific review, *JAMA* 288(7), 872–881.

Nørskov, N. P. and Knudsen, K. E. B. (2016). Validated LC-MS/MS Method for the Quantification of free and bound lignans in cereal-based diets and feces, *J. Agric. Food Chem.* 64(44), 8343–8351.

Oduwole, O. O., Mäkinen, M. J., Isomaa, V. V., Pulkka, A., Jernvall, P., Karttunen, T. J. and Vihko, P. T. (2003). 17β-Hydroxysteroid dehydrogenase type 2: independent prognostic significance and evidence of estrogen protection in female patients with colon cancer, *J. Steroid. Biochem. Mol. Biol.* 87(2–3), 133–140.

Okarter, N. and Liu, R. H. (2010). Health benefits of whole grain phytochemicals, *Crit. Rev. Food Sci. Nutr.* 50(3), 193–208.

Peñalvo, J. L., Adlercreutz, H., Uehara, M., Ristimaki, A. and Watanabe, S. (2008). Lignan content of selected foods from Japan, *J. Agric. Food Chem.* 56(2), 401–409.

Peñalvo, J. L., Heinonen, S. M., Aura, A. M. and Adlercreutz, H. (2005a). Dietary sesamin is converted to enterolactone in humans, *J. Nutr.* 135(5), 1056–1062.

Peñalvo, J. L., Haajanen, K. M., Botting, N. and Adlercreutz, H. (2005b). Quantification of lignans in food using isotope dilution gas chromatography/mass spectrometry, *J. Agric. Food Chem.* 53(24), 9342–9347.

Peterson, J., Dwyer, J., Adlercreutz, H., Scalbert, A., Jacques, P. and McCullough, M. L. (2010). Dietary lignans: physiology and potential for cardiovascular disease risk reduction, *Nutr. Rev.* 68(10), 571–603.

Pietrofesa, R. A., Velalopoulou, A., Arguiri, E., Menges, C. W., Testa, J. R., Hwang, W. T., Albelda, S. M. and Christofidou-Solomidou, M. (2016). Flaxseed lignans enriched in secoisolariciresinol diglucoside prevent acute asbestos-induced peritoneal inflammation in mice, *Carcinogenesis* 37(2), 177–187.

Possemiers, S., Bolca, S., Eeckhaut, E., Depypere, H. and Verstraete, W. (2007). Metabolism of isoflavones, lignans and prenylflavonoids by intestinal bacteria: producer phenotyping and relation with intestinal community, *FEMS Microb. Ecol.* 61(2), 372–383.

Prasad, K. (2000). Antioxidant activity of secoisolariciresinol diglucoside-derived metabolites, secoisolariciresinol, enterodiol, and enterolactone, *Int. J. Angiol.* 9(4), 220–225.doi:10.1007/BF01623898.

Qu, H., Madl, R. L., Takemoto, D. J., Baybutt, R. C. and Wang, W. (2005). Lignans are involved involved in the antitumor activity of wheat bran in colon cancer SW480 cells, *J. Nutr.* 135(3), 598–602.

Reddy, B., Engle, A., Katsifis, S., Simi, B., Bartram, H. P., Perrino, P. and Mahan, C. (1989). Biochemical epidemiology of colon cancer: effect of types of dietary fiber on fecal mutagens, acid, and neutral sterols in healthy subjects, *Cancer Res.* 49(16), 4629–4635.

Reddy, B. S., Engle, A., Simi, B. and Goldman, M. (1992). Effect of dietary fiber on colonic bacterial enzymes and bile acids in relation to colon cancer, *Gastroenterology* 102(5), 1475-1482.

Rodríguez-García, C., Sánchez-Quesada, C., Toledo, E., Delgado-Rodríguez, M. and Gaforio, J. J. (2019). Naturally lignan-rich foods: a dietary tool for health promotion?, *Molecules* 24(5), 917.

Roncaglia, L., Amaretti, A., Raimondi, S., Leonardi, A. and Rossi, M. (2011). Role of bifidobacteria in the activation of the lignan secoisolariciresinol diglucoside, *Appl. Microb. Biot.* 92(1), 159–168.

Seifried, H. E., Anderson, D. E., Fisher, E. I. and Milner, J. A. (2007). A review of the interaction among dietary antioxidants and reactive oxygen species, *J. Nutr. Biochem.* 18(9), 567–579. doi:10.1016/j.jnutbio.2006.10.007.

Serraino, M. and Thompson, L. U. (1992). Flaxseed supplementation and early markers of colon carcinogenesis, *Cancer Let.* 63(2), 159–165.

Shin, M. K., Jeon, Y. D. and Jin, J. S. (2019). Apoptotic effect of enterodiol, the final metabolite of edible lignans, in colorectal cancer cells, *J. Sci. Food Agric.* 99(5), 2411–2419.

Siegel, R. L., Miller, K. D. and Jemal, A. (2020). Cancer statistics 2020, *CA Cancer J. Clin.*, 70, 7–30.

Slavova-Kazakova, A., Karamać, M., Kancheva, V. and Amarowicz, R. (2015). Antixoidant activity of flaxseed extracts in lipid systems, *Mol* 21(1), 17. doi:10.3390/molecules21010017.

Smeds, A. I., Eklund, P. C., Sjöholm, R. E., Willför, S. M., Nishibe, S., Deyama, T. and Holmbom, B. R. (2007). Quantification of a broad spectrum of lignans in cereals, oilseeds, and nuts, *J. Agric. Food Chem.* 55(4), 1337–1346.

Smeds, A. I., Jauhiainen, L., Tuomola, E. and Peltonen-Sainio, P. (2009). Characterization of variation in the lignan content and composition of winter rye, spring wheat, and spring oat, *J. Agric. Food Chem.* 57(13), 5837–5842.

Su, X., Jia, Z., Tao, F., Shen, J., Xu, J., Griffin, J. and Wang, W. (2019). Characterization of anthocyanins in sweet potato leaves grown in various stages and conditions. *Eur. J. Nutr. Food Saf.* 10(4), 253–262.

Szewczyk, M., Abarzua, S., Schlichting, A., Nebe, B., Piechulla, B., Marlen, S., Sibylle, A., Andr, S., Barbara, N., Birgit, P., Volker, B. and Dagmar-Ulrike, R. (2014). Effects of extracts from Linum usitatissimum on cell vitality, proliferation and cytotoxicity in human breast cancer cell lines, *J. Med. Plants Res.* 8(5), 237–245.

Tetens, I., Turrini, A., Tapanainen, H., Christensen, T., Lampe, J. W., Fagt, S., Håkansson, N., Lundquist, A., Hallund, J., Valsta, L. M. and Phytohealth WP1 Working Group (2013). Dietary intake and main sources of plant lignans in five European countries, *Food Nutr. Res.* 57(1), 19805.

Thompson, L. U., Boucher, B. A., Liu, Z., Cotterchio, M. and Kreiger, N. (2006). Phytoestrogen content of foods consumed in Canada, including isoflavones, lignans, and coumestan, *Nutr. Cancer* 54(2), 184–201.

Touré, A. and Xu, X. (2010). Flaxseed lignans: source, biosynthesis, metabolism, antioxidant activity, bio-active components, and health benefits, *Compr. Rev. Food Sci. Food Saf.* 9(3), 261–269.

Tresserra-Rimbau, A., Rimm, E. B., Medina-Remón, A., Martínez-González, M. A., López-Sabater, M. C., Covas, M. I., Corella, D., Salas-Salvadó, J., Gómez-Gracia, E., Lapetra, J., Arós, F., Fiol, M., Ros, E., Serra-Majem, L., Pintó, X., Muñoz, M. A., Gea, A., Ruiz-Gutiérrez, V., Estruch, R., Lamuela-Raventós, R. M. and PREDIMED Study Investigators (2014). Polyphenol intake and mortality risk: a re-analysis of the PREDIMED trial, *BMC Med.* 12(1), 77.

Wang, C. Z., Ma, X. Q., Yang, D. H., Guo, Z. R., Liu, G. R., Zhao, G. X., Tang, J., Zhang, Y. N., Ma, M., Cai, S. Q., Ku, B. S. and Liu, S. L. (2010). Production of enterodiol from defatted flaxseeds through biotransformation by human intestinal bacteria, *BMC Micro.* 10(1), 115.

Wang, W., Ayella, A., Jiang, Y., Ouyang, P. and Qu, H. (2008). Wheat lignans: promising cancer preventive agents. In: Yu, L. (Ed.), *Wheat Antioxidants*, John Wiley & Sons, Inc, Manhattan, KS, pp. 264–272.

Wang, W. and Goodman, M. T. (1999). Antioxidant properties of dietary phenolic agents in a human LDL-oxidation ex vivo model: interaction of protein binding activity, *Nutr. Res.* 19(2), 191–202.

Wang, W., Liu, L. Q., Higuchi, C. M. and Chen, H. (1998). Induction of NADPH: quinone reductase by dietary phytoestrogens in colonic colo205 cells, *Biochem. Pharmacol.* 56(2), 189–195.

Ward, H. A., Kuhnle, G. G., Mulligan, A. A., Lentjes, M. A., Luben, R. N. and Khaw, K. T. (2010). Breast, colorectal, and prostate cancer risk in the European Prospective Investigation into Cancer and Nutrition-Norfolk in relation to phytoestrogen intake derived from an improved database, *Am. J. Clin. Nutr.* 91(2), 440–448. doi:10.3945/ajcn.2009.28282.

Woting, A., Clavel, T., Loh, G. and Blaut, M. (2010). Bacterial transformation of dietary lignans in gnotobiotic rats, *FEMS Microb. Ecol.* 72(3), 507–514.

Wu, W. H., Kang, Y. P., Wang, N. H., Jou, H. J. and Wang, T. A. (2006). Sesame ingestion affects sex hormones, antioxidant status, and blood lipids in postmenopausal women, *J. Nutr.* 136(5), 1270–1275.

Xu, J., Li, Y., Wang, W. (2019). Corn. In: Wang, J., Sun, B. and Tsao, R. (Eds), *Bioactive Factors and Processing Technology for Cereal Foods*, Springer Publishing, Singapore, pp. 33–53.

Xu, J., Tian, G., Ma, C., Gao, H., Chen, C., Yang, W., Deng, Q., Huang, Q., Ma, Z. and Huang, F. (2016). Flaxseed lignan secoisolariciresinol diglucoside ameliorates experimental colitis induced by dextran sulphate sodium in mice, *J. Funct. Foods* 26, 187–195.

Yoder, S. C., Lancaster, S. M., Hullar, M. A. and Lampe, J. W. (2015). Gut microbial metabolism of plant lignans: influence on human health. In: Tuohy, K. and Rio, D. (Eds), *Diet-Microbe Interactions in the Gut*, Academic Press, London, UK, pp. 103–117.

Zamora-Ros, R., Guinó, E., Alonso, M. H., Vidal, C., Barenys, M., Soriano, A. and Moreno, V. (2015). Dietary flavonoids, lignans and colorectal cancer prognosis, *Sci. Rep.* 5(1), 14148.

Zamora-Ros, R., Not, C., Guinó, E., Luján-Barroso, L., García, R. M., Biondo, S., Salazar, R. and Moreno, V. (2012). Association between habitual dietary flavonoid and lignan intake and colorectal cancer in a Spanish case-control study (the Bellvitge Colorectal Cancer Study), *Cancer Causes Control* 24(3), 549–557.

Zhang, W., Wang, X., Liu, Y., Tian, H., Flickinger, B., Empie, M. W., and Sun, S. Z.(2008). Dietary flaxseed lignan extract lowers plasma cholesterol and glucose concentrations in hypercholesterolaemic subjects, *Br. J. Nut.* 99(6), 1301-1309.

Žilić, S. (2016). Phenolic compounds of wheat. 'Their content, antioxidant capacity and bioaccessibility, *MOJ Food Process. Technol.* 2(3), 00037.

Part 2

Developing nutritionally-enhanced cereal products

Chapter 6

Advances in understanding the genetics of the nutritional properties of cereals: maize and oat proteins

Sachin Rustgi, Salman Naveed and Prasanna Shekar, Pee Dee Research and Education Center, Clemson University, USA

1 Introduction

There is a great disparity when we talk about the food; at one end, there are overnourished people (1.9 billion children and adults) (WHO, 2020), and on the other, there are millions who are undernourished (over 820 million people worldwide) (Roser and Ritchie, 2013). Food matters are complex; for instance, meeting the calorific needs is insufficient to ascertain healthy living or alleviate malnutrition. It often requires a balanced nutritious diet. In other words, it is important to feed the hungry, but it is equally important to take care of the hidden hunger, that is, nutritional deficiencies (Global Hunger Index, 2014).

In particular, beta-carotene (250 million children), iron (43% children and 38% pregnant women worldwide), iodine (1.8 billion individuals), and zinc (17.3% of the world population) deficiencies collectively affect millions of individuals around the globe, specifically in the populated and low-income countries (CDC, 2020). Moreover, unhealthy diets were also the leading cause of chronic conditions, such as hyperglycemia and cardiovascular diseases (WHO, 2018). In the last decade, it has been repeatedly shown that a healthy gut assures a healthy body and a brain, and nutritious foods support a healthy gut microbiome, which contributes to a healthy gut (Pennisi, 2019).

http://dx.doi.org/10.19103/AS.2021.0087.12

Several global efforts are underway to produce healthy foods sustainably and educate the public about the importance of diet in healthy living or, in other words, the importance of food in remedying or avoiding various ailments. One such initiative is 'EAT.' It is a nonprofit founded by the Stordalen Foundation, Stockholm Resilience Centre, and the Wellcome Trust, to catalyze a food system transformation. In this mission, EAT has partnered with many institutions and foundations, such as Rockefeller Foundation in the United States, and spreading its message and organizing support in this research area (https://eatforum.org/).

EAT and partners made arrangements to send a delegation to 2020 UN general assembly and to the UN Food Systems Summit 2021 to discuss a five-prong action plan which includes 'ensuring access to safe and nutritious food for all,' as its first target (https://www.unccd.int/actions/food-systems-summit-2021). The National Institutes of Health (NIH) is also gathering public opinion to launch a program titled 'Challenges and Opportunities in Precision Nutrition Research' to support research in this area (https://grants.nih.gov/).

Cereals, specifically wheat, rice, corn, sorghum, barley, and millets, are the major staples that feed a significant portion of the global demand for plant proteins (about 25%) and calories (about 50%) (Sands et al., 2009). The cereals grow in diverse climatic and edaphic conditions, from the Equator to Scandinavia and from well-watered to semi-arid regions, and yield reasonably. Cereals and human civilization have a long connection, as cereal production formed the basis for human civilization (Awika, 2011). At one time point in human history, cereals have also served as currency.

Cereals offer several nutritional components, such as carbohydrates, lipids, proteins, and micronutrients. At the same time, cereals are sources of several allergies and intolerances, such as the allergies caused by wheat, barley, oat, rice, and corn proteins in different individuals and celiac disease caused by wheat and barley seed storage proteins (prolamins) (https://farrp.unl.edu/informallcerealsgrains). These grain constituents not only impart a profound effect on the nutritional properties of the grains but also on their end-use quality. This knowledge has sparked interest in understanding the genetics of the nutritional properties of cereals.

These studies have yielded a massive amount of literature, which is almost impossible to summarize in a chapter, indeed in a book; therefore, in this book chapter, an attempt has made to briefly cover the genetics of corn and oat nutritional properties, whereas the widely covered nutritional aspects of wheat and barley were largely skipped from the discussion. These cereals are also the topic of discussion in several recent reviews and also the subject of several recent books, *Barley: Chemistry and Technology*, *Wheat: Chemistry and Technology*, *Durum wheat: Chemistry and Technology*, *Wheat Quality for improving processing and human health*, *Sorghum Biochemistry: An Industrial Perspective*, and *Wheat and Rice in Disease Prevention and Health*.

2 Maize (*Zea mays* L.)

Maize, after rice and wheat, is the most important cereal crop, which is globally produced for food and feed. Among all cereals, maize provides 62% of dietary proteins in Meso America, 43% in Eastern and Southern Africa, 28% in the Andean Region, 22% in West and Central Africa, and 4% in South Asia (Hossain et al., 2018). Besides proteins, maize kernels are also a good source of carbohydrates, lipids (oil), vitamins, and micronutrients. Interestingly, the maize germplasm exhibits considerable genetic variation for all of these nutritional compounds.

For instance, maize genotypes vary for starch content in kernels from 59.60 to 74.40% (Cook et al., 2012; Guo et al., 2013; Yangcheng et al., 2013), pro-vitamin A or β-carotene content from 3.4 mg/kg to 21.7 mg/kg (Muthusamy et al., 2014; Pillay et al., 2014), seed oil content from 1.4% to 6.0%, zinc (Zn) content from 3.81 mg/kg to 52.95 mg/kg (Chakraborti et al., 2011a; Prasanna et al., 2011; Guleria et al., 2013; Mallikarjuna et al., 2014), and iron (Fe) content from 11.28 mg/kg to 83.35 mg/kg (Agrawal et al., 2012; Prasanna et al., 2011; Mallikarjuna et al., 2014; Chakraborti et al., 2011b). This great level of genetic variability in maize germplasm for these biomolecules and micronutrients offers a unique opportunity to study the genetics of these traits and also the possibility to breed for them.

2.1 Quality protein maize (QPM)

As mentioned earlier, maize is an important source of dietary protein, specifically in low-income countries, where it provides 70% of the protein consumed (Gibbon and Larkins, 2005). Hence, it is popularly known as 'poor man's cereal crop.' In Africa and Latin America, animal proteins are expensive, hence are less readily available to the large fraction of the population, and the maize grains remain the primary source of dietary proteins and daily dietary calories (15–56%) (Prasanna et al., 2001).

Unfortunately, a significant proportion of grain proteins in maize has imbalanced composition and carry very low levels of essential amino acids, lysine (1.81%), and tryptophan (0.35%), and do not provide the Food and Agriculture Organization (FAO) recommended levels of these amino acids. Therefore, conventional maize fed to pregnant/lactating women and babies causes malnutrition and leads to Kwashiorkor disease (May et al., 2018). Given the nutritional importance of maize grain storage proteins in human nutrition, a massive effort was undertaken to improve maize protein quality.

As an outcome of this effort, a quality protein maize (QPM) was developed by incorporating a mutation in the *opaque-2* (*o2*) gene. QPM genotypes possess a more balanced protein composition with high lysine and tryptophan content and lower leucine and isoleucine content than their conventional maize counterparts. Besides the balanced protein phenotype, the *o2* maize mutations (source of QPM) exhibit some unfavorable characteristics, such as opaque endosperm, susceptibility to diseases and pests, and grain breakage at milling.

The International Maize and Wheat Improvement Center (CIMMYT) took the initiative under the leadership of Drs. Surinder Vasal and Evangelina Villegas, who later won the World Food Prize in 2000, to rectify the problems associated with the *o2* maize mutant by breeding in modifiers. These efforts later led to the development of QPM cultivars with hard endosperm (popularly known as 'miracle maize'), which are both nutritionally superior and suitable (4% lysine and >0.8 % tryptophan) for production (Vasal et al., 1980; Villegas et al., 1992; Krivanek et al., 2007).

Research institutions worldwide, including CIMMYT and the International Institute of Tropical Agriculture (IITA), are developing varieties with improved nutritional protein contents by incorporating *o2* and modifier genes in different market classes of corn. Maize cultivars with high yield, improved lysine, and tryptophan contents, and conventional kernel structure are now grown roughly over 9 million acres worldwide and already impacting the lives of millions in Latin America, Africa, Europe, and Asia (Badu-Apraku et al., 2015).

2.1.1 Maize **Zein** *genes and their regulation*

Germ (embryo), endosperm, and pericarp are the main parts of a mature maize kernel (Fig. 1). Maize endosperm comprises 90% starch and 10% protein that consists of albumins, globulins, glutelins, and prolamins dubbed zeins. The starch-rich endosperm cells nourish the embryo at the germination time and contribute to the grain's end-use quality. Out of all proteins, 50–70% consist of proline- and glutamine-rich zeins (Gibbon and Larkins, 2005). Zeins exist exclusively in maize endosperm and play a vital role in storing and supplying nitrogen, carbon, and sulfur to the germinating seeds (Wu et al., 2010; Prasanna et al., 2001).

Synthesis of zeins in starchy endosperm cells takes place on the rough endoplasmic reticulum. After synthesis, zeins accumulate in protein bodies (PBs), which are membrane-bound structures that start as evaginations of the rough endoplasmic reticulum (Larkins and Hurkman, 1978; Wu et al., 2010). PBs, during kernel maturation, become compactly packed between starch grains embedded in desiccated cytoplasm (Gayral et al., 2016) and giving the endosperm its vitreous appearance (Larkins et al., 2017).

Zeins, based on their electrophoretic mobility, can be classified into α-, β-, γ-, and δ-zeins, which are structurally different alcohol-soluble proteins. Among these seed storage proteins, α-zeins are most abundant and exist as polypeptides of 22 kDa and 19 kDa. The second most abundant proteins are

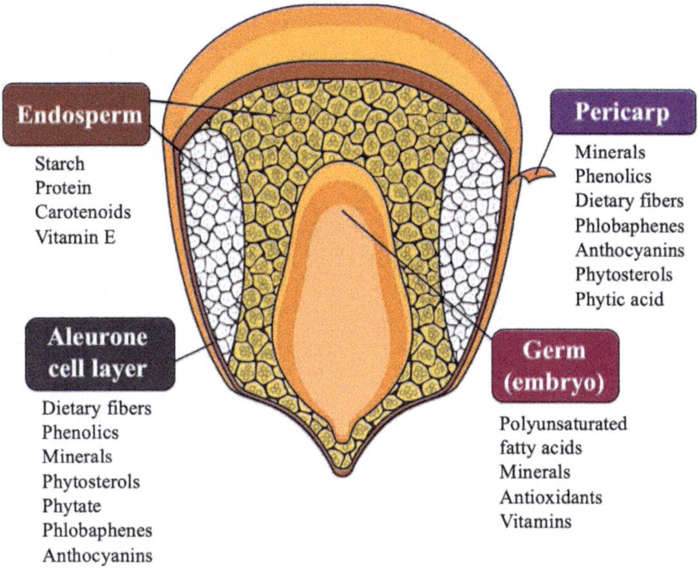

Figure 1 Nutritional contents of mature maize kernel (Modified from Prasanna et al., 2020).

the γ-zeins, which exist as polypeptides of 50 kDa, 27 kDa, and 16 kDa. The β-zein (now renamed as 15 kDa γ-zein) exists as a polypeptide of 15 kDa (Woo et al., 2001). The δ-zeins, which are the least abundant, consist of two related polypeptides of 18 kDa and 10 kDa (Swarup et al., 1995).

Zein polypeptides are coded by different structural genes (Tripathy et al., 2017). In maize cultivar B73, the α-zein polypeptides are encoded by a small gene family of 41–48 genes organized in small clusters on chromosomes 1, 4, and 7 (Miclaus et al., 2011a). Some of the 19 kDa α-zein genes (12 genes) map to the short arm of chromosome 4 in two clusters, *z1A1* (9 genes) and *z1A2* (3 genes). The remaining 19 kD α-zein genes map in two clusters, *z1B* (9 genes) and *z1D* (5 genes) on the short arms of chromosome 7 and chromosome 1, respectively (Song and Messing, 2002). The 22-kD α-zein genes map in two clusters, *z1C1* (14 genes) and *z1C2* (one gene) on the short arm of chromosome 4 (Song and Messing, 2003). Contrary to the α-zeins, β-, γ-, and δ-zeins are typically encoded by the single-copy genes. However, in some cultivars, the 27 kDa γ-zein locus consists of two nearly identical genes in tandem (Das et al., 1991). The genes encoding 50 kDa and 27 kDa γ-zeins exist in proximity on the short arm of chromosome 7 (Yuan et al., 2014), and that of the 16-kD γ-zein on the long arm of chromosome 2. The genes encoding 15 kDa β-zein map on the short arm of chromosome 6, and that of the 10 kDa and 18 kDa δ-zeins, respectively, on chromosomes 9 and 6 (Larkins et al., 2017).

The zein genes are temporally and spatially regulated. But the mechanism is not yet fully understood. In large, their activity is controlled by the transcription factors (TFs) and *cis* nucleotide sequences in the promoters of zein genes. None of the zein genes contain introns; therefore, the post-transcriptional processing of mRNAs does not seem to play a major role in gene expression. So far, three different types of TFs have been recognized to control zein gene expression: bZIP proteins (encoded by *O2*), zinc finger protein (encoded by *pbf1*), and a MADS-box protein (encoded by *ZmMADS47*) (Vicente-Carbajosa et al., 1997; Zhang et al., 2015; Qiao et al., 2016).

2.1.2 High-lysine maize mutants

Different mutants were identified where the synthesis of zeins is altered, leading to unusual size, shape, and count of PBs that result in kernels with a starchy and soft texture. These mutations broadly belong to four categories: (i) affecting zein protein structure; (ii) affecting cytoskeletal proteins; (iii) influencing metabolic processes; and (iv) controlling gene transcription (Table 1). Some of these mutants are spontaneous mutations, whereas others are induced via chemical mutagenesis or transposon tagging (Thompson and Larkins, 1993).

These mutants which were identified over the past century, all exhibit opaque or floury endosperm phenotype and are termed 'opaque' [e.g. *opaque1-17* (*o1*, *o2*, *o5-o7*, *o9-o11*, *o13-o17*), recessive] and 'floury' [e.g. *floury1-4* (*fl1-fl4*), semidominant] (Gibbon and Larkins, 2005; Larkins et al., 2017; Zhang et al., 2018). As stated earlier, these mutants' opaque phenotype is generally associated with the altered protein bodies (PBs) and/or starch granules. Many opaque mutants also exhibited a defect in the expression or accumulation of prolamins (Wu and Messing, 2014).

Among these mutants, *o2* is a high-lysine maize mutant extensively used in maize breeding. The maize *O2* gene encodes a bZIP-family transcription factor that specifically expresses in the endosperm as early as 6 days after pollination (Li et al., 2014; Schmidt et al., 1990). Earlier studies have shown that O2 directly regulates many genes associated with storage functions in the endosperm, particularly zeins, by binding to several conserved cis-motifs known as the O2 box.

Opaque 16 (*o16*) is another important mutant identified in China. Among other high-lysine mutants, only *o16* confers lysine and tryptophan content as high as the *o2* and exhibit the characteristic opaque endosperm phenotype. High-lysine mutant lines, QCL3021 and QCL3024 with homozygous recessive *o16o16* genotype and the opaque endosperm were obtained from a self-crossed population derived from Robertson's Mutator stocks. Later, to study the genetics of *o16* locus, two populations were evaluated for lysine content,

Table 1 List of maize *Opaque* mutants

Mutant	Zein phenotype	Type	Physiological effect	Reference
i) Mutations affecting zein protein structure				
fl2[2]	22-kD α-zein	defective signal peptides	unfolded protein response	Boston and Larkins (2008)
De-B30*[3]	19-kD α-zein	defective signal peptides	unfolded protein response	Boston and Larkins (2008)
fl4[2]	19-kD α-zein	defective signal peptides	unfolded protein response	Wang et al. (2014a)
Mucronate1 (*Mc*)[3]	16-kD γ-zein	frame-shift mutation	unfolded protein response	Kim et al. (2006)
ii) Mutations affecting cytoskeletal proteins				
o1[1]	-	defective myosin or myosin-related proteins	zein translocation into RER	Wang et al. (2012)
fl1[2]	-	defective myosin or myosin-related proteins	zein organization within PB	Holding et al. (2007)
o10[1]	-	defective PB internal protein	Disturbed interaction with Fl1, 22-kD and 16-kD zeins	Yao et al. (2016)
iii) Mutations influencing metabolic processes				
o7[1]	19 kDa zein	defective oxalyl-CoA synthetase	asparagine, lysine and glutamine synthesis	Miclaus et al. (2011b)
mto140[1]	-	defective arogenate dehydrogenase	tyrosine synthesis with pleiotropic effects on other amino acids	Holding et al. (2010)
o6 (*pro1*)[1]	all zeins	defective Δ1-pyrroline-5-carboxylate synthetase	proline synthesis and general reduction in protein synthesis	Wang et al. (2014b)
o5[1]	-	defective monogalactosyldiacylglyerolsynthase (MGD1)	phospholipid composition of amyloplast membranes	Myers et al. (2011)

(*Continued*)

Table 1 *(Continued)*

Mutant	Zein phenotype	Type	Physiological effect	Reference
iv) Mutations controlling gene transcription				
o2[1]	22-kD α-zein	defective bZIP transcription factor	compensatory increase in many non-zein proteins	Schmidt et al. (1990); Aukerman et al. (1991)
fl3[2]	-	PLATZ TF	-	Li et al. (2017)
o11[1]	-	bHLH TF	-	Feng et al. (2018)
v) Mutations involved in unknown processes				
o16	-	-	-	Yang et al. (2005)
o15	27 kDa zein	-	High concentration of 19 and 22 kDa zeins	Dannenhoffer et al. (1995)

Note: [1]Recessive; [2]Semidominant; [3]Dominant.
RER = Rough Endoplasmic Reticulum; PB = Protein Bodies; TF = Transcription Factor.

one was a cross between wild type line QCL3010 and QCL3024, and another was a cross between Qi205 (*o2*) and QCL3021 (Yang et al., 2005).

Pyramiding of both mutants (*o16* and *o2*) led to the development of genotypes with higher lysine and tryptophan content (Yang et al., 2005; Sarika et al., 2016, 2018). Other high-lysine mutants like *floury2* have been identified where zein synthesis is less severely affected than *o2*, and the synthesis of proteins with a higher content of basic amino acids (soluble in acids) was recognized as the factor contributing to high-lysine content.

In addition to recessive mutations, a dominant mutation *defective endosperm-B30* (*de*-B30*) with high-lysine content was identified (Salamini et al., 1979). *De*-B30* carries 50% more lysine than a conventional maize genotype but is also associated with reduced seed weight, hence, minimizing its utilization for quality improvement in breeding programs. A few other high-lysine maize mutants, such as *mucronate* (*mc*) and *o6* with reduced content of all zeins, *o7* with reduced content of 19 kDa zein, and *o15* with reduced content of 27 kDa zein were identified, but these mutants are less well characterized and hence were barely used in breeding activities to improve maize grain protein quality.

The low lysine content is a common problem of cereal proteins; the other grain cereals such as wheat and barley offer lysine content way below the FAO recommended dose for this essential amino acid. Efforts to increase the seed lysine content have remained a major area of research in feed barley since 1960 and have led to the identification of several high-lysine barley mutants (Munck, 1992; Cook et al., 2018). One of the mutants *lys3a* that confers the highest level of the lysine content to barley grains was recently cloned and identified to carry a missense mutation in a domain of one finger (DOF) zinc-finger transcription factor known as Barley Prolamin-box Binding Factor (BPBF) (Moehs et al., 2019; Orman-Ligeza et al., 2020). Indeed, an orthologous gene, *pbf1*, earlier identified in maize, encodes an endosperm-specific transcription factor, which regulates several of the zein genes (27 kDa γ-zein and 22 kDa α-zein) by interacting with O2 (Vicente-Carbajosa et al., 1997). Also, mutations in a wheat homolog of this gene have shown to reduce the grain prolamin content and, conversely, increased the content of more balanced proteins leading to increased lysine content (Moehs et al., 2019). It indicates a similar regulatory mechanism contributing to high-lysine content in cereals and opens up the possibilities to increase the lysine content in other cereals, via inducing mutations in the *O2* and *Pbf1* gene orthologs.

2.1.3 Genetics of quality protein maize (QPM)

As mentioned earlier, QPM has a mutation at the *o2* locus (Mertz et al., 1964), which alters its endosperm protein composition and contributes to the increased lysine and tryptophan content. In comparison with conventional maize (40%–57%), the biological value of QPM is increased (~80%) due to an increase in concentration (~60%–100%) of vital amino acids, lysine, and tryptophan (Bressani, 1992). QPM development was based on the introgression of modifiers in the *o2* background, which changed the endosperm of *o2* mutant from soft and chalky to hard and vitreous (Paez et al., 1969).

Association mapping of the *o2* modifiers in QPM lines revealed that it encodes a 27 kDa zein protein that maps to the long arm of maize chromosome 7 (Lopes and Larkins, 1995). It was also identified that this protein is involved in the vitreous endosperm formation and accumulates three folds more in QPM lines than in the soft *o2* mutant (Wallace et al., 1990). The *o2* modifiers showed complex inheritance (Prasanna et al., 2001), where different loci were identified to control the QPM's vitreous kernel phenotype.

Proteomic analysis of non-zein proteins was performed to identify other factors associated with the endosperm modification in QPM (Gibbon et al., 2003). The analysis unveiled that in QPM the quantity of a starch branching enzyme, involved in amylopectin production, was altered, and it commensurate with the higher starch expansion observed in QPM than a conventional maize genotype. It also revealed that the opaque endosperm suppression in QPM is linked to the type of starch grains accumulated in the endosperm. Later, it was discovered that protein quantity could be improved in the *o2* mutant background while maintaining the low-zein content and high lysine and tryptophan content.

2.1.4 Nutritional aspects of QPM genotypes

Maize is taken as a protein source, but at such places, people are at high risk for protein deficiency, known as kwashiorkor, because of the lower lysine and tryptophan content of seed proteins. In contrast, QPM does not suffer from this problem and serves as a good protein source due to its balanced amino acid profile. The quality of proteins is generally determined by their amino acid composition and bioavailability or digestibility.

Since QPM has high lysine and tryptophan levels, it increases the protein's biological value close to the nutritional value of skim milk (90%), the gold standard for determining the adequacy of nutritional value. Thus, QPM lines offer great hope for efforts to enhance human nutrition in low-income countries (Omosebi et al., 2018). Kernel hardness (measured as the floating index) for conventional maize is 19.7%, whereas it is 57% for QPM. The total kernel protein content for conventional maize is about 9.25% (endosperm 7.9% and germ 1.28%). On the other hand, it is 13.15% (8.6% endosperm and 13.88% germ) in QPM (Gayral et al., 2016).

As mentioned earlier, the protein quality improvement observed in QPM is associated with the *o2* mutant gene (Hossain et al., 2018). In conventional maize genotypes, crude protein content is lower than QPM, and an insignificant proportion of it is contributed by the germ, whereas in QPM, the germ's contribution to the protein content is substantial. These modifications in the protein profile in QPM resulted from biochemical and structural changes to the kernel induced by the modifiers, which have also changed the functionality of extracted proteins (Omosebi et al., 2018; Ortiz-Martinez et al., 2017).

2.1.5 Challenges associated with the development of QPM lines

Breeding QPM cultivars is a significant challenge for plant breeders, specifically with improved protein content, high yield, and other agronomical characteristics of interest. High-lysine opaque mutations due to several pleiotropic effects make QPM development difficult. For instance, the *o2* and *fl2* mutants' soft and chalky kernels are susceptible to damage during harvesting, making them unsuitable for commercial cultivation. Also, flour made from these mutants exhibits poor functional properties, making it unsuitable for food products (Table 2).

Consequently, "high-lysine" corn did not gain the deserved popularity after its first appearance (Gibbon and Larkins, 2005). Moreover, the high-lysine content is primarily regulated by recessive genes in maize; therefore, strict measures for pollination are required, and the selections need to be made using the molecular/biochemical markers, not phenotypic traits. Several genetic *o2* suppressors or modifiers that could restore the normal kernel phenotype in *o2* background were identified (Nelson, 2001). However, breeding them together

Table 2 Comparison of conventional maize with quality protein maize for different nutritional contents (Modified from Damtie, 2019)

Parameters	Conventional maize	Quality protein maize
Moisture content (%)	7.15	6.90
Fat content (%)	4.09	4.80
Crude protein (%)	8.96	11.76
Total kernel protein (%)	9.25	13.15
Crude fiber (%)	1.48	1.09
Vitamin A (μg/100g)	0.06	0.13
Thiamin (mg/100g)	0.35	0.39
Ascorbic acid (mg/100g)	4.27	4.07
Iron (mg/100g)	2.70	2.83
Calcium (mg/100g)	11.47	12.37

with the *o2/fl2* mutant remains challenging due to its recessive nature and the sheer number of modifying genes.

Therefore, breeding QPM lines via conventional approaches require about 10–15 years. However, the conversion of normal elite hybrids to a QPM hybrid may take less time due to the knowledge of combining ability, adaptability, and heterosis of released hybrids (Prasanna et al., 2010). In such cases, the introgression of *o2* takes almost 6–7 years via conventional breeding though the duration could be significantly reduced via advanced marker-assisted backcross breeding (Babu et al., 2005; Muthusamy et al., 2014). Pyramiding of more than one opaque mutant in a single genetic background and with other quality traits could be an effective strategy for further quality enhancement. Molecular marker-assisted selection could be used for this purpose. The high-lysine mutants, such as *o2* and *o16*, have been mapped respectively to chromosomes 7S and 8L (Babu et al., 2005; Yang et al., 2005), and the associated markers were used by Zhang et al. (2013) to introgress *o2* and *o16* genes in waxy maize line QCL 5019. Recombinant inbred lines (RILs) and F_2 population produced by crossing QPM and *o2* mutant showed that the *o2* modifier gene is positively linked with 27kDa γ-zein (Lopes et al., 1995) and silencing of the 27-kDa γ-zein genes eliminates the formation of vitreous endosperm (Yuan et al., 2014). This study confirmed that the 27kDa γ-zein genes could be used as molecular markers in breeding for QPM.

Globally, several QPM varieties have been released in Latin American, African, and Asian countries (Krivanek et al., 2007). However, QPM varieties are yet to gain popularity in developed countries. It may be due to the recessive nature of *o2* combined with the need for the numerous modifier genes required to achieve a hard endosperm phenotype, which increases the difficulty in introgressing them all together into a breeding line.

2.1.6 Alternative ways (transgenic) to develop QPM genotypes

The challenges associated with breeding QPM genotypes using conventional breeding approaches and the increasing availability of data on the maize zein gene regulation, and the modifiers, inspired adaptation of the biotechnological approach to developing high-lysine maize genotypes (Table 3). Three different kinds of transgenics were developed - first, with decreased zein expression by RNA interference. This task was achieved via transgenic silencing of the genes encoding 22 kDa zeins and 19 kDa zeins exclusively or in combination. This approach increased the grain lysine content by 43.8% relative to the wild type, but it was lower than the *o2* mutant. Second, with increased free lysine content. This quest was accomplished by expressing a bacterial lysine-feedback insensitive dihydrodipicolinate synthase (DHDPS) in maize grains. DHDPS is involved in the first committed step in the lysine biosynthesis pathway. This approach increased the free lysine content in the embryo by 66.6% relative to wild type

Table 3 High-lysine maize developed using transgenic approach (Modified from Huang et al., 2009)

Approach	Percentage lysine increase	Reference
i) Decreasing zein synthesis		
22 kDa α-zein reduction via RNAi	11.1–18.5	Segal et al. (2003)
19 kDa α-zein reduction via RNAi	16.3–43.8	Huang et al. (2004)
22 and 19 kDa α-zein reduction via RNAi	85.9–105.2	Huang et al. (2006)
ii) Increasing free lysine content		
Expressing bacterial lysine feedback insensitive DHDPS	66.6	Huang et al. (2005)
LKR/SDH silencing via RNAi	20.7–26.0	Houmard et al. (2007)
DHDPS expression and LKR/SDH silencing	153.3–188.8	Frizzi et al. (2008)
iii) Increasing high-lysine (balanced) protein		
Ectopic expression of sb401	16.1–54.8	Yu et al. (2004)
Ectopic expression of α-lactalbumin	29.4–47.1	Bicar et al. (2008)
Ectopic expression of engineered high-lysine zeins	-	Huang et al. (2009)
Expression of plant lysyl tRNA synthetase	8.3–26.0	Wu et al. (2007)

but did not affect the endosperm's lysine content due to lysine catabolism. This problem was dealt with by silencing the gene for lysine catabolism enzyme, lysine-ketoglutarate reductase-saccharophine dehydrogenase (LKR/SDH). The LKR/SDH RNAi lines showed a 26.0% increase in the total grain lysine content. A novel bifunctional transgenic line, expressing bacterial DHDPS and suppressing LKR/SDH, was developed. This genotype showed the highest (188.8%) increase in maize kernels' total free lysine content ever. Third, with an increased amount of balanced or high-lysine proteins. This approach was accomplished in three different ways – first, by expressing naturally high-lysine proteins, as was done by expressing the potato pollen-specific protein, sb401 or the milk protein, and α-lactalbumin in maize grains. The former approach has resulted in a 16.1–54.8% increase, and the latter approach resulted in a 29–47% increase in the grain lysine content. Second, by engineering high-lysine proteins. Attempts were made to engineer 19 kDa α-zeins, 22 kDa α-zeins, and 27 kDa γ- zeins. Third, by overexpressing the lysyl tRNA synthetase (KRS). The ectopic KRS expression leads to tRNAs' mis-acylation with lysine, causing the translational recording of lysine into zeins at different codons. This novel approach has led to a 26% increase in the lysine content. However, promising the transgenically produced high-lysine corns are less likely to be accepted by the general public. Therefore, despite significant early progress, the research in this area has stalled, and not much further progress in terms of field trials or release of such genotypes has been made.

2.2 Other nutritional aspects of maize

Other than the imbalanced protein profile, lacking the essential amino acids, lysine, and tryptophan, maize niacin (vitamin B3) is bound in an indigestible complex. The ancient civilizations, such as Maya and Aztec, were using the cultural practices of boiling maize in alkaline limewater, known as 'nixtamalization,' to break down the complex and release niacin (Flint-Garcia, 2017). However, this practice is not widely adopted, leading to the deficiency known as Pellagra. Another vital nutrient element of the human diet is pro-vitamin A, for which significant variation exists in maize germplasm. Association mapping and expression analysis have established that favorable alleles at the *lcyE* locus change the flux of metabolites toward the β-carotene branches of the carotenoid pathway leading to a three-fold increase in the pro-vitamin A content of the maize endosperm (Harjes et al., 2008). It suggests that the content of bioavailable vitamin B3 and pro-vitamin A could be increased by adopting the correct food processing practices or the germplasm for breeding.

Other than zeins, maize seeds accumulate a large quantity of the α-amylase inhibitors (ATI) and lipid transfer proteins (LTP), which cause allergy in sensitive individuals (Pastorello et al., 2000, 2009). Allergic reactions to maize are not widespread but have been reported in southern Europe, Mexico, and the United States. However, its exact frequency has not been established. Lipid transfer proteins from other cereals like wheat are also known to cause allergies, but maize LTPs are unique as these are known to induce allergic cross-reactivity with the LTPs found in other cereals (barley, rice) and fruits (cherry, peach). Therefore, the maize-allergic individuals are different from typical cereal-allergic patients and show allergic cross-reactivity to fruits like cherry and peach (Scibilia et al., 2008; Venter et al., 2008). The symptoms of maize allergy include respiratory symptoms, occupational asthma, dermatitis, and in rare cases, could also involve anaphylaxis (Pastorello et al., 2009). Maize thioredoxins serve as minor allergens and are known to exhibit cross-reactivity to cereal allergens related to bakers' asthma and grass pollen allergy (Weichel et al., 2006). Maize germplasm with great genetic diversity for ATIs, LTPs, and thioredoxins exist, which could be exploited to develop reduced-/non-allergenic maize lines for maize allergenic individuals.

Starch constitutes approximately 70% of the maize kernel, provides energy for the germinating seed, and supports embryo growth and development until the seedling starts to photosynthesize actively. Maize carbohydrates generally have a 4:1 ratio of amylopectin to amylose. More amylopectins are desirable as they reduce the food's glycemic index, but may not have a desired impact on the end-use quality. Mutants with high amylopectin (waxy) and high amylose (non-waxy) contents were identified in maize and are being bred in elite lines for specific health or quality-related targets in mind (cf. Flint-Garcia, 2017; Hannah and Boehlein, 2017). However, more research is needed to develop maize lines with desired health benefits without compromising end-use quality.

Another major molecular class is lipids. Maize seeds typically contain 4.3–4.5% oil (Shen and Roesler, 2017). The embryo by mass is 10% of the final maize kernel but contributes about 85% to the total kernel lipids (primarily triacylglycerols). The survey of kernel traits in the maize germplasm revealed a significant decrease (–26%) in kernel oil content between teosinte and maize inbred lines/landraces, but no change was observed in the endosperm-to-embryo ratios of teosinte and maize (Flint-Garcia, 2017). It suggested a possibility of increasing oil content by using teosinte alleles without altering the embryo size in maize. Indeed, a major QTL for kernel oil content was identified on chromosome 6 (Laurie et al., 2004). It encompassed the *DGAT1-2* gene (Zheng et al., 2008), where a 3-bp ancestral insertion (teosinte allele) was found as the causative factor conferring high oil content to grains. *DGAT1-2* was later found to be associated with both oil and starch contents (Cook et al., 2012); thus, it is a plausible target for carbon cycling within the kernels. It suggests that the increase in maize oil content is possible, but it could have an unpredictable effect on end-use quality, which needs to be studied via further research.

3 Oats (*Avena sativa* L.)

Oats, commonly known as white oat (White, 1995), are among six major cereal grains grown worldwide with its production largely focused in the cooler parts of the world, such as the European Union (EU), Russian Federation, Canada, and the United States. Oat production ranks six among the cereals and accounts for less than 2% of total grains produced (Strychar, 2011). Seventy-four percent of the world's oat production is used for feed and 25% as human food (Strychar, 2011).

Despite being the highest-protein cereal-grain with a high content of soluble fiber, mineral, and phenolics, initially, due to the high-fat content (21% saturated, 37% unsaturated, and 43% polyunsaturated) and a problem to quickly develop rancidity, Oats were primarily used as cattle and horse feed. Due to the several recent health claims associated with oats, it is now gaining popularity as a functional food (Ahmad et al., 2010).

Besides being used in food and feed, oats are also used for various clinical and industrial purposes. For instance, oat husk is used for industrial methane production for use as biofuel. Clinically, it is proven to reduce gastrointestinal problems (Anderson and Bridges, 1993; Wrick, 1993, 1994; Stark and Madar, 1994) and the blood cholesterol level and has been officially declared by the

European Commission as gluten-free, and hence suitable for patients with celiac disease. Oats contain a small fraction of prolamins, called 'avenins,' at a much lower quantity than gluten proteins in wheat, barley, and rye (Arendt and Zannini, 2013).

Moreover, oat avenins do not contain any known celiac disease epitopes. Therefore, from 2009 in the European Union, 2013 in the United States, and 2015 in Canada, oat products could be sold as gluten-free if the gluten contamination level stays below 20 ppm. However, this matter remains controversial and an active field of research. Similarly, the studies on the nutritional properties of oat are increasing in number, further catching the attention of the food industry and hence increasing oats use as an ingredient in food products, such as bread (Zhang et al., 1998), breakfast cereals (Ryan et al., 2011), biscuits (Ballabio et al., 2011), and oat milk (Onning et al., 1999).

3.1 Oat kernel structure and composition

Oat kernels consist of the groat and the hull, and the groat comprises germ, bran, and starchy endosperm (Tosh et al., 2010) (Fig. 2). The bran forms the outer layer, which is coarse and rich in vitamins, minerals, and cell wall polysaccharides. Below the bran is the aleurone layer and the endosperm. The oat proteins are mainly present in the endosperm and the aleurone cells. The oat is a good source of micro- and macronutrients consisting of starch, proteins (globulins and prolamins), vitamin E, and phenolic compounds (Table 4). The

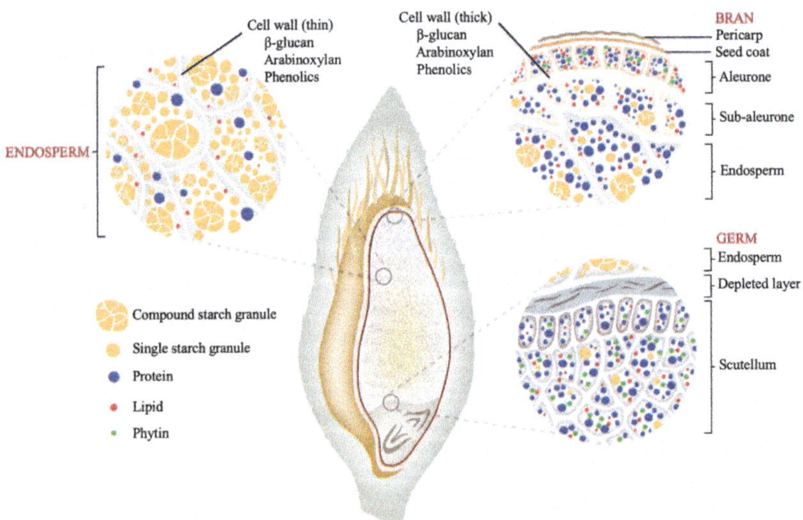

Figure 2 Structural representation of the oat grain presenting different oat tissues (Modified from Grundy et al., 2018).

Table 4 Nutritional composition of oat grains (Modified from Klose and Arendt, 2012)

Nutrient	Amount (per 100 g)
Energy	374 kcal
Moisture	12.2 g
Protein	12.6 g
Carbohydrate	61.5 g
Lipids	7.1 g
Fibre	5.6 g
Sodium	10.0 mg
Potassium	350.0 mg
Calcium	80.0 mg
Phosphorus	340.0 mg
Iron	5.8 mg
Magnesium	-
Zinc	4.5 mg
Thiamine	0.52 mg
Riboflavin	0.17 mg
Vitamin B6	1.0 mg
Vitamin B11	-
Vitamin C	0.0 mg
Vitamin D	0.0 mg

distribution of these nutrients varies based on the cell type and tissue (Klose and Arendt, 2012).

3.1.1 Oat proteins

The major storage proteins of oat are salt-soluble globulins compared with other cereals such as rye, barley, and wheat, where the alcohol-soluble prolamins constitute the major fraction of seed storage proteins (Fig. 3). The globulins account for up to 80% of the storage proteins, and the remaining is made up of the avenins, the prolamin fraction of oat protein (Lásztity, 1998), and other storage proteins.

According to Boyer et al. (1992), the mass ratio of globulins and avenins in oat kernels is 26:1. The globulins consist of 12S, 7S, and 3S fractions, of which the 12S fraction is mainly found in the endosperm and the 7S fraction in the embryo (Burgess and Miflin, 1985; Draper, 1973). The thermal denaturation temperature of oat proteins is 110°C (Harwalkar and Ma, 1987; Molina et al., 2004), which could be due to the strong interactions between the protein subunits (Gorinstein et al., 1996).

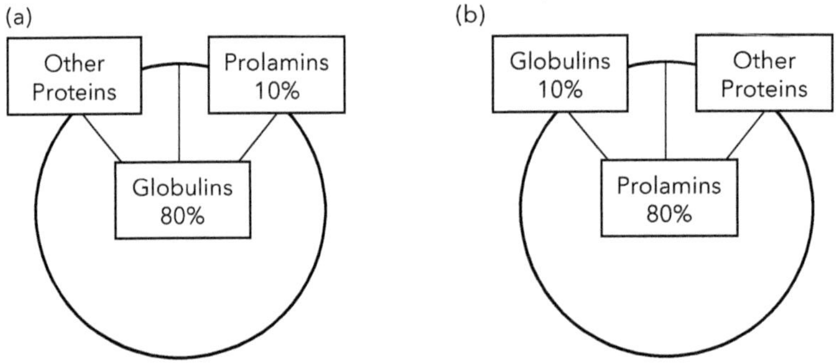

Figure 3 Comparison of oat (a) and wheat (b) seed storage protein fractions.

3.1.1.1 Globulins

The globulins in oat are hexamers of 320 kDa with each subunit being 54 kDa, which crosslinks to a 32 kDa polypeptide through disulfide bonding (Burgess et al., 1983; Casey, 1999; Shotwell et al., 1988). These polypeptides are the product of post-translational processing. In oats, there are 20 different large globulin subunits and 10 different small globulin subunits. The subunits can be separated using ion-exchange chromatography due to differences in their charges and can be divided into the acidic and the basic subunits, respectively, known as α and β subunits (Burgess et al., 1983).

The α subunit has a pH in the range of 5.9–7.2, whereas the β subunits have pH in the range of 8.7–9.2. The β subunits had 27% more basic amino acids, such as arginine, lysine, and histidine, compared to the α subunit. In contrast, the α subunit had a higher proportion of the glutamate residues. Further, the result of α and β globulin subunits' amino acid sequence analysis was consistent with subunit classification based on the differences in charges and the pH values. Later, the two-dimensional gel patterns were also used to reveal the heterogeneity of the α and β subunits.

Shotwell et al. (1990) determined the complete sequence of 518 amino acids for the globulin proteins - 293 amino acids in the α-chain, 201 amino acids in the β-subunit, and 24 amino acids in the signal peptide. Further analysis of the amino acid sequence revealed that they resemble other seed storage proteins having a high percentage of glutamine and asparagine and a low percentage of sulfur-containing amino acids, such as methionine and cysteine. For instance, the N-terminal sequence of the β subunit shared homology with the 11S globulins of *Vicia faba* and *Glycine max*.

According to the FT-IR studies by Liu et al. (2009), the secondary structure of the globulins has more percentage of β-sheets (74%) than α-helices (19%), and β- turns (7%). Earlier, Marcone (1999) determined that the increase or decrease in the pH alters the secondary and the tertiary structure of the globulins by altering the association/dissociation property of subunits (Marcone, 1999). The 12S globulin structure was predicted using quasielastic light scattering and found to be an oblate cylinder in shape and height around 8.5 nm. The two annular trimeric rings stacked one above the other, forms the cylindrical structure (Zhao et al., 2004). Similar observations were also made by Liu et al. (2009) using atomic force microscopy. The oat globulin proteins share similarities with other legume globulin proteins. For instance, the 12S oat globulin protein shares similarity with the 11S legume globulins in their secondary and tertiary structures and the charge difference.

In the oat endosperm, globulins mainly consist of 12S globulins compared to embryo where 7S and 3S globulin fractions are predominant. The 7S globulins have molecular weights around 55 kDa, which resembles the molecular weight of vicilins in legumes. 7S globulins also share the similarity in their amino acid sequence with vicilins. Likewise, the 3S globulins also share similarities with the counterparts in legumes.

About 11 genes encode globulins that are rich in glutamine and poor in sulfur containing amino acids. The globulin genes have three short introns, positioned in the same region as other globulin genes in dicots and other cereals and monocots. This conservation in the globulin gene structure shows that the globulin genes existed before the separation of angiosperms into monocots and dicots. However, the upstream regulatory sequences in the oat proteins are not analogous to dicot seeds as the oat globulin genes express in the endosperm, whereas the dicot genes express in the cotyledons. The globulin gene expression happens to be coordinated with that of the avenin genes even though they do not share any sequence similarity in the regulatory region. The cDNA clones of oat globulins show 31% similarity with glycinin in soybeans, 38% similarity with legumins in peas, and 70% similarity with globulins in rice.

3.1.1.2 Avenins (oat prolamins)

Avenins, belonging to protein class prolamins, are the low-molecular-weight fraction of oat seed storage proteins, which are soluble in 50–70% ethanol or 40% 2-propanol. These proteins have a high percentage of proline and glutamine residues than lysine, which is relatively more prevalent in other proteins (Klose and Arendt, 2012; Shewry et al., 1999). The percentage of charged amino acids in avenins is less than the oat globulins, which result in low charge at any pH (Shewry et al., 1999; Shotwell, 1999). Avenins are mainly found in the oat grain endosperm (Klose and Arendt, 2012; Lending et al., 1989; Shewry, 1995).

The avenins fall in the 17–34 kDa range, and exist as monomers or disulfide aggregates (Shewry, 1995). Avenins are highly polymorphic. The electrophoresis of avenins revealed inter- and intraspecific heterogeneity among the diploid, tetraploid, and hexaploid oat species. The starch gel electrophoresis of avenin

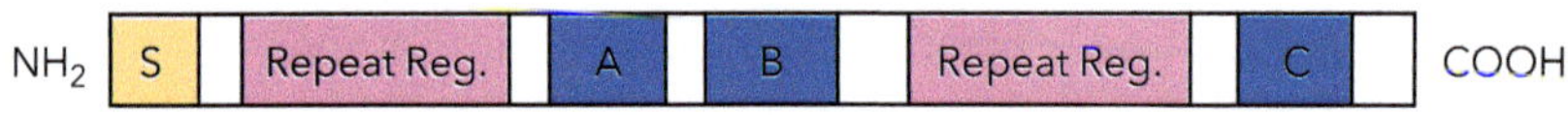

Figure 4 Typical structure of oat A, B, and C avenins. S = signal peptide; A, B, and C = conserved regions, Repeat Reg. = glutamine- and proline-rich repeats (Modified from Balakireva and Zamyatnin, 2016).

proteins indicated they comprise three subunits: α, β, and γ (Fig. 4). The avenins were also separated and characterized by reversed-phase high-performance liquid chromatography and ion-exchange chromatography (Pernollet et al., 1987).

The N terminal of the separated proteins showed homology and had a common repeat sequence of Pro-Phe-Val-Gln-Gln-Gln. Eleven fractions were extracted by Egorov et al. (1994), of which five had blocks of amino acid repeats at the N-terminal end. All avenins carried 8 cysteine residues forming four disulfide bonds. Chesnut et al. (1989) isolated three cDNA clones coding avenins. These genes, pAv122, pAv31, and pAv10, encode polypeptides of 203, 201, and 162 residues, respectively. The polypeptides share similarity in terms of their molecular weights and amino acid sequences. All open reading frames carried an N-terminal signal sequence. Sequence comparisons of the avenin polypeptides showed 72% homology between pAv122 and pAv31, and 54% homology with pAv10 (Chesnut et al., 1989). The sequence also shared similarity with wheat α/β- gliadins and γ-gliadins, rye γ-secalins, and barley B-hordeins. At the same time, the avenins showed differences with barley, rye, and wheat prolamins, specifically concerning molecular weight, number and location of amino acid repeats, and cysteine residues.

3.1.2 β-glucan

Oat seed β-glucans comply with the American Association of Cereal Chemists' definition of dietary fibers released in 2000. The β-glucans are the major component of the endosperm cell walls and the bran. The oat endosperm is made up of parenchymatous cells surrounded by a single aleurone layer. The β-glucan is a linear polysaccharide of 4′ O-linked-β-D-glucopyranosyl (70%) and 3′ O-linked-β-D-glucopyranosyl (30%) units. Daily intake of β-glucans helps reduce the risk of coronary heart disease. As the β-glucans cannot be digested and absorbed in the small intestine, they are low-glycemic index and do not increase consumer's blood cholesterol level (Ripsin et al., 1992; Amundsen et al., 2003; Chen et al., 2006).

According to Schneeman (2001), β-glucans serve as a substrate for gut microflora for fermentation and promote laxation. Indeed, based on this knowledge, a specific product called 'Oatrim' (with a β-glucan content up to 10%) was approved by the FDA (Webster and Wood, 2011). The oat β-glucans were also shown to exhibit anti-cancerous properties and reduce colon cancer incidence (Sadiq Butt et al., 2008; Bode and Dong, 2009; Hsueh et al., 2011). Some reports have also shown it to reduce hypertension in susceptible individuals (He et al., 2004).

The significant variation for the β-glucan content exists in the oat germplasm, which could be exploited to develop varieties to serve specific health-related and quality needs. The genetic of seed β-glucan content is also being studied in oats; the knowledge from these studies is also being exploited in developing high/low β-glucan oat varieties to serve different purposes.

3.1.3 Phenolics and antioxidants

Oat is the major source of compounds that shows antioxidant properties. The most abundant antioxidants in oats are vitamin E, phytic acid, avenanthramides, and the phenolic compound conjugates (Dimberg et al., 1993; Gray et al., 2002). These compounds are primarily present in the outer layers of the kernel. Therefore, the pearling fractions of the oat have high antioxidant activity compared to the flour. About 32 phytosterols, 19 phenolics, 22 alkaloids, and 6 tocols are present in the oat seeds. Some of the phenolics are vanillic acids, vanillin, p-hydroxybenzoic acid, ferulic acid, p-coumaric acid, and caffeic acid.

The phenolic compounds contribute to the nutritional property of the grain. According to the studies performed by Peterson (2001) and Shewry et al. (2008), the phenolics in the oat exhibit antioxidant properties in vitro and in vivo (Ryan et al., 2007). The avenanthramides are a distinct class of antioxidants that exhibit 10–30 times more antioxidant activity than other antioxidants, such as caffeic acid and vanillin (Dimberg et al., 1993). Avenanthramides also show anti-inflammatory properties and antiatherogenic properties (Liu et al., 2004) and help control hypertension (Nie et al., 2006). Indeed, a commercial product under the trade name 'Tranilast' was marketed as an oral antihistamine, antiallergy, and anti-irritant drug in Australia, Europe, Japan, and the United Kingdom (Webster and Wood, 2011; Smulders et al., 2018). The higher content of phytochemicals in oat relative to other cereals also makes it more resistant to harsh weather conditions than other cereals (Grundy et al., 2018), making it an attractive choice for cultivation under changing environmental conditions.

The genetics of phenolic compounds is not well studied in oats; however, significant variation exists for these compounds in the germplasm. Therefore, more studies are needed to decipher the genetics of these biochemical/metabolic traits to use this knowledge to breed genotypes enriched in these compounds for use in industrial purposes.

3.2 Nutritional value of oat proteins

The nutritional value of any protein depends on its essential amino acid content (Table 5). The amino acid composition analysis of oat proteins by Zarkadas et al.

Table 5 Essential amino acid content in oat proteins (Modified from Zarkadas et al., 1995)

Essential amino acids	Amount (per 100 g protein)
Histidine	2.2 g
Isoleucine	3.9 g
Leucine	7.4 g
Lysine	4.0 g
Cysteine	1.6 g
Methionine	2.5 g
Phenylalanine	5.3 g
Tyrosine	3.1 g
Threonine	3.3 g
Tryptophan	-
Valine	5.3 g

(1995) revealed that the oat proteins have all the essential amino acids, except for threonine, and is deficient in lysine. The amount of all the essential amino acids in oat proteins sums up to 44%, which is more than the required nutritional value FAO recommended for preschool children. According to Bressani et al. (1963), the rolled oat proteins serve as a balanced source of required amino acids if supplemented with threonine. Also, the low prolamin content of oat proteins compared to its cereal counterparts, such as wheat, rye, and barley makes it an attractive choice for people with celiac disease (Comino et al., 2011).

3.3 Oat avenins and celiac disease

Despite oats exhibiting important nutritional and pharmacological properties, their safety for celiac patients always remains controversial. An earlier study performed in Norway, which involved 19 adult celiac patients who were challenged with 50 g daily dose of oats for 12 weeks (Lundin et al., 2003), one patient exhibited subtotal villous atrophy. This patient also developed mucosal T-cells that showed avenin-reactivity in vitro. Using these T-cells, two avenin epitopes (PYPEQEEPF and PYPEQEQPF) were also identified. This study provided the first evidence that oat intolerance may exist (Arentz-Hansen et al., 2004).

The later studies found a direct correlation between the immunogenicity of the different oat varieties and the presence of specific peptides with a higher or lower potential immunotoxicity. Real et al. (2012) suggested a wide variation in the potential immunotoxicity of oat cultivars due to differences in the occurrence of high and low immunogenicity epitopes (Real et al., 2012). RP-HPLC was also used to select oats varieties with reduced immunogenicity to develop new non-toxic oat cultivars (Giménez et al., 2017).

More recently, to study the impact of oat varieties on the circulating dendritic cells (DC), several proteins in the oat prolamin fraction were characterized using immunological and proteomic tools. These protein analogs to wheat α- and γ-gliadins showed reactivity to anti-33-mer antibody and differential stimulatory capacity on circulating DC from celiac patients, further confirming their immunogenic properties (Comino et al., 2016).

Collectively, the oat varieties suffer two problems: intrinsic immunoreactivity and wheat, barley, or rye contamination, which is quite common to occur. As mentioned, there exists great genetic diversity for the reduced immunogenicity in oat germplasm, which could be exploited to develop non-immunogenic oat varieties. And the second problem of contamination is being handled in different countries (e.g. Finland, The Netherlands, Ireland, Canada, and the United States) by developing gluten-free oat production chains (Smulders et al., 2018).

4 Conclusion

'Hidden hunger' is an issue, which affects both industrialized and low-income countries. Biofortification of staple grains is the most economical and effective way to impact the lives of many individuals. The development of QPM is one such effort, which could feed the nutritional demands of the populations residing in low-income countries without exerting the financial burden. To expedite the efforts to develop nutritionally enhanced and safe (non-immunogenic) cultivars, it is crucial to decipher the genetics of the nutrients and anti-nutrients produced by staple crops and use this knowledge to breed genotypes with desired attributes. Outstanding genetic resources, such as the oat-maize addition line, a vast collection of mutations, radiation-hybrids, the Nested Association Mapping (NAM) lines with introgression spanning almost all parts of the corn genome, and the constitutively Cas9 expression lines in maize and a vast collection of inbreeds and landraces in oats, are available as germplasm to the research community. Another resource is the availability of the reference maize and oat genome sequences and haplotype maps, which will facilitate analysis of the complex transcriptional regulation of the equally complex gene families encoding allergenic and antigenic proteins or genes encoding enzymes involved in starch and lipid biosynthesis or metabolism, and production of various phenolics/antioxidants. This resource and knowledge, combined with technologies, such as genomic selection, genome-editing, genotype and tissue-culture independent genetic transformation procedures, advances in biochemical and immunological detection procedures, and more sensitive and more accurate non-invasive phenotyping methods, will allow the development of maize and oat genotypes with the desired balance of health-benefiting and quality attributes with a consistent yield to feed the demand for calories and micro-nutrients. All of these approaches are currently being developed and deployed. The challenge will be to bring them together.

5 Where to look for future information

- Bennetzen, J. L. and Hake, S. C. (2009), *Handbook of Maize: Genetics and Genomics*, New York: Springer-Verlag.
- Igrejas, G., Ikeda, T. M. and Guzman, C. (2020), *Wheat Quality for Improving Processing and Human Health*, Cham Switzerland: Springer Nature.
- Khan, K. and Shewry, P. R. (2009), *Wheat: Chemistry and Technology* (4th Edition), St. Paul: AACC International.
- Larkins, B. A. (2017), *Maize Kernel Development*, Wallingford: CAB International.
- Ratnavathi, C. V., Patil, J. V. and Chavan, U. D. (2016), *Sorghum Biochemistry: An Industrial Perspective*, Tokyo: Academic Press.
- Shewry, P. R. and Ullrich, S. E. (2014), *Barley: Chemistry and Technology* (2nd Edition), St. Paul: AACC International.
- Sissons, M., Vbecassis, J., Marchylo, B. and Carcea, M. (2012), *Durum Wheat: Chemistry and Technology* (2nd edition), St. Paul: AACC International.
- Watson, R., Preedy, V. and Zibadi, S. (2014), *Wheat and Rice in Disease Prevention and Health*, Tokyo: Academic Press.
- Webster, F. H. and Wood, P. J. (2011), *Oats: Chemistry and Technology*, St. Paul: AACC International.

6 Acknowledgements

This work was supported by the NIFA Hatch/Multi-state grant (S009).

7 References

Agrawal, P. K., Jaiswal, S. K., Prasanna, B. M., Hossain, F., Saha, S., Guleria, S. K. and Gupta, H. S. (2012). Genetic variability and stability for kernel iron and zinc concentration in maize (*Zea mays* L.) genotypes, *Indian J. Genet.* 72, 421–428.

Ahmad, A., Anjum, F. M., Zahoor, T., Nawaz, H. and Ahmed, Z. (2010). Extraction and characterization of β-d-glucan from oat for industrial utilization, *Int. J. Biol. Macromol.* 46(3), 304–309.

Amundsen, Å. L., Haugum, B. and Andersson, H. (2003). Changes in serum cholesterol and sterol metabolites after intake of products enriched with an oat bran concentrate within a controlled diet, *Scand. J. Nutr.* 47(2), 68–74.

Anderson, J. W. and Bridges, S. R. (1993). Hypocholesterolemic effects of oat bran in humans. In: Wood, P. J. (Ed.), *Oat Bran*, St. Paul: American Association of Cereal Chemists International, pp. 139–157.

Arendt, E. and Zannini, E. (2013). *Cereal Grains for the Food and Beverage Industries*, Cambridge: Woodhead Publishing.

Arentz-Hansen, H., Fleckenstein, B., Molberg, Ø., Scott, H., Koning, F., Jung, G., Roepstorff, P., Lundin, K. E. A. and Sollid, L. M. (2004). The molecular basis for oat intolerance in patients with celiac disease, *PLoS Med.* 1(1), e1.

Aukerman, M. J., Schmidt, R. J., Burr, B. and Burr, F. A. (1991). An arginine to lysine substitution in the bZIP domain of an *opaque-2* mutant in maize abolishes specific DNA binding, *Genes Devel.* 5(2), 310–320.

Awika, J. M. (2011). Major cereal grains production and use around the world. In: Awika, J. M., Piironen, V. and Bean, S. (Eds), *Advances in Cereal Science: Implications to Food Processing and Health Promotion*, Washington, DC: American Chemical Society, pp. 1–13.

Babu, R., Nair, S. K., Kumar, A., Venkatesh, S., Sekhar, J. C., Singh, N. N., Srinivasan, G. and Gupta, H. S. (2005). Two generation marker aided backcrossing for rapid conversion of normal maize lines to quality protein maize (QPM), *Theor. Appl. Genet.* 111(5), 888–897.

Badu-Apraku, B., Annor, B., Oyekunle, M., Akinwale, R. O., Fakorede, M. A. B., Talabi, A. O., Akaogu, I. C., Melaku, G. and Fasanmade, Y. (2015). Grouping of early maturing quality protein maize inbreds based on SNP markers and combining ability under multiple environments, *Field Crops Res.* 183, 169–183.

Balakireva, A. V. and Zamyatnin Jr., A. A. (2016). Properties of gluten intolerance: gluten structure, evolution, pathogenicity and detoxification capabilities, *Nutrients* 8(10), 644.

Ballabio, C., Uberti, F., Manferdelli, S., Vacca, E., Boggini, G., Redaelli, R., Catassi, C., Lionetti, E., Peñas, E. and Restani, P. (2011). Molecular characterisation of 36 oat varieties and in vitro assessment of their suitability for celiac's diet, *J. Cereal Sci.* 54(1), 110–115.

Bicar, E. H., Woodman-Clikeman, W., Sangtong, V., Peterson, J. M., Yang, S. S., Lee, M. and Scott, M. P. (2008). Transgenic maize endosperm containing a milk protein has improved amino acid balance, *Transgen. Res.* 17(1), 59–71.

Bode, A. M. and Dong, Z. (2009). Cancer prevention research–then and now, *Nat. Rev Cancer* 9(7), 508–516.

Boston, R. S. and Larkins, B. A. (2008). The genetics and biochemistry of maize zein proteins. In: Bennetzen, J. L. and Hake, S. C. (Eds), *The Maize Handbook (Vol. II): History and Practice of Genetics, Genomics and Improvement*, New York: Springer, pp. 715–730.

Boyer, S. K., Shotwell, M. A. and Larkins, B. A. (1992). Evidence for the translational control of storage protein gene expression in oat seeds, *J. Biol. Chem.* 267(24), 17449–17457.

Bressani, R. (1992). Nutritional value of high-lysine maize in humans. In: Mertz, E. T. (Ed.), *Quality Protein Maize*, St. Paul: American Association of Cereal Chemists, pp. 205–224.

Bressani, R., Wilson, D. L., Chung, M., Behar, M. and Scrimshaw, N. S. (1963). Supplementation of cereal proteins with amino acids: VI. Effect of amino acid supplementation of rolled oats as measured by nitrogen retention in young children, *J. Nutr.* 81, 399–404.

Burgess, S. R. and Miflin, B. J. (1985). The localization of oat (*Avena sativa* L.) seed globulins in protein bodies, *J. Exp. Bot.* 36(6), 945–954.

Burgess, S. R., Shewry, P. R., Matlashewski, G. J., Aaltosaar, I. and Miflin, B. J. (1983). Characteristics of oat (*Avena sativa* L.) seed globulins, *J. Exp. Bot.* 34(10), 1320–1332.

Casey, R. (1999). Distribution and some properties of seed globulins. In: Shewry, P. R. and Casey, R. (Eds), *Seed Proteins*, Dordrecht: Springer Netherlands, pp. 159–169.

CDC (2020). Micronutrient facts. Center for Disease Control and Prevention, https://www.cdc.gov/nutrition/micronutrient-malnutrition/micronutrients/index.html. Accessed November 7, 2020.

Chakraborti, M., Prasanna, B. M., Hossain, F., Mazumdar, S., Singh, A. M., Guleria, S. and Gupta, H. S. (2011b). Identification of kernel iron and zinc-rich maize inbreds and analysis of genetic diversity using microsatellite markers, *J. Plant Biochem. Biotechnol.* 20(2), 224–233.

Chakraborti, M., Prasanna, B. M., Hossain, F. and Singh, A. (2011a). Evaluation of single cross quality protein maize (QPM) hybrids for kernel iron and zinc concentrations, *Indian J. Genet.* 71, 312–319.

Chen, J., He, J., Wildman, R. P., Reynolds, K., Streiffer, R. H. and Whelton, P. K. (2006). A randomized controlled trial of dietary fiber intake on serum lipids, *Eur. J. Clin. Nutr.* 60(1), 62–68.

Chesnut, R. S., Shotwell, M. A., Boyer, S. K. and Larkins, B. A. (1989). Analysis of avenin proteins and the expression of their mRNAs in developing oat seeds, *Plant Cell* 1(9), 913–924.

Comino, I., Bernardo, D., Bancel, E., de Lourdes Moreno, M., Sánchez, B., Barro, F., Šuligoj, T., Ciclitira, P. J., Cebolla, Á, Knight, S. C., Branlard, G. and Sousa, C. (2016). Identification and molecular characterization of oat peptides implicated on coeliac immune response, *Food Nutr. Res.* 60, 30324.

Comino, I., Real, A., de Lorenzo, L., Cornell, H., López-Casado, M. Á., Barro, F., Lorite, P., Torres, M. I., Cebolla, A. and Sousa, C. (2011). Diversity in oat potential immunogenicity: basis for the selection of oat varieties with no toxicity in coeliac disease, *Gut* 60(7), 915–922.

Cook, F., Hughes, N., Nibau, C., Orman-Ligeza, B., Schatlowski, N., Uauy, C. and Trafford, K. (2018). Barley *lys3* mutants are unique amongst shrunken-endosperm mutants in having abnormally large embryos, *J. Cereal Sci.* 82, 16–24.

Cook, J. P., McMullen, M. D., Holland, J. B., Tain, F., Bradbury, P., Ross-Ibarra, J., Buckler, E. S. and Flint-Garcia, S. A. (2012). Genetic architecture of maize kernel composition in the nested association mapping and inbred association panels, *Plant Physiol.* 158(2), 824–834.

Damtie, Y. (2019). A review of improving the protein quality with the composition of quality traits of maize (*Zea mays* L.) grain and possible breeding techniques, *Asian J. Plant Sci. Res.* 9, 1–12.

Dannenhoffer, J. M., Bostwick, D. E., Or, E. and Larkins, B. A. (1995). Opaque-15, a maize mutant with properties of a defective *opaque-2* modifier, *Proc. Natl. Acad. Sci. U.S.A.* 92(6), 1931–1935.

Das, O. P., Ward, K., Ray, S. and Messing, J. (1991). Sequence variation between alleles reveals two types of copy correction at the 27-kDa zein locus of maize, *Genomics* 11(4), 849–856.

Dimberg, L. H., Theander, O. and Lingnert, H. (1993). Avenanthramides, A group of phenolic antioxidants in oats, *Cereal Chem.* 70, 637–641.

Draper, S. R. (1973). Amino acid profiles of chemical and anatomical fractions of oat grains, *J. Sci. Food Agric.* 24(10), 1241-1250.

Egorov, T. A., Musolyamov, A. K., Andersen, J. S. and Roepstorff, P. (1994). The complete amino acid sequence and disulphide bond arrangement of oat alcohol-soluble avenin-3, *Eur. J. Biochem.* 224(2), 631-638.

Feng, F., Qi, W., Lv, Y., Yan, S., Xu, L., Yang, W., Yuan, Y., Chen, Y., Zhao, H. and Song, R. (2018). OPAQUE11 is a central hub of the regulatory network for maize endosperm development and nutrient metabolism, *Plant Cell* 30(2), 375-396.

Flint-Garcia, S. A. (2017). Kernel evolution: from Teosinte to Maize. In: Larkins, B. A. (Ed.), *Maize Kernel Development*, Wallingford: CAB International, pp. 1-15.

Frizzi, A., Huang, S., Gilbertson, L. A., Armstrong, T. A., Luethy, M. H. and Malvar, T. M. (2008). Modifying lysine biosynthesis and catabolism in corn with a single bifunctional expression/silencing transgene cassette, *Plant Biotechnol. J.* 6(1), 13-21.

Gayral, M., Gaillard, C., Bakan, B., Dalgalarrondo, M., Elmorjani, K., Delluc, C., Brunet, S., Linossier, L., Morel, M. and Marion, D. (2016). Transition from vitreous to floury endosperm in maize (*Zea mays* L.) kernels is related to protein and starch gradients, *J. Cereal Sci.* 68, 148-154.

Gibbon, B. C. and Larkins, B. A. (2005). Molecular genetic approaches to developing quality protein maize, *Trends Genet.* 21(4), 227-233.

Gibbon, B. C., Wang, X. and Larkins, B. A. (2003). Altered starch structure is associated with endosperm modification in quality protein maize, *Proc. Natl. Acad. Sci. U.S.A.* 100(26), 15329-15334.

Giménez, M. J., Real, A., García-Molina, M. D., Sousa, C. and Barro, F. (2017). Characterization of celiac disease related oat proteins: bases for the development of high-quality oat varieties suitable for celiac patients, *Sci Rep* 7, 42588.

Global Hunger Index (2014). Addressing the challenge of hidden hunger. 2014 Global Hunger Index: the challenge of hidden hunger, https://www.ifpri.org/sites/default/files/ghi/2014/index.html. Accessed November 7, 2020.

Gorinstein, S., Zemser, M., Friedman, M., Rodrigues, W. A., Martins, P. S., Vello, N. A., Tosello, G. A. and Paredes-López, O. (1996). Physicochemical characterization of the structural stability of some plant globulins, *Food Chem.* 56(2), 131-138.

Gray, D. A., Clarke, M. J., Baux, C., Bunting, J. P. and Salter, A. M. (2002). Antioxidant activity of oat extracts added to human LDL particles and in free radical trapping assays, *J. Cereal Sci.* 36(2), 209-218.

Grundy, M. M., Fardet, A., Tosh, S. M., Rich, G. T. and Wilde, P. J. (2018). Processing of oat: the impact on oat's cholesterol lowering effect, *Food Funct.* 9(3), 1328-1343.

Guleria, S. K., Chahota, R. K., Kumar, P., Kumar, A., Prasanna, B. M., Hossain, F., Agrawal, P. K. and Gupta, H. S. (2013). Analysis of genetic variability and genotype × year interactions on kernel zinc concentration in selected Indian and exotic maize (*Zea mays*) genotypes, *Indian J. Agr. Sci.* 83, 836-841.

Guo, Y., Yang, X., Chander, S., Yan, J., Zhang, J., Song, T. and Li, J. (2013). Identification of unconditional and conditional QTL for oil, protein and starch content in maize, *Crop J.* 1(1), 34-42.

Hannah, L. C. and Boehlein, S. (2017). Starch biosynthesis in maize endosperm. In: Larkins, B. A. (Ed.), *Maize Kernel Development*, Wallingford: CAB International, pp. 149-159.

Harjes, C. E., Rocheford, T. R., Bai, L., Brutnell, T. P., Kandianis, C. B., Sowinski, S. G., Stapleton, A. E., Vallabhaneni, R., Williams, M., Wurtzel, E. T., Yan, J. and Buckler, E.

S. (2008). Natural genetic variation in lycopene epsilon cyclase tapped for maize biofortification, *Science* 319(5861), 330–333.

Harwalkar, V. R. and Ma, C. -Y. (1987). Study of thermal properties of oat globulin by differential scanning calorimetry, *J. Food Sci.* 52(2), 394–398.

He, J., Streiffer, R. H., Muntner, P., Krousel-Wood, M. A. and Whelton, P. K. (2004). Effect of dietary fiber intake on blood pressure: a randomized, double-blind, placebo-controlled trial, *J. Hypertens.* 22(1), 73–80.

Holding, D. R., Meeley, R. B., Hazebroek, J., Selinger, D., Gruis, F., Jung, R. and Larkins, B. A. (2010). Identification and characterization of the maize arogenate dehydrogenase gene family, *J. Exp. Bot.* 61(13), 3663–3673.

Holding, D. R., Otegui, M. S., Li, B.-L., Meeley, R. B., Dam, T., Hunter, B. G., Jung, R. and Larkins, B. A. (2007). The maize *floury1* gene encodes a novel ER protein involved in zein protein body formation, *Plant Cell* 19(8), 2569–2582.

Hossain, F., Muthusamy, V., Pandey, N., Vishwakarma, A. K., Baveja, A., Zunjare, R. U., Thirunavukkarasu, N., Saha, S., Manjaiah, K. M. M., Prasanna, B. M. and Gupta, H. S. (2018). Marker-assisted introgression of *opaque2* allele for rapid conversion of elite hybrids into quality protein maize, *J. Genet.* 97(1), 287–298.

Houmard, N. M., Mainville, J. L., Bonin, C. P., Huang, S., Luethy, M. H. and Malvar, T. M. (2007). High-lysine corn generated by endosperm-specific suppression of lysine catabolism using RNAi, *Plant Biotech. J.* 5(5), 605–614.

Hsueh, C. W., Chia, H. H., Jeng, D. H., Mon, Y. Y., Shing, J. W., Chau, J. W. (2011). Inhibitory effect of whole oat on aberrant crypt foci formation and colon tumor growth in ICR and BALB/c mice, *J. Cereal Sci.* 53(1), 73–77.

Huang, S. S., Adams, W. R., Zhou, Q., Malloy, K. P., Voyles, D. A., Anthony, J., Kriz, A. L. and Luethy, M. H. (2004). Improving nutritional quality of maize proteins by expressing sense and antisense zein genes, *J. Agric. Food Chem.* 52(7), 1958–1964.

Huang, S., Frizzi, A., Florida, C. A., Kruger, D. E. and Luethy, M. H. (2006). High lysine and high tryptophan transgenic maize resulting from the reduction of both 19- and 22-kD a-zeins, *Plant Mol. Biol.* 61(3), 525–535.

Huang, S., Frizzi, A. and Malvar, T. M. (2009). Engineering high lysine corn, In: H. Krishnan (Ed.), *Modification of Seed Composition to Promote Health and Nutrition*, Madison: American Society of Agronomy, Crop Science Society of America, Soil Science Society of America, pp. 233-248.

Huang, S. S., Kruger, D. E., Frizzi, A., D'Ordine, R. L., Florida, C. A., Adams, W. R., Brown, W. E. and Luethy, M. H. (2005). High-lysine corn produced by the combination of enhanced lysine biosynthesis and reduced zein accumulation, *Plant Biotech. J.* 3(6), 555–569.

Kim, C. S., Gibbon, B. C., Gilikin, J. W., Larkins, B. A., Boston, R. S. and Jung, R. (2006). The maize Mucronate mutation is a deletion in the 16-kD γ-zein gene that induces the unfolded protein response, *Plant J.* 48(3), 440–451.

Klose, C. and Arendt, E. K. (2012). Proteins in oats; their synthesis and changes during germination: a review, *Crit. Rev. Food. Sci. Nutr.* 52(7), 629–639.

Krivanek, A. F., De-Groote, H., Gunaratna, N. S., Diallo, A. O. and Friesen, D. (2007). Breeding and disseminating quality protein maize (QPM) for Africa, *Afr. J. Biotechnol.* 6, 312–324.

Larkins, B. A. and Hurkman, W. J. (1978). Synthesis and deposition of zein in protein bodies of maize endosperm, *Plant Physiol.* 62(2), 256–263.

Larkins, B. A., Wu, Y., Song, R. and Messing, J. (2017). Maize seed storage proteins. In: Larkins, B. A. (Ed.), *Maize Kernel Development*, Wallingford: CAB International, pp. 175–189.

Lásztity, R. (1998). Oat grain–A wonderful reservoir of natural nutrients and biologically active substances, *Food Rev. Int.* 14(1), 99–119.

Laurie, C. C., Chasalow, S. D., LeDeaux, J. R., McCarroll, R., Bush, D., Hauge, B., Lai, C., Clark, D., Rocheford, T. R. and Dudley, J. W. (2004). The genetic architecture of response to long-term artificial selection for oil concentration in the maize kernel, *Genetics* 168(4), 2141–2155.

Lending, C. R., Chesnut, R. S., Shaw, K. L. and Larkins, B. A. (1989). Immunolocalization of avenin and globulin storage proteins in developing endosperm of *Avena sativa* L, *Planta* 178(3), 315-324.

Li, G., Wang, D., Yang, R., Logan, K., Chen, H., Zhang, S., Skaggs, M. I., Lloyd, A., Burnett, W. J., Laurie, J. D., Hunter, B. G., Dannenhoffer, J. M., Larkins, B. A., Drews, G. N., Wang, X. and Yadegari, R. (2014). Temporal patterns of gene expression in developing maize endosperm identified through transcriptome sequencing, *Proc. Natl. Acad. Sci. U.S.A.* 111(21), 7582–7587.

Li, Q., Wang, J., Ye, J., Zheng, X., Xiang, X., Li, C., Fu, M., Wang, Q., Zhang, Z. and Wu, Y. (2017). The maize imprinted gene Floury3 Encodes a PLATZ protein required for tRNA and 5S rRNA transcription through interaction with RNA polymerase III, *Plant Cell* 29(10), 2661–2675.

Liu, G., Li, J., Shi, K., Wang, S., Chen, J., Liu, Y. and Huang, Q. (2009). Composition, secondary structure, and self-assembly of oat protein isolate, *J. Agric. Food Chem.* 57(11), 4552-4558.

Liu, L., Zubik, L., Collins, F. W., Marko, M. and Meydani, M. (2004). The antiatherogenic potential of oat phenolic compounds, *Atherosclerosis* 175(1), 39–49.

Lopes, M. A. and Larkins, B. A. (1995). Genetic analysis of *opaque2* modifier gene activity in maize endosperm, *Theor. Appl. Genet.* 91(2), 274–281.

Lopes, M. A., Takasaki, K., Bostwick, D. E., Helentjaris, T. and Larkins, B. A. (1995). Identification of two *opaque2* modifier loci in quality protein maize, *Mol. Gen. Genet.* 247(5), 603–613.

Lundin, K. E. A., Nilsen, E. M., Scott, H. G., Løberg, E. M., Gjøen, A., Bratlie, J., Skar, V., Mendez, E., Løvik, A. and Kett, K. (2003). Oats induced villous atrophy in coeliac disease, *Gut* 52(11), 1649–1652.

Mallikarjuna, M. G., Nepolean, T., Hossain, F., Manjaiah, K. M., Singh, A. M. and Gupta, H. S. (2014). Genetic variability and correlation of kernel micronutrient among exotic quality protein maize inbreds and their utility in breeding programme, *Ind. Jrnl. Gen. Plnt. Bree.* 74(2), 166–173.

Marcone, M. F. (1999). Biochemical and biophysical properties of plant storage proteins, *Food R. Int.* 32(2), 79–92.

May, T., Klatt, K. C., Smith, J., Castro, E., Manary, M., Caudill, M. A., Jahoor, F. and Fiorotto, M. L. (2018). Supplementation prevents a hallmark disturbance of kwashiorkor in weanling mice fed a maize vegetable diet: hepatic steatosis of undernutrition, *Nutrients* 10(5), 653.

Mertz, E. T., Bates, L. S. and Nelson, O. E. (1964). Mutant gene that changes protein composition and increases lysine content of maize endosperm, *Science* 145(3629), 279–280.

Miclaus, M., Wu, Y., Xu, J. H., Dooner, H. K. and Messing, J. (2011b). The maize high-lysine mutant *opaque7* is defective in an acyl-coA synthetase-like protein, *Genetics* 189(4), 1271–1280.

Miclaus, M., Xu, J. H. and Messing, J. (2011a). Differential gene expression and epiregulation of alpha zein gene copies in maize haplotypes, *PLoS Genet.* 7(6), e1002131.

Moehs, C. P., Austill, W. J., Holm, A., Large, T. A. G., Loeffler, D., Mullenberg, J., Schnable, P. S., Skinner, W., van Boxtel, J., Wu, L. and McGuire, C. (2019). Development of reduced gluten wheat enabled by determination of the genetic basis of the *lys3a* low hordein barley mutant, *Plant Physiol.* 179(4), 1692–1703.

Molina, M. I., Petruccelli, S. and Añón, M. C. (2004). Effect of pH and ionic strength modifications on thermal denaturation of the 11S globulin of sunflower (*Helianthus annuus*), *J. Agric. Food Chem.* 52(19), 6023–6029.

Munck, L. (1992). The case of high-lysine barley breeding. In: Shewry, P. R. (Ed.), *Barley: Genetics, Biochemistry, Molecular Biology and Biotechnology*, Wallingford, OX: CAB International, pp. 573–601.

Muthusamy, V., Hossain, F., Thirunavukkarasu, N., Choudhary, M., Saha, S., Bhat, J. S., Prasanna, B. M. and Gupta, H. S. (2014). Development of β-carotene rich maize hybrids through marker assisted introgression of β-carotene hydroxylase allele, *PLoS ONE* 9(12), e113583.

Myers, A. M., James, M. G., Lin, Q., Yi, G., Stinard, P. S., Hennen-Bierwagen, T. A. and Becraft, P. W. (2011). Maize *opaque5* encodes monogalactosyldiacylglycerol synthase and specifically affects C18:3/C18:2 galactolipids necessary for amyloplast and chloroplast function, *Plant Cell* 23(6), 2331–2347.

Nelson, O. E. (2001). Maize: the long trail to QPM. In: Reeve, E. C. R. and Black, I. (Eds), *Encyclopedia of Genetics*, London/Chicago: Fitzroy Dearborn Publishers, pp. 657–660.

Nie, L., Wise, M. L., Peterson, D. M. and Meydani, M. (2006). Avenanthramide, a polyphenol from oats, inhibits vascular smooth muscle cell proliferation and enhances nitric oxide production, *Atherosclerosis* 186(2), 260–266.

Omosebi, M. O., Osundahunsi, O. F. and Fagbemi, T. N. (2018). Effect of extrusion on protein quality, antinutritional factors, and digestibility of complementary diet from quality protein maize and soybean protein concentrate, *J. Food Biochem.* 42(4), e12508.

Onning, G., Wallmark, A., Persson, M., Akesson, B., Elmståhl, S. and Oste, R. (1999). Consumption of oat milk for 5 weeks lowers serum cholesterol and LDL cholesterol in free-living men with moderate hypercholesterolemia, *Ann. Nutr. Metab.* 43(5), 301–309.

Orman-Ligeza, B., Borrill, P., Chia, T., Chirico, M., Dolezel, J., Drea, S., Karafiátová, M., Schatlowski, N., Solomon, C. U., Steuernagel, B., Wulff, B. B. H., Uauy, C. and Trafford, K. (2020). *LYS3* encodes a prolamin-box-binding transcription factor that controls embryo growth in barley and wheat, *J. Cereal Sci.* 93, 102965.

Ortiz-Martinez, M., Otero-Pappatheodorou, J. T., Serna-Saldívar, S. O. and García-Lara, S. (2017). Antioxidant activity and characterization of protein fractions and hydrolysates from normal and quality protein maize kernels, *J. Cereal Sci.* 76, 85–91.

Paez, A. V., Helm, J. L. and Zuber, M. S. (1969). Lysine content of *opaque-2* maize kernels having different phenotypes 1, *Crop Sci.* 9(2), 251–252.

Pastorello, E. A., Farioli, L., Pravettoni, V., Ispano, M., Scibola, E., Trambaioli, C., Giuffrida, M. G., Ansaloni, R., Godovac-Zimmermann, J., Conti, A., Fortunato, D. and Ortolani, C. (2000). The maize major allergen, which is responsible for food induced allergic reactions, is a lipid transfer protein, *J. Allergy Clin. Immunol.* 106(4), 744-751.

Pastorello, E. A., Farioli, L., Pravettoni, V., Scibilia, J., Conti, A., Fortunato, D., Borgonovo, L., Bonomi, S., Primavesi, L. and Ballmer-Weber, B. (2009). Maize food allergy: lipid-transfer proteins, endochitinases, and alpha-zein precursor are relevant maize allergens in double-blind placebo-contrlled maize-challenge positive patients, *Anal. Bioanal. Chem.* 395(1), 93-102.

Pennisi, E. (2019). Gut bacteria linked to mental well-being and depression, *Science* 363(6427), 569.

Pernollet, J. C., Huet, J. C., Galle, A. M. and Sallantin, M. (1987). N-terminal sequences of oat avenins compared to other cereal prolamines, *Biochimie* 69(6-7), 683-689.

Peterson, D. M. (2001). Oat antioxidants, *J. Cereal Sci.* 33(2), 115-129.

Pillay, K., Siwela, M., Derera, J. and Veldman, F. J. (2014). Provitamin A carotenoids in biofortified maize and their retention during processing and preparation of South African maize foods, *J. Food Sci. Technol.* 51(4), 634-644.

Prasanna, B. M., Mazumdar, S., Chakraborti, M., Hossain, F., Manjaiaha, K. M., Agrawal, P. K., Guleria, S. K. and Gupta, H. S. (2011). Genetic variability and genotype × environment interactions for kernel iron and zinc concentrations in maize (*Zea mays* L.) genotypes, *Indian J. Agric. Sci.* 81, 704-711.

Prasanna, B. M., Palacios-Rojas, N., Hossain, F., Muthusamy, V., Menkir, A., Dhliwayo, T., Ndhlela, T., San Vicente, F., Nair, S. K., Vivek, B. S., Zhang, X., Olsen, M. and Fan, X. (2020). Molecular breeding for nutritionally enriched maize: status and prospects, *Front. Genet.* 10, 1392.

Prasanna, B. M., Pixley, K. V., Warburton, M. L. and Xie, C. (2010). Molecular marker-assisted breeding for maize improvement in Asia, *Mol. Breed.* 26(2), 339-356.

Prasanna, B. M., Vasal, S. K., Kassahun, B. and Singh, N. N. (2001). Quality protein maize, *Curr. Sci.* 81, 1308-1319.

Qiao, Z., Qi, W., Wang, Q., Feng, Y., Yang, Q., Zhang, N., Wang, S., Tang, Y. and Song, R. (2016). ZmMADS47 regulates zein gene transcription through interaction with Opaque2, *PLoS Genet.* 12(4), e1005991.

Real, A., Comino, I., de Lorenzo, L., Merchán, F., Gil-Humanes, J., Giménez, M. J., López-Casado, M. Á, Torres, M. I., Cebolla, Á, Sousa, C., Barro, F. and Pistón, F. (2012). Molecular and immunological characterization of gluten proteins isolated from oat cultivars that differ in toxicity for celiac disease, *PLoS ONE* 7(12), e48365.

Ripsin, C. M., Keenan, J. M., Jacobs, D. R. Jr., Elmer, P. J., Welch, R. R., Van Horn, L., Liu, K., Turnbull, W. H., Thye, F. W. and Kestin, M. (1992). Oat products and lipid lowering. A Meta-analysis', *JAMA* 267(24), 3317-3325.

Roser, M. and Ritchie, H. (2013). Hunger and undernourishment. Ourworld.InData.org, https://ourworldindata.org/hunger-and-undernourishment. Accessed November 7, 2020.

Ryan, D., Kendall, M. and Robards, K. (2007). Bioactivity of oats as it relates to cardiovascular disease, *Nutr. Res. Rev.* 20(2), 147-162.

Ryan, L., Thondre, P. S. and Henry, C. J. K. (2011). Oat-based breakfast cereals are a rich source of polyphenols and high in antioxidant potential, *J. Food Compos. Anal.* 24(7), 929-934.

Sadiq Butt, M., Tahir-Nadeem, M., Khan, M. K., Shabir, R. and Butt, M. S. (2008). Oat: unique among the cereals, *Eur. J. Nutr.* 47(2), 68-79.

Salamini, F., Di Fonzo, N. and Gentinetta, E. (1979). A dominant mutation interfering with protein accumulation in maize seeds. In: Proceedings of an International symposium on seed protein improvement in cereals and grain legumes, Vol I., Vienna: Int. Atomic Energy Agency, pp. 97–108.

Sands, D. C., Morris, C. E., Dratz, E. A. and Pilgeram, A. (2009),. Elevating optimal human nutrition to a central goal of plant breeding and production of plant-based foods, *Plant Sci.* 177(5), 377–389.

Sarika, K., Hossain, F., Muthusamy, V., Baveja, A., Zunjare, R., Goswami, R., Thirunavukkarasu, N., Saha, S. and Gupta, H. S. (2016). Exploration of novel *opaque16* mutation as a source for high-lysine and tryptophan in maize endosperm, *Ind. Jrnl. Gen. Plnt. Bree.* 77(1), 59–64.

Sarika, K., Hossain, F., Muthusamy, V., Zunjare, R. U., Baveja, A., Goswami, R., Thirunavukkarasu, N., Jha, S. K. and Gupta, H. S. (2018). *Opaque16*, a high lysine and tryptophan mutant, does not influence the key physicobiochemical characteristics in maize kernel, *PLOS ONE* 13(1), e0190945.

Schmidt, R. J., Burr, F. A., Aukerman, M. J. and Burr, B. (1990). Maize regulatory gene opaque-2 encodes a protein with a leucine-zipper motif that binds to zein DNA, *Proc. Natl. Acad. Sci. U.S.A.* 87(1), 46–50.

Schneeman, B. O. (2001). Linking agricultural production and human nutrition, *J. Sci. Food Agric.* 81(1), 3–9.

Scibilia, J., Pastorello, E. A., Zisa, G., Ottolenghi, A., Ballmer-Weber, B., Pravettoni, V., Scovena, E., Robino, A. and Ortolani, C. (2008). Maize food allergy: a double-blind placebo-controlled study, *Clin. Exp. Allergy* 38(12), 1943–1949.

Segal, G., Song, R. and Messing, J. (2003). A new opaque variant of maize by a single dominant RNA interference-inducing transgene, *Genetics* 165(1), 387–397.

Shen, B. and Roesler, K. (2017). Maize kernel oil content. In: Larkins, B. A. (Ed.), *Maize Kernel Development*, Wallingford: CAB International, pp. 160–174.

Shewry, P. R. (1995). Plant storage proteins, *Biol. Rev. Camb. Philos. Soc.* 70(3), 375–426.

Shewry, P. R., Tatham, A. S. and Halford, N. G. (1999). Avenins : the prolamins of oats. In: Shewry, P. R. and Casey, R. (Eds), *Seed Proteins*, Netherlands, Dordrecht: Springer, pp. 79–92.

Shewry, P. R., Piironen, V., Lampi, A. M., Nystrom, L., Li, L., Rakszegi, M., Fras, A., Boros, D., Gebruers, K., Courtin, C. M., Delcour, J. A., Andersson, A. A. M., Dimberg, L., Bedo, Z. and Ward, J. L. (2008). Phytochemical and fiber components in oat varieties in the health grain diversity screen, *J. Agric. Food Chem.* 56(21), 9777–9784.

Shotwell, M. A. (1999). Oat globulins. In: Shewry, P. R. and Casey, R. (Eds) *Seed Proteins*, Dordrecht: Kluwer Academic Publishers, pp. 389–400.

Shotwell, M. A., Afonso, C., Davies, E., Chesnut, R. S. and Larkins, B. A. (1988). Molecular characterization of oat seed globulins, *Plant Physiol.* 87(3), 698–704.

Shotwell, M. A., Boyer, S. K., Chesnut, R. S. and Larkins, B. A. (1990). Analysis of seed storage protein genes of oats, *J. Biol. Chem*. 265(17), 9652–9658.

Smulders, M. J. M., van de Wiel, C. C. M., van den Broeck, H. C., van der Meer, I. M., Israel-Hoevelaken, T. P. M., Timmer, R. D., van Dinter, B. J., Braun, S. and Gilissen, L. J. W. J. (2018). Oats in healthy gluten-free and regular diets: a perspective, *Food Res. Int.* 110, 3–10.

Song, R. and Messing, J. (2002). Contiguous genomic DNA sequence comprising the 19-kDa zein gene family from *Zea mays*, *Plant Physiol.* 130(4), 1626–1635.

Song, R. and Messing, J. (2003). Gene expression of a gene family in maize based on noncollinear haplotypes, *Proc. Natl. Acad. Sci. U.S.A.* 100(15), 9055–9060.

Stark, A. and Madar, Z. (1994). Dietary fiber. In: Goldberg, I. (Ed.), *Functional Foods: Designer Foods, Pharma Foods, Nutraceuticals*, New York: Chapman & Hall, pp. 183–201.

Strychar, R. (2011). World oat production, trade, and usage. In: Webster, F. H. and Wood, P. J. (Eds), *Oats: Chemistry and Technology*, AACC International, pp. 1–10.

Swarup, S., Timmermans, M. C. P., Chaudhuri, S. and Messing, J. (1995). Determinants of the high methionine trait in wild and exotic germplasm may have escaped selection during early cultivation of maize, *Plant J.* 8(3), 359–368.

Thompson, G. A. and Larkins, B. A. (1993). Characterization of the zein genes and their regulation in maize endosperm. In: Walbot, V. and Freeling, M. (Eds), *The Maize Handbook*, Heidelberg, Germany: Springer, pp. 639–647.

Tosh, S. M., Brummer, Y., Miller, S. S., Regand, A., Defelice, C., Duss, R., Wolever, T. M. and Wood, P. J. (2010). Processing affects physicochemical properties of β-glucan in oat bran cereal, *J. Agric. Food. Chem.* 58(13), 7723–7730.

Tripathy, S. K., Ithape, D. M., Maharana, M. and Prusty, A. M. (2017). Quality protein maize (QPM): genetic basis and breeding perspective, *Trop. Plant Res.* 4(1), 145–152.

Vasal, S. K., Villegas, E., Bjarnason, M., Gelaw, B. and Goertz, P. (1980). Genetic modifiers and breeding strategies in developing hard endosperm opaque-2 materials. In: Pollmer, W. G. and Phillips, R. H. (Eds), *Improvement of Quality Traits of Maize for Grain and Silage Use*, London: Martinus Nijhoff, pp. 37–73.

Venter, C., Skypala, I. and Dean, T. (2008). Maize allergy: what we have learned so far, *Clin. Exp. Allergy* 38(12), 1844–1846.

Vicente-Carbajosa, J., Moose, S. P., Parsons, R. L. and Schmidt, R. J. (1997). A maize zinc-finger protein binds the prolamin box in zein promoters and interacts with the basic leucine zipper transcriptional activator Opaque2, *Proc. Natl. Acad. Sci. U.S.A.* 94(14), 7685–7690.

Villegas, E., Vasal, S. K. and Bjarnason, M. (1992). Quality protein maize-what is it and how was it developed. In: Mertz, E. T. (Ed.), *Quality Protein Maize*, St. Paul: Am. Soc. Cereal Chem., pp. 27–48.

Wallace, J. C., Lopes, M. A., Paiva, E. and Larkins, B. A. (1990). New methods for extraction and quantitation of zeins reveal a high content of γ-zein in modified *opaque-2* maize, *Plant Physiol.* 92(1), 191–196.

Wang, G., Qi, W. W., Wu, Q., Yao, D. S., Zhang, J. S., Zhu, J., Wang, G., Wang, G., Tang, Y. and Song, R. (2014a). Identification and characterization of maize *floury4* as a novel semi-dominant opaque mutant that disrupts protein body assembly, *Plant Physiol.* 165(2), 582–594.

Wang, G., Wang, F., Wang, G., Wang, F., Zhang, X., Zhong, M., Zhang, J., Lin, D., Tang, Y., Xu, Z. and Song, R. (2012). *Opaque1* encodes a myosin XI motor protein that is required for endoplasmic reticulum motility and protein body formation in maize endosperm, *Plant Cell* 24(8), 3447–3462.

Wang, G., Zhang, J., Wang, G., Fan, X., Sun, X., Qin, H., Xu, N., Zhong, M., Qiao, Z., Tang, Y. and Song, R. (2014b). Proline responding1 plays a critical role in regulating general protein synthesis and cell cycle in maize, *Plant Cell* 26(6), 2582–2600.

Webster, F. H. and Wood, P. J. (2011). *Oats: Chemistry and Technology*, St. Paul: AACC International.

Weichel, M., Glaser, A. G., Ballmer-Weber, B. K., Schmid-Grendelmeier, P. and Crameri, R. (2006). Wheat and maize thioredoxins: a novel cross-reactive cereal allergen family related to Baker's asthma, *J. Allergy Clin. Immunol.* 117(3), 676–681.

White, E. M. (1995). Structure and development of oats. In: Welch, R. W. (Ed.), *The Oat Crop: Production and Utilization*, London, UK: Chapman and Hall, pp. 88–119.

WHO (2018). Noncommunicable diseases. World Health Organization, https://www.who.int/news-room/fact-sheets/detail/noncommunicable-diseases. Accessed November 7, 2020.

WHO (2020). Obesity and overweight. World Health Organization, https://www.who.int/news-room/fact-sheets/detail/obesity-and-overweight. Accessed November 7, 2020.

Woo, Y. M., Hu, D. W., Larkins, B. A. and Jung, R. (2001). Genomics analysis of genes expressed in maize endosperm identifies novel seed proteins and clarifies patterns of zein gene expression, *Plant Cell* 13(10), 2297–2317.

Wrick, K. L. (1993). Functional foods: cereal products at the food-drug interface, *Cereal Foods World* 38, 205–214.

Wrick, K. L. (1994) The Potential Role of Functional Foods in Medicine and Public Health. In: Goldberg I. (Eds), Functional Foods, *Boston*: Springer, pp.480–494.

Wu, Y., Holding, D. R. and Messing, J. (2010). Gamma-zeins are essential for endosperm modification in quality protein maize, *Proc. Natl. Acad. Sci. U.S.A.* 107(29), 12810–12815.

Wu, Y. and Messing, J. (2014). Proteome balancing of the maize seed for higher nutritional value, *Front. Plant Sci.* 5, 240.

Wu, X. R., Kenzior, A., Willmot, D., Scanlon, S., Chen, Z., Topin, A., He, S. H., Acevedo, A. and Folk, W. R. (2007). Altered expression of plant lysyl tRNA synthetase promotes tRNA misacylation and translational recording of lysine, *Plant J.* 50(4), 627–636.

Yang, W., Zheng, Y., Zheng, W. and Feng, R. (2005). Molecular genetic mapping of a high-lysine mutant gene (opaque-16) and the double recessive effect with *opaque-2* in maize, *Mol. Breed.* 15(3), 257–269.

Yangcheng, H., Jiang, H., Blanco, M. and Jane, J. L. (2013). Characterization of normal and waxy corn starch for bioethanol production, *J. Agric. Food Chem.* 61(2), 379–386.

Yao, D., Qi, W., Li, X., Yang, Q., Yan, S., Ling, H., Wang, G., Wang, G. and Song, R. (2016). Maize *opaque10* encodes a cereal specific protein that is essential for the proper distribution of zeins in endosperm protein bodies, *PLoS Genet.* 12(8), e1006270.

Yu, J., Peng, P., Zhang, X., Zhao, Q., Zhy, D., Sun, X., Liu, J. and Ao, G. (2004). Seed-specific expression of a lysine rich protein sb401 gene significantly increases both lysine and total protein content in maize seeds, *Mol. Breed.* 14(1), 1–7.

Yuan, L., Dou, Y., Kianian, S. F., Zhang, C. and Holding, D. R. (2014). Deletion mutagenesis identifies a haploinsufficient role for γ-zein in *opaque2* endosperm modification, *Plant Physiol.* 164(1), 119–130.

Zarkadas, C. G., Yu, Z. and Burrows, V. D. (1995). Protein quality of three new Canadian-developed naked oat cultivars using amino acid compositional data, *J. Agric. Food Chem.* 43(2), 415–421.

Zhang, D. C., Moore, W. R. and Doehlert, D. C. (1998). Effects of oat grain hydrothermal treatments on wheat-oat flour dough properties and bread baking quality, *Cereal Chem.* 75(5), 602–605.

Zhang, S., Zhan, J. and Yadegari, R. (2018). Maize opaque mutants are no longer so opaque, *Plant Reprod.* 31(3), 319–326.

Zhang, W., Yang, W., Wang, M., Wang, W., Zeng, G., Chen, Z. and Cai, Y. (2013). Increasing lysine content of waxy maize through introgression of *Opaque-2* and *Opaque-16* genes using molecular assisted and biochemical development, *PLoS ONE* 8(2), e56227.

Zhang, Z., Yang, J. and Wu, Y. (2015). Transcriptional regulation of zein gene expression in maize through the additive and synergistic action of *opaque2*, prolamine-box binding factor and O2 heterodimerizing proteins, *Plant Cell* 27(4), 1162–1172.

Zhao, Y., Mine, Y. and Ma, C. Y. (2004). Study of thermal aggregation of oat globulin by laser light scattering, *J. Agric. Food Chem.* 52(10), 3089–3096.

Zheng, P., Allen, W. B., Roesler, K., Williams, M. E., Zhang, S., Li, J., Glassman, K., Ranch, J., Nubel, D., Solawetz, W., Bhattramakki, D., Llaca, V., Deschamps, S., Zhong, G. Y., Tarczynski, M. C. and Shen, B. (2008). A phenylalanine in DGAT is a key determinant of oil content and composition in maize, *Nat. Genet.* 40(3), 367–372.

Chapter 7

Developments in fractionation methods to improve extraction of aleurone or its beneficial compounds from wheat grain

V. Lullien-Pellerin, IATE, Univ Montpellier, INRAE, Institut Agro, Montpellier, France

1 Introduction

In wheat grain, the aleurone tissue constitutes a single row of cubical cells with a mean thickness of around 50–65 μm, displaying 15% variation depending on the cultivar (Peyron et al., 2002; Stevens, 1973) and is located between the starchy endosperm and the outer layers (Evers et al., 1999, Fig. 1). It originates from the endosperm during grain development (Becraft and Yi, 2011) and represents around 9% of the grain dry mass (Barron et al., 2011). Aleurone cells have been found to store soluble proteins, minerals, lipids and vitamins and a number of micronutrients enclosed within thick cell walls mainly composed of relatively linear arabinoxylans and β-glucans (Brouns et al., 2011; Antoine et al., 2002; Pomeranz, 1988; Bacic and Stone, 1981). Together with the embryo it is the only remaining live tissue once the grain matures and is known to play a key role in grain germination producing hydrolytic enzymes under the control of the hormone gibberelic acid (Bethke et al., 1998).

The aleurone layer is not generally found after classical milling into fractions (flours, semolina) used for food making but is recovered in bran fractions used for animal feed where it represents around 50% in dry weight of the bran particles (Barron et al., 2011). However, its presence in flours after milling appeared to depend on the wheat grain hardness, being present more in flours from hard

http://dx.doi.org/10.19103/AS.2021.0087.13

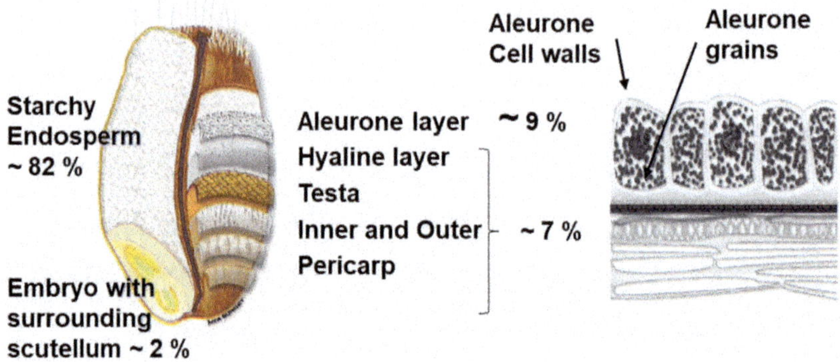

Figure 1 Schematic representation of wheat grain tissues and details of the aleurone layer and its upper surrounding tissues. Adapted from original images presented in Surget and Barron (2005).

common wheat grains than in those obtained from soft common wheat grains (Greffeuille et al., 2005). This appears to be the result of different breaking behaviour between grains of different hardness. Grains from soft cultivars appear to break mostly in the sub-aleuronic region, resulting in an increased amount of residual starchy endosperm in the bran. This distinctive behaviour was recently confirmed to be linked with the presence of wild-type alleles of specific genes encoding both puroindolines A and B, two proteins suspected to be involved in the strength of cohesion between starch granules and the protein matrix in the starchy endosperm and responsible for grain endosperm softening (Lullien-Pellerin et al., 2020; Giroux et al., 2003). The introduction of wild-type alleles of both puroindolines in a durum wheat background, which does not contain these genes carried by the D genome, led to a softer grain structure (Murray et al., 2016; Morris et al., 2011). This introduction also led, after milling, to a bran fraction containing more aleurone content and starch from the residual endosperm than bran obtained from grains which do not carry these wild-type puroindolines (Heinze et al., 2016). Flours from softer endosperm thus contain less aleurone content which is mainly recovered in the bran fraction.

As the aleurone layer contains a number of compounds with beneficial properties, research over the last 25 years has tried to increase its presence in food products. This has mainly been illustrated through the HealthGrain project (FP6 EU 2005-10), which aimed to improve its isolation and study its effect on food quality and human health. Isolation of the aleurone layer has gained an increasing interest which has motivated scientists to better define this tissue in order to harmonize studies relating to its separation (Atwell et al., 2010).

It is important to note that the location, metabolic activity and role in grain germination of the aleurone layer present challenges that need to be addressed.

An example is wheat grains contaminated with mycotoxins like deoxynivalenol (DON), depending on cultivation conditions and cultivar resistance to infection by *Fusarium graminearum* (Culze et al., 2020). The toxins may be present in the aleurone layer, even if external tissues like the outer and inner pericarp are more contaminated (Rios et al., 2009). The safety of products containing the aleurone layer must therefore be assessed to avoid potentially negative effects on health (Cimbalo et al., 2020).

The fibre and enzyme content of the aleurone layer can also have negative effects on sensory and technological properties. Bread volume can, for example, be significantly reduced after addition of the aleurone layer (Bagdi et al., 2016; Noort et al., 2010), with a need to adjust water content and particle size. Indeed, aleurone-enriched flour for making bread has been found to have distinct hydration and rheological properties (Li et al., 2019; Bucsella et al., 2016).

It is also important to note that some nutritionally-important compounds do not appear available in the aleurone layer such as minerals which are complexed with phytic acid (O'Dell et al., 1972) or phenolic acids covalently linked to cell walls (Rhodes et al., 2002).

Two general approaches have been developed to isolate the aleurone layer from wheat grains, either from bran fractions or from grains after sequential debranning (Fig. 2).

2 Aleurone layer isolation from bran fractions

Either wet or dry processes have been developed which have been mainly described in patents. They exploit the differences in properties between the

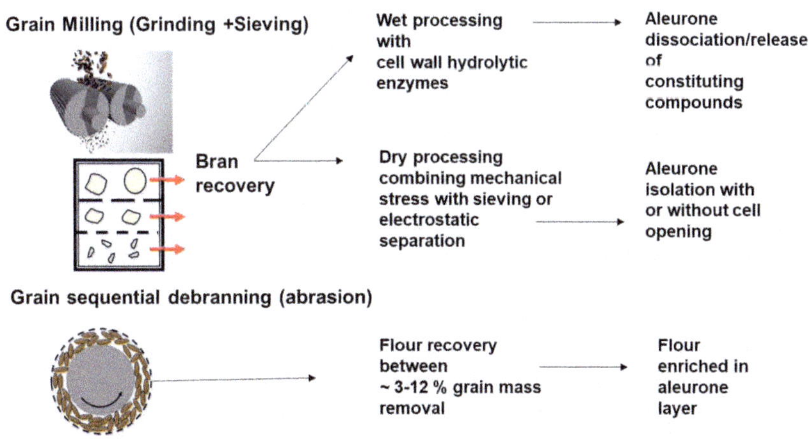

Figure 2 Schematic representation of distinct strategies for wheat aleurone layer isolation.

different wheat grain tissues due to distinct biochemical composition and structure (Antoine et al., 2003). Different approaches include: disassociation of tissues at cell wall interfaces, which depends on tissue adhesion; reducing particle size in order to recover each tissue as individual cells; or damaging aleurone tissues to recover the cellular content. These different approaches involve different techniques and parameters.

2.1 Wet processing

The first description of aleurone purification with sufficient yield was described in Stevens (1970). Bran was first ground with a hammer mill and then disintegrated in a mixture of benzene and carbon tetrachloride and the sediment retreated in similar mixture after centrifugation to recover different parts of the aleurone tissue after several similar steps of suspension, disintegration and centrifugation.

Bohm et al. (2003) developed different examples of possible wet isolation but do not give many details on their process. They used a cocktail of four different cell-wall hydrolyzing enzymes or only xylanase, in the presence of water and the wheat bran fraction (particles between 400–800 μm), at a temperature between 45–55°C, to respectively obtain the aleurone cellular content and aleurone cell-wall fragments after sieving, or to disrupt the aleurone layer from the surrounding tissues. In this process, the addition of water and increase in temperature were used for the appropriate working of added exogenous hydrolytic enzymes, and one can also expect a concomitant activation of endogenous enzymes. These endogenous activities can either come from possible contaminating residual microorganisms which are known to shelter in the crease or to be present at the grain surface (Laca et al., 2006) or from remaining active tissues as in the aleurone layer or the embryo. Therefore, other modifications of the present molecules than those expected can occur.

Kvist et al. (2010) also described wet processing which was used to isolate the different bran-composing tissues or interesting compounds (protein, lipid, non-starch poly- and oligo-saccharides) through a combination of enzymatic treatment and wet milling followed by sequential centrifugation and ultrafiltration.

2.2 Dry processing

2.2.1 Using impact mill coupled or not with electrostatic separation

The first successful attempt to isolate the aleurone layer from bran was described by Stone and Minifie (1988). It was based on recovering particles of diameters between 130 μm and 290 μm, which were ground with a hammer-mill with a tip speed in the range of 76–100 m/s to reduce the bran particle sizes which was further classified to remove particles under 40 μm. The sized bran particles were electrostatically charged by rubbing them against each other and/or against the walls of an elutriator column. Then the charged particles, convoyed with air stream, were separated through a uniform electrical field (with a high potential difference of around 10 000 V), depending on their differences in surface charges to be collected at the end by deflection into distinct receiver bins. The study showed that the aleurone particles were deflected towards the negative electrode, therefore carrying a positive charge but unfortunately, material used to charge the particles was not described. This patented process allowed recovery of a very pure 'aleurone' fraction (95%), even if its yield was rather low (10%).

Explanation for the ability to charge differently the aleurone layer and the other surrounding bran tissues was revealed afterwards from studies on hand-isolated tissues (Antoine et al., 2004). The hand-isolated aleurone layer was found to exhibit more capacitive ability than the outer pericarp and the tissues in-between.

It was also found that the specific mechanical properties of each composing tissues in bran play an important role in their separation and are clearly influenced by the temperature and water content (Hemery et al., 2010). Therefore, impact grinding at ambient temperature was found to be more efficient to dissociate the bran-composing tissues than cryogrinding (Hemery et al., 2011a). It was also found that opening the aleurone cells by grinding can change the charging behaviour of the particles as a result of modification in the surface composition. Thus, the grinding step before separation has a non-negligible influence on their charging properties. Indeed, size and moisture content of the generated particles were shown to play a role on their charging properties and delay of discharging (Hemery et al., 2009a). A simple ball milling of bran with sufficient time to reduce the size of the particles under the cell size was found to allow the release of the aleurone cellular content (Antoine et al., 2004b).

A series of patents (Bohm and Kratzer, 2008; Laux et al., 2006; Behrens and Bohm, 2004; Bohm and Buri, 2004; Bohm et al., 2003) from the Bühler SA company was released to protect the production of an aleurone-enriched fraction. The process involves grinding bran classically with an impact mill at a speed between 70 m/s and 120 m/s and with a moisture content of 10% to 20%. The ground bran was then sieved and the aleurone layer isolated using either air classification or electrostatic separation after particle charging within a stainless steel pipe.

Two different aleurone-enriched fractions can be obtained with 95% (LEURON® 1) or 99% purity (LEURON® 2) depending on the grinding process. Both fractions include aleurone cell clusters with lower cell number with improved purity. LEURON® 2 had a mean particle size around 200 µm corresponding to 5-40 cells still attached to the hyaline layer and containing low amount of pericarp and starch granules as the non-aleurone material (Buri et al., 2004). Fractions were characterized by microscopy and by their content in compounds known to be present in this wheat tissue.

Another approach coupling both an ultrafine grinding, charging particles in a teflon tube and separation in a uniform electric field (Fig. 3a) was described in Hemery et al. (2011b) using a slightly modified pilot electrostatic separator (TEP System, Tribo Flow Separations, Lexington, USA). Obtained fractions were finely characterized to identify and quantify composing tissues through the use of previously defined biochemical markers (Hemery et al., 2009b). The described process allowed isolating a positively charged aleurone-enriched fraction after three cycles of separation with a 34% yield in comparison to the initial bran. This aleurone-enriched fraction was mostly enriched with aleurone cell walls, which was multiplied by a two-fold factor in comparison to the initial bran. Indeed part of the corresponding aleurone

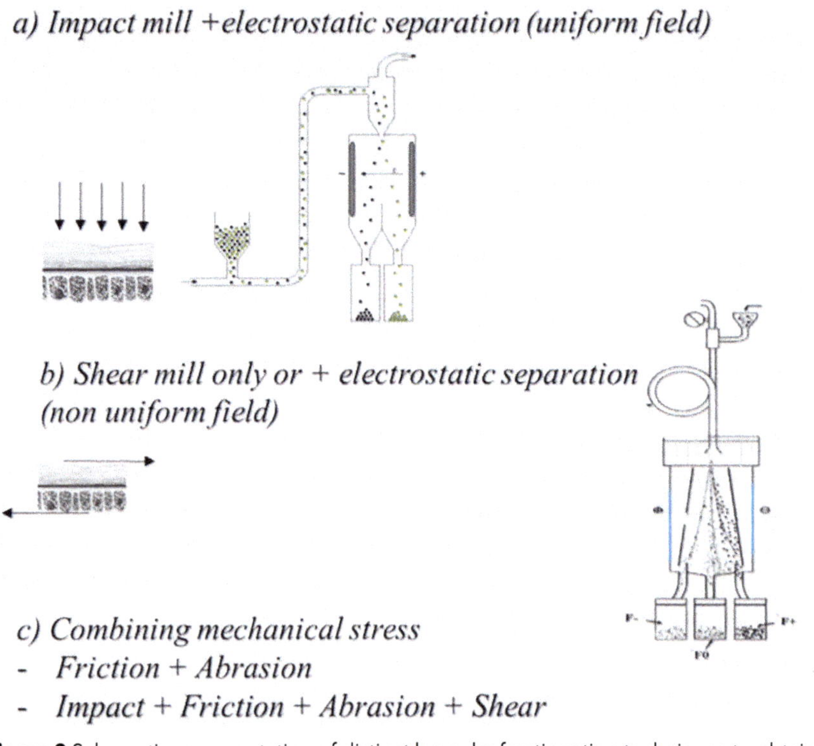

Figure 3 Schematic representation of distinct bran dry fractionation techniques to obtain fraction enriched with aleurone tissue.

cell content was found to remain stuck to the charging pipe and thus cannot be easily isolated.

2.2.2 Using shear milling coupled or not with electrostatic separation

A coarse bran, brushed to reduce potential residual attached endosperm, was reduced in size using a pin mill and the fractions under 200 µm air-classified into two fractions according to their size and density differences (Antoine et al., 2004c). The remaining grain endosperm in wheat bran as well as the pericarp tissue were mainly recovered in the higher yield obtained fraction displaying larger size particles (with a median size around 83 µm). This fraction also contained 43% of the cell walls from the aleurone or the intermediate (=inner pericarp+testa+hyaline) layers and 31% of the aleurone cellular content meaning that the aleurone and the intermediate cell walls appeared difficult to dissociate. The finest fraction (with 50% of the overall particles under 7 µm) was found to mostly contain the aleurone cellular content (66% of the fraction).

Chen et al. (2013a) described a distinct grinding protocol to dissociate wheat bran constitutive tissues using a centrifugal impact mill in which shear stress mainly predominates (Mayer-Laigle et al., 2018). The generated particles were transported out of the crushing chamber through an air classifier, and classified with different sieves with openings between 100 and 420 µm. The outer pericarp was found to be detached from the bran fragments when the particle size reached around 460 µm, whereas the aleurone was released as fragments of 20–30 intact cells as the particle size was around 200 µm. These authors concluded that their results fully appeared in agreement with their operating constraints (high-speed rotating rotor-tooth with a tip speed of around 70 m/s) and the already reported mechanical properties of each of the bran constituting tissues (Greffeuille et al., 2007; Antoine et al., 2003) and peeling properties between the aleurone and surrounding tissues (Martelli et al., 2010). In comparison with Hemery et al. (2011b), this process appeared to increase both the aleurone cell walls and cellular content from initial bran for a similar total enrichment between 55% and 62%.

This enriched fraction was further used, after tribo-charging, for triboelectric separation between the aleurone layer and the other residual surrounding tissues and progress were made to optimize parameters studying the effect of different charging materials (Teflon, Nylon or stainless) as well as the air flow rate and water content (Chen et al., 2014). In contrast with Hemery et al. (2011b), particles to wall pipe collisions were favoured and were then separated in a non-uniform electric field (Fig. 3b) to collect three fractions at the end of the triboelectric separation. The positioning of the composing bran tissues was established in the known triboelectric material series: starch displayed the most positive particles superior to polyamide followed with the aleurone and intermediate tissues showing similar charging properties and followed with the outer pericarp being more positive than polyethylene (Chen et al., 2020). The authors related these properties to differences in the phenolic compounds at the particle surface observed by Raman microscopy. They also improved the efficiency and selectivity of the electrostatic separation between tissues after particle drying between 7 and 13% of water content whereas the yield can be increased with the particle size decreasing without any improvement of the separation selectivity.

2.2.3 Other processing methods (Fig. 3c)

In 2014, Pape and Galindo developed new equipment for bran finishing that is able to apply friction and abrasive constraints using an annular milling chamber equipped with an eccentric cylindrical rotor at the inner side and a frame assembly including abrasive surfaces and perforated screens at the outer side and displaying longitudinal lobes protruding radially from the rotor. This process has advantages of being able to better recover the aleurone without fragments from the other tissues as reported by their authors (Pape and Galindo, 2014).

Different companies have developed aleurone enriched flour products though production methods that are not well described. For example, the Shandong ZhiShiFang Food Science and Technology Company Ltd., established in 2012, describes the production of aleurone flour which was found to be possibly added to wheat dough without degradation of its hydration and rheological properties (Li et al., 2019). It was used by Xu et al. (2020) to increase the nutritional properties of instant fried noodles. The process was patented in China by Wu and Qi (2013). McIntosh et al. (2001) also described the use of aleurone-enriched flour produced in Australia by Goodman Fielder Pty Limited and which appears to display benefits to reduce induced colon adenoma in rats. Although not well described, the process involves isolating bran followed by shearing aleurone cells in order to release their content and concomitantly recovered part of the germ which resulted in a flour significantly enriched with folate (Fenech et al., 1999).

Trying to reduce the phytic acid content in wheat bran, Guo et al. (2015) developed a protocol using a teeth roller mill to open the aleurone cells combined with the ultrasonic washing of the aleurone cell content. They efficiently cleaned the residual bran from the aleurone cell content, and thus one can expect to recover this latter from the washing water. Indeed using a similar ultrasonic washing, Ndolo et al. (2015) demonstrates its ability to release niacin from the aleurone cells.

3 Aleurone isolation by abrasion from wheat grain

Harris at al. (2005) prepared fractions enriched either with the aleurone layer or the outer pericarp respectively, after grain tempering at a final moisture of 17% and debranning onto a PeriTec machine (Satake Co., Hiroshima, Japan). They obtained an enriched pericarp fraction, also containing part of the germ, after debranning to 2.5% of the grain mass. Then, the debranned grains were milled in a pilot mill mimicking the traditional commercial processing and the bran fractions (coarse and fine) were recovered as an enriched aleurone fraction. It was further characterized by both scanning electron microscopy and chemical analysis of the cell walls and demonstrated that the aleurone-rich fraction also contain the testa and hyaline layer.

Rios et al. (2009) studying the mycotoxin distribution in wheat grain and concomitantly using specific biochemical markers to monitor each of the outer layers from grain demonstrated that the aleurone layer can be reached between 5 and 12% of grain abrasion using a tangential abrasive dehulling system (Satake model TM-05C, Stockport, UK) even if co-extraction of the surrounding tissues also partly occurred. Thus, one can expect to obtain a fraction enriched with the aleurone layer by grain abrasion. Indeed, Delcour et al. (2012) described the possible preparation of such a fraction between 3 and 15% of grain, even if no complete characterization was described in their publication.

Chen et al. (2013b) also using a tangential abrasive dehulling machine but with different specification (model no. 4F-10/220, Veenables Machine Works, Saskatoon, Canada) isolated a fraction with similar antioxidant properties to a commercial aleurone flour developed by Cargill (GrainWise® wheat aleurone, undescribed process, Minneapolis, USA). This fraction was obtained after 1 min abrasion of wheat grain displaying 11% water content and corresponded to around 10% of grain mass removal.

Other aleurone-enriched fraction appeared in different publications without detailed explanation of the developed process. Indeed, Zaupa et al. (2014) described durum wheat aleurone-enriched fractions, provided by Barilla (Parma, Italy), which were obtained after a three debranning step processing, followed with a micronizing step, and subjected to air-separation to release two distinctive fractions. One of the fraction (IN) corresponded to the inner part of the aleurone close to the starchy endosperm and was recovered with an extraction yield of 0.75% and displayed particles <100 μm. The other one (OUT) included the outer tissue next to the aleurone layer (i.e. at least the hyaline layer) and was extracted with a yield of 1.25% and displayed particles ranging from 300 to 425 μm.

The aleurone-enriched fraction either coming from bran or grain abrasion was generally directly added in different products (as some examples, in bread: Bagdi et al., 2016; Blandino et al., 2013; Price et al., 2012 and 2010, in pasta: Bagdi et al., 2014, or in noodles: Xu et al., 2020) to test its effect either on safety, rheological, sensory, nutritional and health properties. However, the used fraction is rarely well described in terms of composing tissues or detailed biochemical composition, making comparison between fractions and their effects difficult.

Indeed, depending on the process applied, different percentage of isolation and purity was obtained as well as distinct level of cell opening (Fig. 4). As an example, process described and patented by Bühler led to intact sheet of aleurone tissues (Buri et al., 2004), whereas fine grinding and electrostatic separation described in Hemery et al. (2011b) appeared to contribute to isolate aleurone cell content from their cell walls in distinguished fractions. However, the former fraction can be reground in finer particles <50 um in order to release the cell content. It can also be treated with specific enzymes in order, for example, to hydrolyse the cell-wall material either to obtain oligomers of the constituting cell wall arabinoxylans (AXOS) or other-bound compounds (i.e. the phenolic acids) as illustrated in Delcour et al. (2012) and Rosa et al. (2013). However, this enzyme treatment implied to shift to wet processing.

There are thus different strategies , depending on whether the aim is to isolate cell-wall constituents, aleurone cellular content, insoluble or soluble compounds.

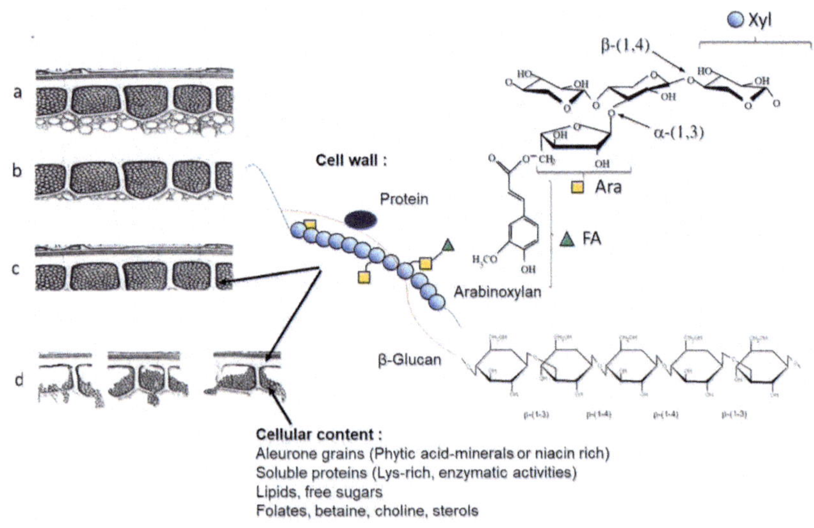

Figure 4 Schematic representation of distinct tissue composition and structure of aleurone enriched fractions (a, b, c, d) depending on the processing steps and defined target to concentrate, that is, the aleurone cell walls or intracellular compounds or the overall tissue.

4 Aleurone cell-wall components release: fibres and antioxidant properties

According to Stone (2006), aleurone cell walls are mainly composed of arabinoxylans (65%, AX) and (1,3)(1,4) β- glucans (29%) and also contain a significantly higher level (1.85%) of esterified hydroxycinnamic acids than in the endosperm cells, being mainly ferulic acid (FA; Rhodes et al., 2002). Aleurone cell walls also contain specific proteins (Rhodes and Stone, 2002) and low amount of cellulose and glucomannans (Table 1). Taking into account the arabinoxylan and β-glucan content of the isolated aleurone layer (Antoine et al., 2003), the aleurone cell walls were found to represent around 30–50% of the aleurone cell mass. After isolating aleurone from two wheat cultivars, Hemery et al. (2009b) found the aleurone cell wall proportion to be 42.4% and 48.0% respectively, with a standard deviation of 2.5%.

The cell-wall polysaccharide fraction mainly constituted with AX contributes to the dietary fibre content of the aleurone fraction which are mainly insoluble fibres (Buri et al., 2004). AX from the aleurone layer was found to display a low substitution level by arabinose (Ara) of the xylose (Xyl) linear backbone, but a high level of esterified FA (Barron et al., 2007; Parker et al., 2005; Antoine et al., 2003), therefore leading to different properties than AX from the other grain tissues, that is, a better ability for fermentation (Amrein et al., 2003).

Digestion with cell-wall hydrolases (xylanases from *Bacillus subtilis* and β-glucanase and cellulase from *Trichoderma reesei*) of aleurone-enriched fraction was found to produce more AXOS (i.e. yield of 164g/Kg with the purest fraction produced by Bühler) than other bran tissues due to lower AX substitution (Vangsøe et al., 2019), contributing potentially to the increase in prebiotic effects (Grootaert et al., 2007).

Higher yield (62%) of AXOS production from wheat aleurone produced by Cargill was however reported by Malunga and Beta (2015) using only xylanase from *Neocallimastix patriciarum* and further fractionation using ethanol, allowing to collect a fraction which was found to display antioxidant properties.

Table 1 Aleurone cell-wall composition as reported in Stone (2006), Rhodes et al. (2002), Rhodes and Stone (2002).

Arabinoxylans	65
β-glucans	29
Cellulose	2
Gluco-mannans	2
Phenolic acids (90% FA)	1.85
Proteins (Pro, Ser or Gly rich)	1

Same authors pointed out the lower yield of AXOS obtained using another xylanase, from *Trichoderma viride*, with the same aleurone fraction.

As they digest cell walls, use of these enzymes could simultaneously release intracellular compounds but this has not been reviewed in the studies discussed.

Intensive ball-grinding for long duration (5 days) was also found to solubilize AX through the destruction of the treated tissues and to produce AXOS (Van Craeyveld et al., 2009). Aleurone mild grinding was also found to increase its antioxidant capacity by a two-fold factor when its median particle size was reduced under 50 μm due to greater exposure of bound FA, which is the major contributor of the aleurone antioxidant properties (Anson et al., 2008), and to slightly increase (from 1.5 to 2.4%) the FA solubility (Rosa et al., 2013). Moreover, the use of combination between xylanase and feruloyl esterase activities was found to allow an almost total release (up to 86%) of FA in bioaccessible forms and to a significant increase of the aleurone radical scavenging activity by up to four-fold. Rhodes and Stone (2002) pointed out that hydrolysis of the aleurone cell walls by xylanase can also release part of the associated proteins which are rich in glycine, proline or serine amino-acid depending on the considered class.

5 Aleurone cell content release

Aleurone cell content was found to contain two different types (I and II) of small inclusions (0.5–5 μm), which accumulated along grain development into vacuoles that result in aleurone grains or also called protein bodies or globoids (Fulcher et al., 1981; Morrison et al., 1975). Type I inclusion is birefringent using polarized optics and can be solubilized with dilute acid due to its chemical composition (Fulcher et al., 1981). They were found to contain phytic acid, a myoinositol 1,2,3,4,5,6 hexa*kis*phosphate, known as IP6, complexed with cations, notably Mg, and Fe, but also Ca and K (Moore et al., 2012; Regvar et al., 2011; Bethke et al., 1998). Singh et al. (2014) using spatial X-ray fluorescence micro-imaging clearly showed the colocation of P, and thus phytate, with K, Mg and Fe in the aleurone. Moreover isolation of phytate globoids using a mix between cottonseed oil and carbon tetrachloride showed they are almost equally composed of phytic acid and protein and also contains minerals with K and Mg as the most abundant (Bohn et al., 2007). Small lipid bodies (100 nm), rich in triglycerides, also called oleosomes, were clearly found around the aleurone grains at least from type I (Regvar et al., 2011; Bethke et al., 1998; Morrison et al., 1975). Phytic acid is considered to have anti-nutritional properties in reducing mineral bioavailability and is preferentially removed or reduced in food products. But, its recovery may be of interest for phosphorus (P) production to ensure sustainability of this non-renewable resource (Shen et al.,

2019). Phytic acid is also considered as an interesting antioxidant with anti-carcinogenic activities (Silva and Bracarense, 2016; Schlemmer et al., 2009). Its amount in the aleurone layer can be estimated to around 8–15% (Table 2) taking into account the purity of the fraction (Hemery et al., 2009b; Antoine et al., 2004c; Buri et al., 2004). Concomitantly, ash and lipid content can be estimated to be around 12–19% and 7% (Table 2), respectively (Pomeranz, 1988; Buri et al., 2004).

Type II inclusions are electron-dense structures (Fulcher et al., 1981) and were found to contain niacin (vitamin B3) and also appeared associated with proteins (Morrison et al., 1975). Niacin was confirmed to be concentrated in the aleurone cell content which can be released using ultrasonic washing (Ndolo et al., 2015). Depending on authors, the niacin amount in the wheat aleurone (Table 2) varied between 170 and 814 µg/g (Ndolo et al., 2015; Buri et al., 2004; Pomeranz, 1988) and therefore significantly contributes to the recommended daily uptake. However, no real study was undertaken to evaluate its real bio-accessibility from different aleurone preparation. Indeed the bio-accessible niacin amount can be different than the total niacin content (Chamlagain et al., 2020). Zaupa et al. (2014) studying in vitro digestion of debranning fractions from durum wheat estimated niacin bio-accessibility to 55% from a micronized debranning fraction enriched in aleurone cell content whereas the concomitant thiamine content was rather fully bio-accessible.

Taking into account their presence as small micrometric inclusions in the aleurone layer, one can expect the possible purification of aleurone grains at a micro-scale size. Astonishingly, neither attempt to isolate them

Table 2 Aleurone layer composition in % as estimated from Ndolo et al. (2015); Corol et al. (2012); Nurmi et al. (2012); Hemery et al. (2009b); Antoine et al. (2004c); Buri et al. (2004); Fenech et al. (1999); Pomeranz (1988); Stevens (1970). Concentration in percent of each compounds in only the aleurone cellular content may be approximate by multiplying the value in the Table by a 2 fold factor taking into account the proportion of cell walls in the aleurone tissue and assuming that there are no further effect of the material structure on the compound extraction.

Protein	≥ 16–21
Ash	12–19
Phytic acid	8–15
Free sugars	10
Lipids	7
Betaine	1.5
Choline	0.2
Sterols	0.26
Niacin	0.02–0.08
Lignans	0.007–0.009
Folates	0.0002–0.0005

using dry fractionation appeared in the literature. This is possibly due to the presence of lipid bodies that probably play a role in particle agglomeration during size reduction, and which need to be removed to ease separation in dry conditions. Nevertheless, the combination between a teeth roller mill and ultrasonic washing developed by Guo et al. (2015) should probably be a step to recover the aleurone cell content from the washing water. Digestion with cell-wall hydrolytic enzymes (xylanases and feruloyl esterases as in Rosa et al., 2013, with possible addition of β-glucanases) should also help the release of aleurone cellular content. However, using only dry fractionation, Antoine et al. (2004c) was able to obtain a fraction enriched with the aleurone cell content even if part of the starchy endosperm and of the closed surrounding tissue remained.

As the major part of the protein in the aleurone was soluble, the release of the cell content helps their recovery. The protein content in the aleurone layer was estimated at least around 16–21% (Table 2) taking into account the moisture content or the purity of the fraction from published data (Buri et al., 2004; Pomeranz, 1988) with only a negligible amount found in cell walls, that can be evaluated to 2–3% considering both the amount of protein (1%) present in purified cell walls, as evaluated by Rhodes and Stone (2002), and the proportion of cell walls in the tissue (around 40–48% of the tissue mass, Hemery et al., 2009b). They are found to contain a higher amount of lysine and thus appear complementary to the grain storage proteins that are deficient in this amino-acid (Buri et al., 2004). Considering the aleurone physiological role these soluble proteins are mainly enzymes involved in the germination process but their abundance will certainly depend on the hormonal balance between abscisic acid and gibberellin and reactive oxygen species level (Zheng and Wang, 2014) and also the grain environment (climate and microflora). Few studies have tried to further characterize the nature of the different proteins using proteome analysis (Meziani et al., 2012; Jerkovic et al., 2010; Laubin et al., 2008). Around 600–700 different proteins were found by 2-D electrophoresis from which half can be identified in the database in 2010 (Jerkovic et al., 2010), whereas Meziani et al. (2012) separated more than 1000 different proteins. If the main proportion of these proteins was attributed to the globulin-like storage proteins from the cupin superfamily, the others are involved in the metabolic activities such as the carbohydrate metabolism, protein synthesis, protein folding, transport mechanism, and so on or involved in the plant defence against pathogens (Meziani et al., 2012; Jerkovic et al., 2010). Nevertheless, real knowledge of the aleurone layer content in this case can be biased by the necessary grain soaking and endosperm softening to be able to isolate the corresponding tissue by hand-microdissection.

Aleurone flour was also found to be a rich source of folate with amount varying from 1.7 to 5.5 µg/g (Table 2) depending on the processing method to produce the enriched fraction (Buri et al., 2004; Fenech et al., 1999). Nevertheless, the fraction from Fenech and co-authors (1999) may also contain part of the germ explaining its higher content. Interestingly, all of the contained folate appears bioavailable following human consumption (Fenech et al., 1999).

Graham et al. (2009) quantified and demonstrated the aleurone location of betaine and choline, two important lipotrope compounds that may prevent fat deposits when consumed (Fardet et al., 2012). Respective concentrations of betaine and choline in the aleurone fraction, isolated by using the patented Bühler process, were at 15.5 and 2.1 mg/g respectively (Table 2), but may vary depending on the wheat cultivar and environmental conditions (Corol et al., 2012). Concentration of these fractions is twice as high as the corresponding bran fraction, which is consistent with their location in the aleurone tissue which accounts for half the mass of the bran fraction (Barron et al., 2011). Moreover due to its higher amount in whole durum wheat or einkorn grains (Corol et al., 2012), one can expect a higher amount of these compounds in corresponding aleurone fractions. Incorporation of the Bühler aleurone fraction in bread was also found efficient to increase the human plasma betaine concentration after consumption (Keaveney et al., 2015).

Bread enrichment with an aleurone-enriched flour was also found to increase the lignan and corresponding enterolignan content in plasma and urine of fed pigs (Bolvig et al., 2016) proving its presence and absorption by animals. Buri et al. (2004) estimated the total lignan content to be around 7–9 mg/100 g of pure aleurone (Table 2) taking into account the percentage of this tissue in the fraction further determined in Hemery et al. (2009b). Lignans are diphenolic compounds which can be converted by the gut microflora into enterolignans displaying phyto-oestrogene activities with potential beneficial effects on health (Landete, 2012; Adlercreutz, 2002). However, more work is needed to better define their effect depending on the dose, the involved mechanisms and potential conjunction of effects with other molecules. Localization in the aleurone tissue is unknown.

The aleurone layer also contributes to the sterol content of the grain even if it was not found exclusively in this grain tissue as demonstrated in Nurmi et al. (2012). Total sterol content was estimated to be 0.26% of the aleurone layer (Table 2) taking into account the fraction purity and sitosterol was found as the major compound (40% of the total). Around 15% of the sterol content were found bound to FA (Nurmi et al., 2012) leading to steryl-ferulates. Free sugars (around 10%) as sucrose, raffinose and probably stachyose were also present in the aleurone cell content (Table 2) as determined early by Stevens (1970) and further confirmed by Buri et al. (2004). Moreover vitamin E activity was also found in the aleurone-enriched fraction described in Buri et al. (2004) and was probably due to tocotrienols that are identified in wheat bran even if germ was found to contain the highest activity (Morrison et al., 1982). These E-vitamers may be associated with lipid bodies in the aleurone intracellular content as already shown in oat grains (White and Gray, 2006).

Coloured wheat grains were also found to store natural pigments which can be valuably exploited for their sensorial or health properties (Saini et al., 2020). Indeed, anthocyanins, a specific class of flavonoid with antioxidant properties, were found, quantified and characterized in blue aleurone from coloured grains (Tosi et al., 2020; Böhmdorfer et al., 2018; Krüger and Morlock, 2018; Jaafar et al., 2013). But, only the amount of anthocyanins in bran was reported between 0.005 and 0.129% (Tosi et al., 2020; Böhmdorfer et al., 2018). They are present as glycosylated forms in grains that were tentatively determined in Krüger and Morlock (2018). Carotenoids were also found in the aleurone even if found in higher quantity in the germ part (Ndolo and Beta, 2013). The aleurone layer of wheat was shown to have significant content of lutein and zeaxanthin, which display antioxidant properties. Their suggested role was to protect the aleurone lipids from peroxidation.

6 Conclusion and future trends

Aleurone isolation by dry fractionation methods had made progress in recent years and avoids both the loss of components in the liquid fraction and the production of effluents in comparison with wet processing. It also reduced the modification of molecules during fractionation particularly if cryogrinding is used. Depending on the processing steps, the amount of residual tissues close to the aleurone layer was different and either the aleurone layer was recovered as intact cells or the cellular content was separated from the cell walls. Monitoring of new processing methods was aided by the identification in the generated fractions of specific biochemical markers of either the aleurone cellular content or the cell walls as notably characterized by Hemery et al. (2009b). As biochemical measurements need time and specific skills and sufficient quantity of material, spectral methods were also developed to identify and monitor the aleurone layer from the other grain tissues and may be promising in the future for rapidly monitoring its isolation (Wold et al., 2017; Corcel et al., 2016; Barron, 2011).

Aleurone isolation or enrichment is of interest not only to enhance its constituting compounds in food products but also to potentially further isolate AXOS with pre-biotic effect and phenolic compounds with antioxidant activities from its cell walls and proteins, minerals, lipids, lipotropes, sterols and vitamins and antioxidants from its cell content. Moreover one can expect further separation between these components in the near future with the progress of both dry fractionation at a micro-scale level and the use of new green solvents such as the eutectic ones (Cherif et al., 2020; Jablonský et al., 2018). However, the choice of the strategy to exploit a specific aleurone part or a specific compound has to take into account the possible co-purification of a non-desirable molecule, the possible use of the remaining residue and the processing cost. One can also keep in mind the potential interest to isolate a complex of molecules rather than a specific molecule to keep the natural environment or help synergistic effect as claimed in Fardet (2010).

Interestingly, as the aleurone is an alive tissue, it may be possible to modulate its composition depending on the physiological state (as shown, for example, for protease, phytase and xylanase activities along germination, Lemmens et al., 2019; Diaz-Mendoza et al., 2019) and/or using grain post-treatment (e.g. hydro-thermal treatment to lower phytate content, Lemmens et al., 2018; Fredlund et al., 1997) enlarging the possibilities for aleurone exploitation. Finally, the natural or transgenic variability can be explored or constructed to specifically enhance the amount of particular compounds to be obtained.

7 Where to look for further information

Further reading:

- General reviews on the aleurone layer can be found in: Antoine, C. et al. (2002) *Sci. des Aliments* 22, 545–556; Brouns, F. et al. (2011), see reference list.
- A review about the relationship between the aleurone structure and properties can be found in: Rosa-Sibakov et al. (2015), *Trends Food Sci. Tech.*, 41: 118–134.
- A review on dry processing methods to notably enrich products with compounds from the aleurone layer can be found in: Hemery et al. (2007), *J. Cereal Sci.*, 46: 327–347.
- The aleurone fraction from Cargill 'Grainwise aleurone' is described here: https://www.yumpu.com/en/document/view/33458297/grainwise-wheat-aleurone-cargill-foods.

Major international research projects to enrich cereal foods with the aleurone layer or content to improve nutritional and health properties: HealthGrain Project (https://healthgrain.org/).

8 References

Adlercreutz, H. (2002). Phyto-oestrogens and cancer, *Lancet Oncol.* 3(6): 364–373.

Amrein, T. M., Granicher, P., Arrigoni, E. and Amado, R. (2003). In vitro digestibility and colonic fermentability of aleurone isolated from wheat bran, *Lebensm. Wiss. Technol.* 36(4): 451–460.

Anson, N. M., van den Berg, R., Havenaar, R., Bast, A. and Haenen, G. R. (2008). Ferulic acid from aleurone determines the antioxidant potency of wheat grain (Triticum aestivum L.), *J. Agric. Food Chem.* 56(14): 5589–5594.

Antoine, C., Castellon, J., Toureille, A., Rouau, X. and Dissado, L. (2004a). Comparison of constitutive wheat bran tissues by dielectric spectroscopy and effect of their moisture content, *Polym. Int.* 53(12): 2169–2177.

Antoine, C., Lullien-Pellerin, V., Abecassis, J. and Rouau, X. (2002). Nutritional interest of the wheat seed aleurone layer, *Sci. Aliments* 22(5): 545–556.

Antoine, C., Lullien-Pellerin, V., Abecassis, J. and Rouau, X. (2004b). Effect of wheat bran ball-milling on fragmentation and marker extractability of the aleurone layer, *J. Cereal Sci.* 40(3): 275–282.

Antoine, C., Peyron, S., Lullien-Pellerin, V., Abecassis, J. and Rouau, X. (2004c). Wheat bran tissue fractionation using biochemical markers, *J. Cereal Sci.* 39(3): 387–393.

Antoine, C., Peyron, S., Mabille, F., Lapierre, C., Bouchet, B., Abecassis, J. and Rouau, X. (2003). Individual contribution of grain outer layers and their cell wall structure to the mechanical properties of wheat bran, *J. Agric. Food Chem.* 51(7): 2026–2033.

Atwell, B., Fulcher, G., Lullien-Pellerin, V., Marquart, L., Meloan, E., Slade, L., Slavin, J., Von Reding, W., Wellnitz, J. and Kohn, S. (2010). New aleurone definition approval, *Cereal Foods World* 55: 242.

Bacic, A. and Stone, B. A. (1981). Chemistry and organization of aleurone cell wall components from wheat and barley, *Functional Plant Biol.* 8(5): 475–495.

Bagdi, A., Szabo, F., Gere, A., Kokai, Z., Sipos, L. and Tomoskozi, S. (2014). Effect of aleurone-rich flour on composition, cooking, textural, and sensory properties of pasta, *LWT Food Sci. Technol.* 59(2): 996–1002.

Bagdi, A., Toth, B., Lorincz, R., Szendi, Sz., Gere, A., Kokai, Z., Sipos, L.. and Tomoskozi, S. (2016). Effect of aleurone-rich flour on composition, baking, textural, and sensory properties of bread, *LWT Food Sci. Technol.* 65: 762–769.

Barron, C. (2011). Prediction of relative tissue proportions in wheat mill streams by Fourier transform mid-infrared spectroscopy, *J. Agric. Food Chem.* 59(19): 10442–10447.

Barron, C., Samson, M. F., Lullien-Pellerin, V. and Rouau, X. (2011). Wheat grain tissue proportions in milling fractions using biochemical marker measurements: application to different wheat cultivars, *J. Cereal. Sci.* 53(3): 306–311.

Barron, C., Surget, A. and Rouau, X. (2007). Relative amounts of tissue in mature wheat (Triticum aestivum L.) grain and their carbohydrate and phenolic acid composition, *J. Cereal Sci.* 45(1): 88–96.

Becraft, P. W. and Yi, G. (2011). Regulation of aleurone development in cereal grains, *J. Exp. Bot.* 62(5): 1669–1675.

Behrens, R. and Bohm, A. (2004). Method for the electric contact charging of a mixture containing aleurone and shell particles, International Patent WO2004016354, Applicant, Bühler A.G.

Bethke, P. C., Swanson, S. J., Hillmer, S. and Jones, R. L. (1998). From storage compartment to lytic organelle: the metamorphosis of the aleurone protein storage vacuole, *Ann. Bot.* 82(4): 399–412.

Blandino, M., Sovrani, V., Marinaccio, F., Reyneri, A., Rolle, L., Giacosa, S., Locatelli, M., Bordiga, M., Travaglia, F., Coïsson, J. D. and Arlorio, M. (2013). Nutritional and technological quality of bread enriched with an intermediate pearled wheat fraction, *Food Chem.* 141(3): 2549–2557.

Bohm, A., Bogoni, C., Behrens, R. and Otto, T. (2003). Method for the extraction of aleurone from bran, US Patent Application Publication US 2003/0175384 A1, Applicant, Bühler A.G.

Bohm, A. and Buri, R. (2004). Composition comprising aleurone, method of administering the same, and method of manufacturing the same, US Patent Application 2004/0258776 A1, Applicant, Bühler A.G.

Bohm, A. and Kratzer, A. (2008). Method for isolating aleurone particles, United States Patent, US 2008/0257787 A1, Applicant, Bühler A.G.

Böhmdorfer, S., Oberlerchner, J. T., Fuchs, C., Rosenau, T.and Grausgruber, H. (2018). Profiling and quantification of grain anthocyanins in purple pericarp × blue aleurone wheat crosses by high-performance thinlayer chromatography and densitometry, *Plant Methods* 14: 29.

Bohn, L., Josefsen, L., Meyer, A. S. and Rasmussen, S. K. (2007). Quantitative analysis of phytate globoids isolated from wheat bran and characterization of their sequential dephosphorylation by wheat phytase, *J. Agric. Food Chem.* 55(18): 7547–7552.

Bolvig, A. K., Adlercreutz, H., Theil, P. K., Jørgensen, H. and Bach Knudsen, K. E. (2016). Absorption of plant lignans from cereals in an experimental pig model, *Brit. J. Nutr.* 115(10): 1711–1720.

Brouns, F., Hemery, Y., Price, R. and Anson, N. M. (2011). Wheat aleurone: separation, composition, health aspects and potential food use, *Crit. Rev. Food Sci. Nutr.* 52(6): 553–568.

Bucsella, B., Molnar, D., Harasztos, A. H. and Tomoskozi, S. (2016). Comparison of the rheological and end-product properties of an aleurone-rich flour, whole grain wheat and rye flour, *J. Cereal Sci.* 69: 40–48.

Buri, R. C., von Reding, W. and Gavin, M. H. (2004). Description and characterization of wheat aleurone, *Cereal Foods World* 49: 274–282.

Chamlagain, B., Rautio, S., Edelmann, M., Ollilainen, V. and Piironen, V. (2020). Niacin contents of cereal-milling products in food-composition databases need to be updated, *J. Food Compos. Anal.* 91: 103518.

Chen, Y., Dunford, N. T. and Goad, C. (2013b). Evaluation of wheat bran obtained by tangential abrasive dehulling device, *Food Bioprocess Tech.* 6(7): 1655–1663.

Chen, Z., Liu, F., Wang, L., Li, Y., Wang, R. and Chen, Z. (2014). Tribocharging properties of wheat bran fragments in air-solid pipe flow, *Food Res. Int.* 62: 262–271.

Chen, Z., Xia, Q., Zha, B., Sun, J., Xu, B. and Chen, Z. (2020). Triboelectric separation of wheat bran tisues: influence of tribo-material, water content, and particle size, *J. Food Process Eng.* 43(1): e13346. doi:10.1111/jfpe.13346.

Chen, Z., Zha, B., Wang, L., Wang, R., Chen, Z.. and Tian, Y. (2013a). Dissociation of aleurone cell cluster from wheat bran by centrifugal impact milling, *Food Res. Int.* 54(1): 63–71.

Cherif, M. M., Grigorakis, S., Halahlah, A., Loipassaki, S. and Makris, D. P. (2020). High-efficiency extraction of phenolics from wheat waste biomass (bran) by combining deep eutectic solvent, ultrasound-assisted pretreatment and thermal treatment, *Environ. Process.* 7(3): 845–859.

Chulze, S. N., Palazzini, J. M., Lullien-Pellerin, V., Ramirez, M. L., Cuniberti, M. and Magan, N. (2020). Fusarium species infection in wheat: impact on quality and mycotoxin accumulation. In: Igrejas, G., Ikeda, T. and Guzmán, C. (Eds), *Wheat Quality for Improving Processing and Human Health Wheat Quality*, Springer, Cham, pp. 421–445.

Cimbalo, A., Alonso-Garrido, M., Font, G. and Manyes, L. (2020). Toxicity of mycotoxins *in vivo* on vertebrate organisms: a review, *Food Chem. Toxicol.* 137: 111161. doi:10.1016/j.fct.2020.111161.

Corcel, M., Devaux, M.-F., Guillon, F. and Barron, C. (2016). Comparison of UV and visible autofluorescence of wheat grain tissues in macroscopic images of cross-sections and particles, *Comput. Electron. Agr.* 127: 281-288.

Corol, D. I., Ravel, C., Raksegi, M., Bedo, Z., Charmet, G., Beale, M. H., Shewry, P. R. and Ward, J. L. (2012). Effects of genotype and environment on the contents of betaine, choline, and trigonelline in cereal grains, *J. Agric. Food Chem.* 60(21): 5471-5481.

Diaz-Mendoza, M., Diaz, I. and Martinez, M. (2019). Insights on the proteases involved in barley and wheat grain germination, *Int. J. Mol. Sci.* 20(9): 2087. doi:10.3390/ijms20092087.

Delcour, J. A., Rouau, X., Courtin, C. M., Poutanen, K. and Ranieri, R. (2012). Technologies for enhanced exploitation of the health promoting potential of cereals, *Trends Food Sci. Tech.* 25(2): 78-86.

Evers, A. D., O'Brien, L. and Blakeney, A. B. (1999). Cereal structure and composition, *Aust. J. Agric. Res.* 50(5): 629-650.

Fardet, A. (2010). New hypotheses for the health-protective mechanisms of whole-grain cereals: what is beyond fibre?, *Nutrit. Res. Rev.* 23(1): 65-134.

Fardet, A., Martin, J.-F., Touazi, D., Caillavet, F. and Chardigny, J.-M. (2012). Lipotropes from plant-based foods supplied by a standard French diet vs. food guide pyramid recommendations: grain products are the best sources at lower cost, J. Food Comp. Anal. 28(2): 135-148.

Fenech, M., Noakes, M., Clifton, P. and Topping, D. (1999). Aleurone flour is a rich source of bioavailable folate in humans, *J. Nutr.* 129(6): 1114-1119.

Fredlund, K., Asp, N. -G., Larsson, M., Marklinder, I. and Sandberg, A. -S. (1997). Phytate reduction in whole grains of wheat, rye, barley and oats after hydrothermal treatment, *J. Cereal Sci.* 25(1): 83-91.

Fulcher, R. G., O'Brien, T. P. and Wong, S. I. (1981). Microchemical detection of niacin, aromatic amine and Phytin reserves in cereal bran, *Cereal Chem.* 58: 130-135.

Giroux, M. J., Sripo, T., Gerhardt, S. and Sherwood, J. (2003). Puroindolines: their role in grain hardness and plant defence, *Biotech. Gene. Engin. Rev.* 20: 277-290.

Graham, S. F., Hollis, J. H., Migaud, M. and Browne, R. A. (2009). Analysis of betaine and choline content of aleurone, Bran and flour fractions of wheat (Triticum aestivum L.) using 1H nuclear magnetic resonance (NMR) spectroscopy, *J. Agric. Food Chem.* 57(5): 1948-1951.

Greffeuille, V., Abecassis, J., L'Helgouac'h, B. C. and Lullien-Pellerin, V. (2005). Differences in aleurone layer fate between hard and soft common wheats at grain milling, *Cereal Chem.* 82(2): 138-143.

Greffeuille, V., Mabille, F., Rousset, M., Oury, F.-X., Abecassis, J. and Lullien-Pellerin, V. (2007). Mechanical properties of outer layers from near-isogenic lines of common wheat differing in hardness, *J. Cereal Sci.* 45(2): 227-235.

Grootaert, C., Delcour, J. A., Courtin, C. M., Broekaert, W. F., Verstraete, W. and Van de Wiele, T. (2007). Microbial metabolism and prebiotic potency of arabinoxylan oligosaccharides in the human intestine, *Trends Food Sci. Tech.* 18(2): 64-71.

Guo, J., Bian, Y.-Y., Zhu, K.-X., Guo, X.-N., Peng, W. and Zhou, H.-M. (2015). Reducing phytate content in wheat bran by directly removing the aleurone cell content with teeth roller mill and ultrasonic cleaner, *J. Cereal Sci.* 64: 133-138.

Harris, P. J., Chavan, R. R. and Ferguson, L. R. (2005). Production and characterisation of two wheat-bran fractions: an aleurone-rich and a pericarp-rich fraction, *Mol. Nutr. Food Res.* 49(6): 536–545.

Heinze, K., Kiszonas, A. M., Murray, J. C., Morris, C. F. and Lullien-Pellerin, V. (2016). Puroindoline genes introduced into durum wheat reduce milling energy and change milling behavior similar to soft common wheats, *J. Cereal Sci.* 71: 183–189.

Hemery, Y., Chaurand, M., Holopainen, U., Lampi, A.-M., Lehtinen, P., Piironen, V., Sadoudi, A. and Rouau, X. (2011a). Potential of dry fractionation of wheat bran for the development of food ingredients, part I: influence of ultra-fine grinding, *J. Cereal Sci.* 53(1): 1–8.

Hemery, Y., Holopainen, U., Lampi, A.-M., Lehtinen, P., Nurmi, T., Piironen, V., Edelmann, M. and Rouau, X. (2011b). Potential of dry fractionation of wheat bran for the development of food ingredients, part II: electrostatic separation of particles, *J. Cereal Sci.* 53(1): 9–18.

Hemery, Y., Lullien-Pellerin, V., Rouau, X., Abecassis, J., Samson, M.-F., Aman, P., von Reding, W., Spoerndli, C. and Barron, C. (2009b). Biochemical markers: efficient tools for the assessment of wheat grain tissue proportions in milling fractions, *J. Cereal Sci.* 49(1): 55–64.

Hemery, Y. M., Mabille, F., Martelli, M. R. and Rouau, X. (2010). Influence of water content and negative temperatures on the mechanical properties of wheat bran and its constitutive layers, *J. Food Eng.* 98(3): 360–369.

Hemery, Y., Rouau, X., Dragan, C., Bilici, M., Beleca, R. and Dascalescu, L. (2009a). Electrostatic properties of wheat bran and its constitutive layers: influence of particle size, composition, and moisture content, *J. Food Eng.* 93(1): 114–124.

Jaafar, S. N. S., Baron, J., Siebenhandl-Ehn, S., Rosenau, T., Boehmdorfer, S. and Grausgruber, H. (2013). Increased anthocyanin content in purple pericarp x blue aleurone wheat crosses, *Plant Breed.* 132(6): 546–552.

Jablonský, M., Škulcová, A., Malvis, A. and Šima, J. (2018). Extraction of value added components from food industry based and agro-forest biowastes by deep eutectic solvents, *J. Biotechnol.* 282: 46–66.

Jerkovic, A., Kriegel, A. M., Bradner, J. R., Atwell, B. J., Roberts, T. H. and Willows, R. D. (2010). Strategic distribution of protective proteins within bran layers of wheat protects the nutrient-rich endosperm, *Plant Physiol.* 152(3): 1459–1470.

Keaveney, E. M., Price, R. K., Hamill, L. L., Wallace, J. M. W., McNulty, H., Ward, M., Strain, J. J., Ueland, P. M., Molloy, A. M., Piironen, V., von Reding, W., Shewry, P. R., Ward, J. L. and Welch, R. W. (2015). Postprandial plasma betaine and other methyl donor-related responses after consumption of minimally processed wheat bran or wheat aleurone, or wheat aleurone incorporated into bread, *Brit. J. Nutr.* 113(3): 445–453.

Krüger, S. and Morlock, G. E. (2018). Fingerprinting and characterization of anthocyanins in 94 colored wheat varieties and blue aleurone and purple pericarp wheat crosses, *J. Chromatogr. A* 1538: 75–85.

Kvist, S., Carlsson, T., Lawther, J. M. and DeCastro, F. B. (2010). Process for the fractionation of cereal brans, US Patent Application Publication US 2010/7709033 B2, Applicant, Biovelop International B.V.

Laca, A., Mousia, Z., Diaz, M.,Webb, C. and Pandiella, S. S. (2006). Distribution of microbial contamination within cereal grains, *J. Food Eng.* 72(4): 332–338.

Landete, J. M. (2012). Plant and mammalian lignans: a review of source, intake, metabolism, intestinal bacteria and health, *Food Res. Int.* 46(1): 410–424.

Laubin, B., Lullien-Pellerin, V., Nadaud, I., Gaillard-Martinie, B., Chambon, C.. and Branlard, G. (2008). Isolation of the wheat aleurone layer for 2D electrophoresis and proteomics analysis, *J. Cereal Sci.* 48(3): 709-714.

Laux, J., von Reding,W. and Widmer, G. (2006). Aleurone product and corresponding production method, US Patent Application Publication US 2006/0177529 A1, Applicant, Bühler A.G.

Lemmens, E., De Brier, N., Goos, P., Smolders, E. and Delcour, J. A. (2019). Steeping and germination of wheat (Triticum aestivum L.). I. Unlocking the impact of phytate and cell wall hydrolysis on bio-accessibility of iron and zinc elements', *J. Cereal Sci.* 90: 102847.

Lemmens, E., De Brier, N., Spiers, K. M., Ryan, C., Garrevoet, J., Falkenberg, G., Goos, P., Smolders, E. and Delcour, J. A. (2018). The impact of steeping, germination and hydrothermal processing of wheat (Triticum aestivum L.) grains on phytate hydrolysis and the distribution, speciation and bio-accessibility of iron and zinc elements, *Food Chem.* 264: 367-376.

Li, X., Hu, H., Xu, F., Liu, Z., Zhang, L. and Zhang, H. (2019). Effects of aleurone-rich fraction on the hydration and rheological properties attributes of wheat dough, *Int. J. Food Sci. Technol.* 54(5): 1777-1786.

Lullien-Pellerin, V., Haraszi, R., Anderssen, R. S. and Morris, C. F. (2020). Understanding the mechanics of wheat grain fractionation and the impact of puroindolines on milling and product quality. In: Igrejas, G., Ikeda, T. and Guzmán, C. (Eds), *Wheat Quality for Improving Processing and Human Health Wheat Quality*, Springer, Cham, pp. 369-385.

Malunga, L. N. and Beta, T. (2015). Antioxidant capacity of arabinoxylan oligosaccharide fractions prepared from wheat aleurone using Trichoderma viride or neocallimastix patriciarum xylanase, *Food Chem.* 167: 311-319.

Martelli, M. R., Barron, C., Mabille, F., Rouau, X. and Sadoudi, A. (2010). Adherence within biological multilayered systems: development and application of a peel test onwheat grain peripheral tissues, *J. Cereal Sci.* 52(1): 83-89.

Mayer-Laigle, C., Rajaonarivony, R. K., Blanc, N. and Rouau, X. (2018). Comminution of dry lignocellulosic biomass: part II. Technologies, improvement of milling performances, and security issues, *Bioengineering* 5(3): 50. doi:10.3390/bioengineering5030050.

McIntosh, G. H., Royle, P. J. and Pointing, G. (2001). Wheat aleurone flour increases cecal beta-glucuronidase activity and butyrate concentration and reduces colon adenoma burden in azoxymethane-treated rats, *J. Nutr.* 131(1): 127-131.

Meziani, S., Nadaud, I., Gaillard-Martinie, B., Chambon, C., Benali, M. and Branlard, G. (2012). Proteomic analysis of the mature kernel aleurone layer in common and durum wheat, *J. Cereal Sci.* 55(3): 323-330.

Moore, K. L., Zhao, F. J., Gritsch, C. S., Tosi, P., Hawkesford, M. J., McGrath, S. P., Shewry, P. R. and Grovenor, C. R. M. (2012). Localisation of iron in wheat grain using high resolution secondary ion mass spectrometry, *J. Cereal Sci.* 55(2): 183-187.

Morris, C. F., Simeone, M. C., King, G. E. and Lafiandra, D. (2011). Transfer of soft kernel texture from Triticum aestivum to durum wheat, Triticum turgidum ssp. Durum, *Crop Sci.* 51(1): 114-122.

Morrison, W. R., Coventry, A. M. and Barnes, P. J. (1982). The distribution of acyl lipids and tocopherols in flours millstreams, *J. Sci. Food Agric.* 33(9): 925-933.

Morrison, I. N., Kuo, J. and O'Brien, T. P. (1975). Histochemistry and fine structure of developing wheat aleurone cells, *Planta* 123(2): 105-116.

Murray, J. C., Kiszonas, A. M., Wilson, J. and Morris, C. F. (2016). Effect of soft kernel texture on the milling properties of soft durum wheat, *Cereal Chem.* 93(5): 513–517.

Ndolo, V. U. and Beta, T. (2013). Distribution of carotenoids in endosperm, germ, and aleurone fractions of cereal grain kernels' ,*Food Chem.* 139(1-4): 663-671.

Ndolo, V. U., Fulcher, R. G. and Beta, T. (2015). Application of LC-MS-MS to identify niacin in aleurone layers of yellow corn, barley and wheat kernels, *J. Cereal Sci.* 65: 88–95.

Noort, M. W. J., van Haaster, D., Hemery, Y., Schols, H. A. and Hamer, R. J. (2010). The effect of particle size of wheat bran fractions on bread quality–evidence for fibre-protein interactions, *J. Cereal Sci.* 52(1): 59–64.

Nurmi, T., Lampi, A. M., Nyström, L., Hemery, Y., Rouau, X. and Piironen, V. (2012). Distribution and composition of phytosterols and steryl ferulates in wheat grain and bran fractions, *J. Cereal Sci.* 56(2): 379–388.

O'Dell, B. L., de Boland, A. R. and Koirtyohann, S. R. (1972). Distribution of phytate and nutritional important elements among the morphological components of cereal grains, *J. Agr. Food Chem.* 20: 718–723.

Pape, W. T. and Galindo, R. M. G. (2014). Recovery of aleurone-rich flour from bran, Applied Milling Systems Inc., Houston, TX (US), SuperBrix Internacional S.A., Barranquilla (CO) United States Patent, US 2014/8783588 B2.

Parker, M. L., Ng, A. and Waldron, K. W. (2005). The phenolic acid and polysaccharide composition of cell walls of bran layers of mature wheat (Triticum aestivum L. cv. Avalon) grains, *J. Sci. Food Agric.* 85(15): 2539–2547.

Peyron, S., Mabille, F., Devaux, M. F. and Autran, J. C. (2002). Influence of structural characteristics of aleurone layer on milling behavior of durum wheat (Triticum durum Desf.), *Cereal Chem.* 80: 62–67.

Pomeranz, Y. (1988). Chemical composition of kernel structures. In: Pomeranz, Y. (Ed), Wheat: Chemistry and Technology, Vol. 2, Am. Assoc. Cereal Chem, St Paul, MN, pp 97–100.

Price, R. K., Keaveney, E. M., Hamill, L. L., Wallace, J. M. W., Ward, M., Ueland, P. M., McNulty, H., Strain, J. J., Parker, M. J. and Welch, R. W. (2010). Consumption of wheat aleurone-rich foods increases fasting plasma betaine and modestly fasting homocysteine and LDL-cholesterol in adults, *J. Nutr.* 140(12): 2153–2157.

Price, R. K., Wallace, J. M. W., Hamill, L. L., Keaveney, E. M., Strain, J. J., Parker, M. J. and Welch, R. W. (2012). Evaluation of the effect of wheat aleurone-rich foods on markers of antioxidant status, inflammation and endothelial function in apparently healthy men and women, *Brit. J. Nutr.* 108(9): 1644–1651.

Regvar, M., Eichert, D., Kaulich, B., Gianoncelli, A., Pongrac, P., Vogel-Mikus, K. and Kreft, I. (2011). New insights into globoids of protein storage vacuoles in wheat aleurone using synchrotron soft X-ray microscopy, *J. Exp. Bot.* 62(11): 3929–3939.

Rhodes, D. I., Sadek, M. and Stone, B. A. (2002). Hydroxycinnamic acids in walls of wheat aleurone cells, *J. Cereal Sci.* 36(1): 67–81.

Rhodes, D. I. and Stone, B. A. (2002). Proteins in walls of wheat aleurone cells, *J. Cereal Sci.* 36(1): 83–101.

Rios, G., Pinson-Gadais, L., Abecassis, J ., Zakhia-Rozis, N. and Lullien-Pellerin, V. (2009). Assessment of dehulling efficiency to reduce deoxynivalenol and Fusarium level in durum wheat grains, *J. Cereal Sci.* 49(3): 387–392.

Rosa, N. N., Dufour, C., Lullien-Pellerin, V. and Micard, V. (2013). Exposure or release of ferulic acid from wheat aleurone: impact on its antioxidant capacity, *Food Chem.* 141(3): 2355–2362.

Saini, P., Kumar, N., Kumar, S., Mwaurah, P. W., Panghal, A., Attkan, A. K., Singh, V. K., Garg, M. K. and Singh, V. (2020). Bioactive compounds, nutritional benefits and food applications of colored wheat: a comprehensive review, *Crit. Rev. Food Sci.*: 1-14, doi:10.1080/10408398.2020.1793727.

Schlemmer, U., Frølich, W., Prieto, R. M. and Grases, F. (2009). Phytate in foods and significance for humans: food sources, intake, processing, bioavailability, protective role and analysis, *Mol. Nutr. Food Res.* 53(Suppl. 2): S330–S375.

Shen, J., Wang, L., Jiao, X., Meng, F., Zhang, L., Feng, G., Zhang, J., Yuan, L., Ma, L., Hou, Y., Zhang, T., Zhang, W., Li, G., Zhang, K. and Zhang, F. (2019). Innovations of phosphorus sustainability: implications for the whole chain, *Front. Agr. Sci. Eng.* 6(4): 321–331.

Silva, E. O. and Bracarense, A. P. (2016). Phytic acid: from antinutritional to multiple protection factor of organic systems, *J. Food Sci.* 81(6): R1357–R1362.

Singh, S. P., VogelMikuš, K., Vavpetic, P., Jeromel, L., Pelicon, P., Kumar, J. and Tuli, R. (2014).Spatial Xray fluorescence microimaging of minerals in grain tissues of wheat and related genotypes, *Planta* 240(2): 277–289.

Stevens, D. J. (1970). Free sugars of wheat aleurone cells, *J. Sci. Food Agric.* 21(1): 31–34.

Stevens, D. J. (1973). The aleurone layer of wheat. III. Isolation and scanning electron microscope examination of the layer and of its component parts, *J. Sci. Food Agric.* 24(3): 307–317.

Stone, B. A. (2006). Cell walls of cereal grains, *Cereal Foods World* 51(2): 62–65.

Stone, B. A. and Minifie, J. (1988). Recovery of aleurone cells from wheat bran, US patent, 4,746,073, Applicant, La Trobe University, Australia.

Surget, A. and Barron, C. (2005). Histologie du grain de blé, *Ind. Cer.* 145: 3-7.

Tosi, P., Hidalgo, A. and Lullien-Pellerin, V. (2020). 'The impact of processing on potentially beneficial wheat grain components for human health' quality. In: Igrejas, G., Ikeda, T. and Guzmán, C. (Eds), *Wheat Quality for Improving Processing and Human Health Wheat Quality*, Springer, Cham, pp. 387–420.

Van Craeyveld, V., Holopainen, U., Selinheimo, E., Poutanen, K., Delcour, J. A. and Courtin, C. M. (2009). Extensive dry ball milling of wheat and rye bran leads to *in situ* production of arabinoxylan oligosaccharides through nanoscale fragmentation, *J. Agric. Food Chem.* 57(18): 8467–8473.

Vangsøe, C. T., Sørensen, J. F. and Bach Knudsen, K. E. (2019). Aleurone cells are the primary contributor to arabinoxylan oligosaccharide production from wheat bran after treatment with cell wall-degrading enzymes, *Int. J. Food Sci. Tech.* 54(10): 2847–2853.

White, D. A. and Gray, F. D. A. (2006). Characterisation of oat (*Avena sativa* L.) oil bodies and intrinsically associated E-vitamers, *J. Cereal Sci.* 43(2): 244–249.

Wold, J. P., Airado-Rodriguez, D., Holtekjolen, A.-K., Holopainen, U. and Sahlstrom, S. (2017). Rapid and non-destructuve determination of aleurone content in pearling fractions of barley by NIR and fluorescence spectroscopy, *J. Agric. Food Chem.* 65: 1813–1821.

Wu, L. and Qi, J. (2013). Method for extracting content of wheat aleurone layer through pure physical dry process, Shandong Zhishifang Food Technology Co Ltd, CN Patent 103229933(A).

Xu, M., Hou, G. G., Ma, F., Ding, J., Deng, L., Kahraman, O., Niu, M., Trivettea, K., Lee, B., Wu, L. and Baik, B.-K. (2020). Evaluation of aleurone flour on dough, textural, and nutritional properties of instant fried noodles, *LWT Food Sci. Technol.* 126: 109294.

Zaupa, M., Scazzina, F., Dall'Asta, M., Calani, L., Del Rio, D., Bianchi, M. A., Melegari, C., De Albertis, P., Tribuzio, G., Pellegrini, N. and Brighenti, F. (2014). In vitro bioaccessibility of phenolics and vitamins from durum wheat aleurone fractions, *J. Agric. Food Chem.* 62(7): 1543–1549.

Zheng, Y. and Wang, Z. (2014). Differentiation mechanism and function of the cereal aleurone cells and hormone effects on them, *Plant Cell Rep.* 33(11): 1779–1787.

Chapter 8

Wheat flour fortification and human health

Helena Pachón, Food Fortification Initiative and Emory University, USA

1 Introduction

The objective of this chapter is to review evidence of the human health impact of wheat flour fortification.

1.1 What is wheat flour fortification?

Food fortification is the addition of nutrients to foods while they are being processed (WHO and FAO 2006). Also known as enrichment, food fortification is a unique intervention in health circles because it is delivered by the private sector in contrast to most public-health interventions that are implemented by the public sector. For wheat flour, fortification occurs in a mill after parts of the wheat kernel are ground to flour: the endosperm in refined white flour and the bran, germ and endosperm in whole-grain flour (Bauernfeind and DeRitter 1991). Small concentrations of nutrients, usually vitamins and minerals, are added to this flour in the mills. Wheat flour can be fortified in large, industrial-sized mills or in small, non-industrialized mills. This chapter will focus on

http://dx.doi.org/10.19103/AS.2021.0087.23

large-scale industrial fortification of wheat flour where most of the evidence and success with flour fortification is observed.

With a few exceptions, most nutrients can be added to wheat flour through fortification (WHO and FAO 2006). Nutrients in the outer layers of the wheat kernel that were removed during milling can be added to wheat flour through fortification. The B vitamins thiamin, riboflavin and niacin are examples of this (Bauernfeind and DeRitter 1991); adding these nutrients back to flour through fortification is known as restitution or restoration. Nutrients can be added back at the same, lower or higher levels than present in the kernel. Nutrients that are not naturally found in the wheat kernel can also be added to flour through fortification. Vitamin B12 is an example of this (USDA 2020).

Nutrients are usually added through fortification coupled with another ingredient(s). For example, iron can be added in the form of ferrous sulfate or sodium iron ethylenediaminetetraacetate (NaFeEDTA) (WHO 2009); these forms are known as fortificants or fortification compounds. Niacin can be added in the form of niacinamide, nicotinic acid or nicotinamide (WHO and FAO 2006). Because nutrients are added to flour to benefit human health, forms that are better absorbed by the human body (i.e. more bioavailable) are preferred. They may be costlier than less well-absorbed forms (e.g. NaFeEDTA compared with electrolytic iron); however, less needs to be added of the more bioavailable form to have a comparable health benefit (Hurrell et al. 2010).

There are different reasons why some nutrients or fortificants are not typically added to wheat flour through fortification. For bioavailability purposes, another food may be a better choice to add the nutrient to; this is the case with vitamin A. Vitamin A requires fat for absorption; oil, margarine and butter are better options for fortifying because they are lipid rich. Vitamin A can be added to flour; however, it is a costly ingredient because of the processing required to encapsulate the vitamin so it can be mixed into flour (WHO and FAO 2006). A fortificant may interfere with the technological processing or sensory characteristics of the food made with fortified flour. For example, ferrous sulfate can cause rancidity in high-fat foods (WHO and FAO 2006). For these foods, a less reactive iron compound may be used in flour fortification.

1.2 Why fortify wheat flour?

Many countries with mandatory fortification document their reason for fortifying wheat flour (Marks et al. 2018). It is to address a widespread health problem caused by a nutrient deficiency(ies) in the population, such as iron deficiency, anemia and neural tube defects.

The reason wheat flour is chosen to fortify is because it meets two basic criteria. First, food made with wheat flour (such as bread, noodles, pasta) is consumed by a large proportion of the population trying to be reached.

Published by Burleigh Dodds Science Publishing Limited, 2021.

Second, most of the world's wheat flour is produced in large-scale, industrial mills (FFI 2020). This is important because large-scale fortification is easier for industry to implement and for the government to monitor compared with small-scale fortification (WHO and FAO 2006).

2 Status of wheat flour fortification

The status of wheat flour fortification can be organized by three non-mutually exclusive categories: countries with foundational documents that establish a wheat flour fortification program; countries with documented performance of existing flour fortification programs; and countries without fortification programs which have the potential to benefit from wheat flour fortification. Country statistics on these three categories of flour fortification can be found at the Global Fortification Data Exchange website (FortificationData.org). This chapter will focus on a subset of countries with 'foundational documents' - those with legislation that mandates or allows voluntary fortification and those with standards for wheat flour fortification - and documented performance of their flour fortification programs.

2.1 Countries that mandate wheat flour fortification

As of February 10, 2020, 83 countries have legislation that effectively mandates the fortification of wheat flour (Global Fortification Data Exchange 2020a). This means the country has 'documentation [which] indicates that fortification of all or some of the food is compulsory or required'. These countries are shown in green in Fig. 1.

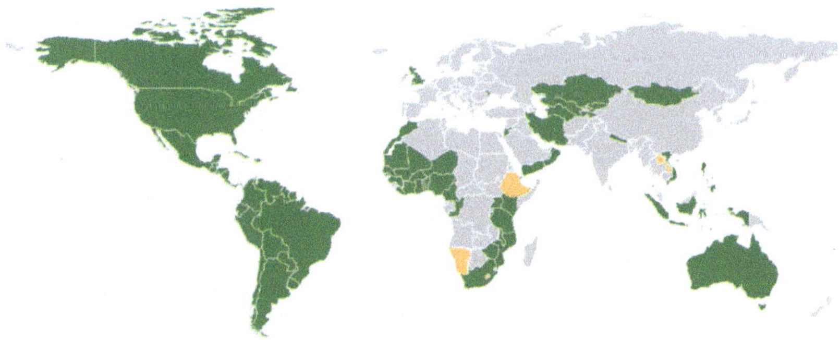

Figure 1 Countries in green are those with legislation that has the effect of mandating wheat flour fortification with one or more nutrients (Global Fortification Data Exchange 2020a). Countries in yellow are those confirmed to not have mandatory legislation of wheat flour. Countries in grey are those unlikely to have mandatory fortification of wheat flour; however, this information has not been confirmed by an in-country contact.

Published by Burleigh Dodds Science Publishing Limited, 2021.

Countries with mandatory wheat flour fortification share the following characteristics:

- Following the United Nations' 2018 regional designation (United Nations Statistics Division 2020), 25 are in the Africa Region, 35 are in the Americas region, 16 in the Asia region, three are in the Europe region and four in Oceania.
- Sixteen are low-income countries, 29 are lower-middle-income countries, 25 are upper-middle-income countries and 13 are high-income countries per the World Bank's 2017 designation (World Bank 2020).
- Sixteen countries also have mandatory fortification of maize flour (Global Fortification Data Exchange 2020a).

2.2 Countries that allow voluntary fortification of wheat flour

As of February 10, 2020, 14 countries have 'official documentation and/or a food standard that provides guidance or regulations for fortification but does not have the effect of mandating or requiring fortification', that is, voluntary fortification of wheat flour (Global Fortification Data Exchange 2020a). These countries are shown in blue in Fig. 2.

2.3 Countries with standards for wheat flour fortification

Standards are documents that 'indicate standardized fortification levels of the food vehicle in question with one or more nutrients' (Global Fortification Data Exchange 2020b). Among the 97 countries with mandatory or voluntary wheat flour fortification, the Global Fortification Data Exchange has standards for 91. Table 1 lists the nutrients that are included in these 91 standards (Global

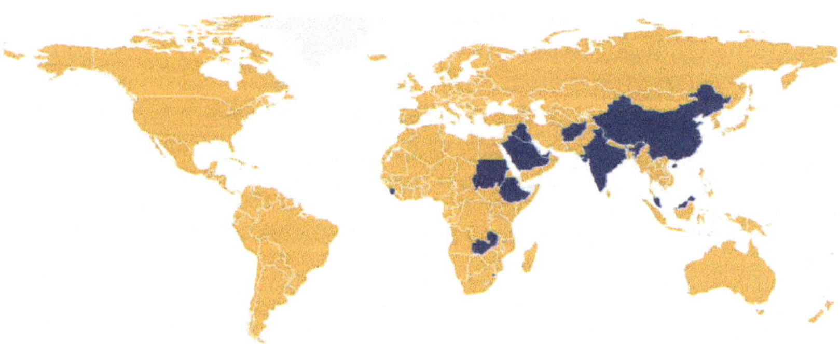

Figure 2 Countries in blue have voluntary wheat flour fortification with one or more nutrients (Global Fortification Data Exchange 2020a). Countries in yellow do not have voluntary fortification of wheat flour.

Table 1 Among 91 countries with mandatory or voluntary wheat flour fortification whose standards are available, the nutrients and amounts that are listed in fortification standards (Global Fortification Data Exchange 2020b)

Nutrient	Number of countries	Amount (mg/kg)
Iron	89	15-120
Folic acid (vitamin B9)	73	0.1-5.11
Thiamin (vitamin B1)	66	1.25-10
Riboflavin (vitamin B2)	64	1.3-6.6
Niacin (vitamin B3)	64	6.7-60
Zinc	31	12.5-101.3
Calcium	23	1.28-2,400
Vitamin A	20	0.62-10
Vitamin B12	20	0-0.04
Vitamin B6	14	2.0-6.5
Vitamin D	7	0.01-0.02
Selenium	1	0.21

Fortification Data Exchange 2020b). Overall, standards include up to four different minerals and eight different vitamins.

Most countries include iron, folic acid, thiamin, riboflavin and niacin in their standards for wheat flour fortification. Thirty one or fewer countries include zinc, calcium, vitamin A, vitamin B12, vitamin B6, vitamin D or selenium in their standards.

3 How the human health impact of wheat flour fortification is measured

The remainder of the chapter will center on health improvements observed from large-scale fortification of wheat flour, either alone or in combination with maize flour.

There are two types of studies to assess if food fortification has a health impact. The first are efficacy studies which tell us the 'extent to which [fortification] produces a beneficial result under ideal conditions' (Samet et al. 2008). Usually, efficacy is based on the results of randomized controlled trials. In comparison, effectiveness studies estimate the extent to which fortification produces a beneficial result 'when deployed in the field in the usual circumstances'. Evidence is needed from efficacy studies to ensure that fortification can have a beneficial impact. Additionally, effectiveness studies report if these benefits are observed when programs are implemented under real-life conditions.

This chapter will focus on effectiveness trials. This is because there is good evidence from efficacy trials that if people deficient in a nutrient consume a food fortified with that nutrient, their nutrient status will improve as will their health (e.g. Muthayya et al. 2012; Black et al. 2012). What is less known is how fortification operates under real-life conditions such as when it is offered through a government's social program or it is provided through the open market.

4 Examples of health outcomes associated with wheat flour fortification that have been studied

Generally speaking, when people consume a nutrient provided from any source (such as a non-fortified food, a fortified food, a supplement), they will experience an increase in their bodies' levels of that nutrient. This improvement in nutritional status can in turn improve functional outcomes: 'nutrient-dependent physiological functions' that can be measured (Solomons and Allen 1983). For example, we expect that when people consume folic acid from any source, it will increase their blood folate levels (Fig. 3). In women of childbearing age, this will lead to a reduction in neural tube defects, a functional outcome.

Specific to fortified wheat flour, researchers have assessed its role in affecting biological markers and functional outcomes; all of these results will be reviewed in this chapter.

4.1 Biological markers of nutritional status

Biological markers of nutritional status can be measured in a minimally invasive way from biological fluids and tissues such as blood, urine and hair (Gibson 1990). In assessing the impact of wheat flour fortification in effectiveness studies, only blood and breast milk samples have been taken (Table 2).

Folate provided through the fortification compound, folic acid, is the most studied nutrient added to fortified flour. Several biological markers have been assessed such as the concentration of plasma or serum folate and red blood cell folate; these were used to quantify the prevalence (or percentage) of folate deficiency and when combined with hemoglobin, to determine the prevalence

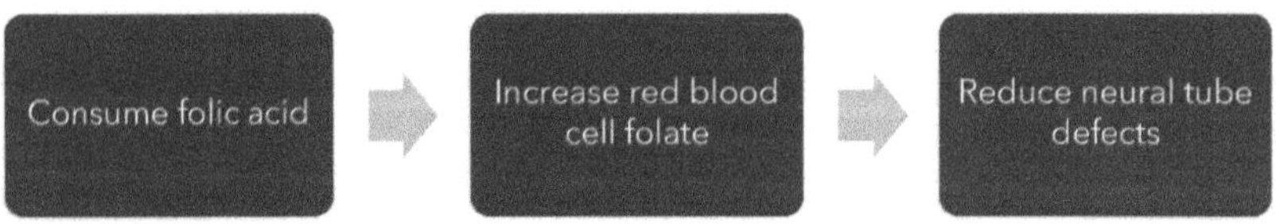

Figure 3 A representation of how fortification with a nutrient such as folic acid can lead to improvement in a biological marker of that nutrient such as red blood cell folate. In turn, an increase in red blood cell folate can lead to improvement in a functional outcome: for example, a reduced risk of neural tube defects.

Published by Burleigh Dodds Science Publishing Limited, 2021.

Table 2 Biological markers of nutritional status and functional outcomes assessed in studies of wheat flour fortification's impact on human health

Nutrient provided through fortification	Biological markers of nutritional status[a]	Functional outcomes[a]
Folate	Folate deficiency Folate-deficiency anemia High homocysteine Plasma/serum folate Plasma/serum homocysteine Red blood cell folate	Anemia Cancer Cognitive function Congenital heart disease Coronary heart disease Hypersensitivity Masking of vitamin B12 deficiency Myocardial infarction Orofacial clefts Neural tube defects Stroke Thyroid- and diabetes-related disorders
Iron	Body iron stores Plasma/serum ferritin Iron deficiency Iron-deficiency anemia Soluble transferrin receptor	Anemia
Zinc	Plasma/serum zinc Zinc deficiency	None assessed
Vitamin B12	Breastmilk vitamin B12 Plasma/serum vitamin B12 Vitamin B12 deficiency	None assessed
Multiple	Hemoglobin	Anemia[b]

[a] Some of the markers of nutritional status measure the concentration of nutrients or other constituents in the blood or breastmilk. Some of the markers of nutritional status and most of the functional outcomes refer to the prevalence (or percentage) of people who have the condition. The functional outcomes refer to the incidence (or number of people newly diagnosed with the condition), prevalence (or percentage of people who have the condition) or deaths (or number of people who die due to the condition).

[b] Some studies assessed the effect of fortification with multiple nutrients on the prevalence of anemia, without evaluating the contribution of fortification in improving nutrient-specific biological markers of nutritional status (for example, Assunção et al. 2007).

of folate-deficiency anemia. For iron, zinc and vitamin B12 added to wheat flour, studies assessed their impact on biological markers, as well (Table 2).

4.2 Functional outcomes

The impact of wheat flour fortification on several functional outcomes was studied (Table 2). For example, positive functional outcomes evaluated were reductions in neural tube defects and anemia while negative functional outcomes assessed were masking of vitamin B12 deficiency and an increase in cancer incidence (Table 2). Some of these outcomes are described further.

4.2.1 Neural tube defects

Neural tube defects are a type of congenital anomaly that affects the development of a baby's spine and brain while *in utero* (Avagliano et al. 2018). It is estimated that between 213 800 and 322 000 babies are born with neural tube defects around the world every year (Blencowe et al. 2018). For healthy spine and brain development, the neural tube must close by 28 days after conception (van Gool et al. 2018); this developmental milestone in the fetus occurs before most women know they are pregnant (Martinez et al. 2018).

The two most common forms of neural tube defects are spina bifida and anencephaly (Avagliano et al. 2018). Spina bifida is when the baby's spine is not formed correctly. Spina bifida can be treated, but it cannot be cured, and individuals with spina bifida have varying degrees of permanent disability for the rest of their lives. Anencephaly is when the brain is not formed correctly. All babies with anencephaly die *in utero* or shortly after birth.

With adequate folate status in women before conception, a healthy neural tube forms in the fetus (Martinez et al. 2018). Folic acid is a form of vitamin B9 that is well absorbed by the body. It can be provided in pill form or as a fortification compound (IOM 1998). Folic acid consumed by women before conception and in the first few weeks after conception prevents around 70% of these birth defects (Czeizel and Dudás 1992; MRC Vitamin Study Research Group 1991). For this reason, women capable of becoming pregnant are recommended to increase their folic acid intake by consuming supplements with folic acid, foods fortified with folic acid and foods rich in food folates (a form of vitamin B9 that the body does not absorb as well as it absorbs folic acid) (Institute of Medicine 1998).

4.2.2 Anemia

Anemia is 'a condition in which the number of red blood cells or the hemoglobin concentration within them is lower than normal' (WHO 2020). An estimated 800 million women and preschool children worldwide have anemia (Stevens et al. 2013). In public-health practice, anemia is determined by measuring hemoglobin levels in blood (Chaparro and Suchdev 2019). If the value is below a cut-off, a person is considered to be anemic. Anemia has multiple causes, both nutritional and non-nutritional in nature. Dietary deficiencies in the nutrients iron, copper, zinc, folate, vitamin B12, riboflavin, vitamin B6, thiamin, vitamin A and vitamin E - which contribute to hemoglobin synthesis - can cause anemia (Kraemer and Zimmermann 2007). Non-nutritional causes of anemia include malaria, hemoglobin disorders such as thalassemia and chronic inflammation (Chaparro and Suchdev 2019).

The prevalence of anemia can only be reduced if the causes of the anemia are addressed. In some world regions, there are both nutritional and non-nutritional causes of anemia (Kassebaum et al. 2014). In these cases, fortification with nutrients involved in hemoglobin synthesis can only reduce the occurrence of anemia if there is a deficiency in these nutrients in the diet.

4.2.3 Masking of vitamin B12 deficiency

A concern emerged in the mid-1900s related to both folate and vitamin B12. Folate deficiency independently causes megaloblastic anemia; that is anemia where the red blood cells are larger than normal (IOM 2000). Vitamin B12 deficiency also independently causes megaloblastic anemia. Additionally, vitamin B12 deficiency causes potentially irreversible neurological conditions such as 'memory loss, disorientation and frank dementia'.

The masking of vitamin B12 deficiency occurs in a specific situation where a person has megaloblastic anemia due to vitamin B12 deficiency only (Berry 2019). This is often observed in older adults who are unable to absorb vitamin B12 from the diet as well as they did when they were younger (Allen et al. 2018). In these individuals, if folic acid is provided, the anemia is corrected. However, if vitamin B12 is not provided, vitamin B12 deficiency can persist and with it, potentially irreversible neurological conditions.

When folic acid corrects megaloblastic anemia while not treating the underlying vitamin B12 deficiency, it is known as 'folic acid masking of vitamin B12 deficiency'. Fortification with folic acid may mask vitamin B12 deficiency.

4.2.4 Cancer

Folic acid is reported to both prevent and cause cancer (Smith et al. 2008). Specifically, folic acid 'may protect against the initiation of cancer, but facilitate the growth of preneoplastic [pre-cancerous] cells'. The cancer research conducted with folic acid has mainly focused on folic acid delivered through large-dose supplements, and not lower-dose food fortification. Evidence of wheat flour fortification with folic acid causing cancer is reviewed in this chapter.

5 Additional considerations when assessing the health impact of wheat flour fortification

There are several challenges with assessing the health impact of food fortification programs through effectiveness studies. The first five of these issues can affect the interpretation of the research results; the last issue is to address the paucity of such data from countries that implement fortification programs. Potential solutions for overcoming these challenges are noted.

5.1 Lack of a control group

Because large-scale fortification is often implemented under a national mandate, it rarely offers an opportunity to have a randomly selected control group that does not get fortification for a period of time. The lack of such a group makes it difficult to infer causality for fortification (Victora et al. 2004). Thus, we cannot state with certainty that fortification causes an improvement in a health outcome.

For example, using two national surveys from Costa Rica, researchers observed there was a lower prevalence of iron deficiency, anemia and iron-deficiency anemia in children in 2008 compared with 1996 (Martorell et al. 2015). Between the two surveys, maize flour and milk were mandated to be fortified with iron, and the iron compound used to fortify wheat flour was changed to a fortificant that the body absorbs well (i.e. ferrous fumarate). Is it plausible that fortification contributed to the health impact observed?

The investigators generated a program-impact pathway of various factors in Costa Rica's food fortification program (Fig. 4) (Martorell et al. 2015). First, they assessed whether there was a potential to benefit from food fortification. They concluded there was a potential to benefit because micronutrient deficiencies were present in 1996 (27% of children were iron deficient). Next, they assessed if a fortification policy had been created and legislation passed. The answer was yes. Then, they assessed if bioavailable fortificants were mandated. The answer was also yes. Next, they determined if foods were fortified at mandated levels.

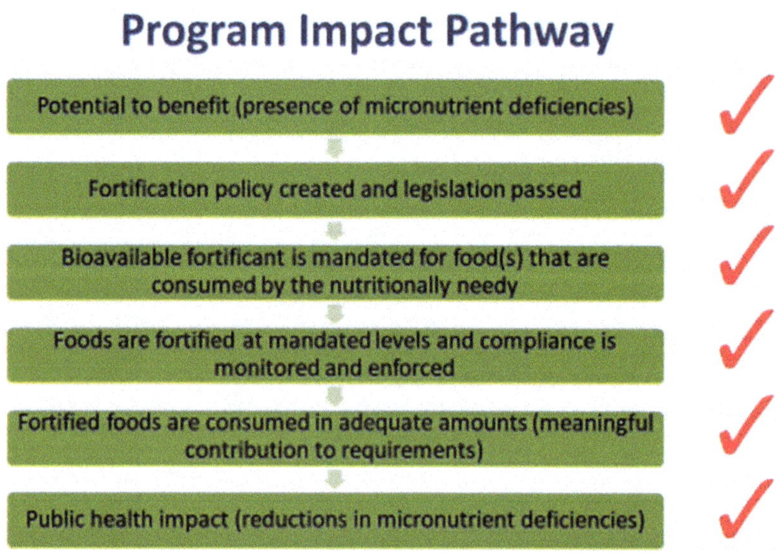

Figure 4 Program impact pathway developed by researchers to determine the plausibility of food fortification improving health outcomes (Martorell et al. 2015).

Published by Burleigh Dodds Science Publishing Limited, 2021.

They obtained data from the government regulatory agency that confirmed that all 246 wheat flour samples obtained in bakeries over a one-year period met or exceeded the iron content required by law. After, they evaluated if fortified foods were consumed in adequate amounts. They analyzed dietary data and estimated that fortified foods contributed 49% of children's dietary iron requirement. Finally, they assessed the public health impact and saw a reduction in biological makers and a functional outcome. The affirmative responses to all of these questions suggest that it is plausible that food fortification with iron in Costa Rica contributed to the health impacts observed.

This type of complementary, program-related information can be presented for any program to argue for fortification's contribution to health impacts. For example, program decision makers can compile and triangulate information generated through government monitoring, such as compliance with fortification (Smarter Futures no date). Unfortunately, this type of information is rarely presented in effectiveness studies (Pachón et al. 2015).

5.2 Challenges of using birth defects registry data

The following experience from Peru highlights the importance of verifying electronic birth defects registry information with a review of clinical records, to minimize misclassification errors. In 2012, Ricks et al. (2012) published an article that evaluated the impact on neural tube defects (NTDs) of wheat flour fortification with folic acid which was decreed in Peru in 2005. Their work showed no reduction in NTDs in a large maternity hospital in Lima, after folic-acid fortification of wheat flour began; the pre-fortification NTD estimates (18.4/10 000 live and still births) were from 2004 to 2005 and the post-fortification NTD estimates (20.0/10 000 live and still births) were from 2007 to 2008. Electronic registry data were used to generate the NTD estimates.

Tarqui-Mamani (2013) wrote a letter to the editor of the journal that published Ricks' paper. Tarqui-Mamani's research team used the same data as Ricks; however, they reviewed clinical charts and found that 32.9% of cases in the electronic registry noted as NTDs were in fact other congenital anomalies. This suggests that the Ricks' paper overestimated the number of NTD cases in both the pre- and post-fortification periods.

In 2013, Tarqui-Mamani was part of a research team that reported their analysis of the same electronic registry data as Ricks, but after having double-checked the clinical charts (Sanabria Rojas et al. 2013). They reported on data collected in longer pre- (2001–2005) and post-fortification periods (2006–2010). The birth prevalence of NTDs that they reported in both the pre-fortification (2005: 13.6/10 000) and post-fortification (2010: 7.6/10 000) periods were lower than what Ricks reported, suggesting again, that Ricks misclassified congenital anomalies as NTDs. The Sanabria results indicate a lower prevalence of NTDs in the post-fortification period compared with the pre-fortification period.

5.3 Selecting an outcome indicator that is only responsive to the nutrients added through fortification

As noted earlier for anemia, there are many factors that together or in isolation can cause low hemoglobin levels; these include nutritional and non-nutritional causes. When anemia is the sole outcome studied to measure the impact of a nutrition intervention such as fortification, one can never be completely sure if the change (or lack of change) was due to the nutrients delivered. It is preferable to select an outcome that is directly and exclusively changed in the human body because of a particular nutrient that is provided through fortification. Examples of such outcomes are the biological markers described previously (Table 2).

5.4 Allowing sufficient time before measuring outcomes

There are two main reasons why a minimum amount of time is needed between the start of fortification program implementation and the measurement of health outcomes. One is that programs need sufficient time to ensure a consistent supply of adequately fortified food reaches the target population (Smarter Futures no date). Mills that have never fortified need time to purchase and install feeders, purchase vitamins and minerals, purchase bags with new nutrient labels, train mill staff in adding nutrients and testing for this addition, and in documenting the fortification process during all shifts. At the same time, governments need time to train inspectors in auditing mill activities and integrating inspections for fortification into existing protocols and schedules. Additionally, there may be several months between the production of flour and its appearance in the market for consumers or in mass-produced products that use flour as an ingredient. Another reason is that some biological markers and most functional indicators require a longer period of time before their presence can be measured in the human body. Neural tube defects are measured at the end of a nine-month pregnancy period and cancers can take decades to manifest (Keum and Giovannucci 2014).

For programs that are evaluated 'too soon' after initiation, a lack of impact can be due to either of the aforementioned reasons or to the program being ineffective (e.g. wrong food was chosen to be fortified; wrong nutrients, levels or fortification compounds selected). In Brazil, the first published studies that assessed hemoglobin levels and the prevalence of neural tube defects (NTDs) showed no difference between the pre- and post-fortification periods (Table 3), whereas later studies did observe higher hemoglobin levels and lower prevalence of NTDs in the post-fortification period.

Table 3 Studies from Brazil that reported hemoglobin levels and neural tube defects before and after fortification of wheat and maize flour

Study	Outcome	Results
Assunção et al. 2007	Hemoglobin	Hemoglobin levels are the same in pre- and post-fortification periods
Fujimori et al. 2011	Hemoglobin	Hemoglobin levels higher in the post- than pre-fortification period
Assunção et al. 2012	Hemoglobin	Hemoglobin levels higher in the post- than pre-fortification period
Pacheco et al. 2009	Neural tube defects	NTD prevalence is the same in pre- and post-fortification periods
López-Camelo et al. 2010	Neural tube defects	Prevalence of one type of NTD is the same in pre- and post-fortification periods: total spina bifida; prevalence of two types of NTDs lower in the post-fortification than pre-fortification period: anencephaly, cephalocele
Pacheco Santos et al. 2016	Neural tube defects	NTD prevalence lower in post-fortification than pre-fortification period

For these reasons, it is prudent to measure program-performance information such as the percentage of flour produced that is adequately fortified (i.e. compliance) and the percentage of people consuming adequately fortified flour (i.e. coverage) before embarking on an impact evaluation. Programs that are not delivering adequately fortified food to most of the target population are unlikely to see a health impact; in those cases, program performance should be improved before assessing impact.

5.5 *Unethical to conduct randomized controlled studies for some outcomes*

A randomized controlled trial would unequivocally answer the question 'does consumption of folic acid-fortified flour by pregnant women cause a reduction in neural tube defects?' However, because it has been established that folic acid delivered to pregnant women (in a supplement) reduces the first occurrence and recurrence of neural tube defects, it would be unethical to conduct such a trial where a group of pregnant women would knowingly be deprived of folic acid (Oakley 2009). For this reason, only observational studies, like those described in this chapter, can be ethically completed. Conclusions from these studies can be strengthened with program-performance data as noted earlier.

5.6 *Impact evaluation surveys can be costly*

Because stand-alone, fortification impact evaluation surveys can be costly, one solution is to use existing data to assess the health impact of flour fortification.

For example, in June 2004, wheat and maize flour fortification with iron and folic acid became mandatory in Brazil (Global Fortification Data Exchange, 2020c). Researchers based in the city of Recife were interested in determining if the fortification mandate had an impact on the number of babies born with neural tube defects (Pacheco et al. 2009).

Brazil has a National Information System on Live Births (Pacheco et al. 2009). The information in this system was used to determine if a child was born in Recife with a neural tube defect. The researchers then counted the number of babies born with neural tube defects before fortification became mandatory and after fortification became mandatory.

This research project did not require primary data collection by the researchers. They were able to use existing data to assess if fortification had a health impact. This and the study from Costa Rica (Martorell et al. 2015) provide a valuable lesson. Existing data, such as national nutrition surveys and live births registries, can be used to estimate the health impact of flour fortification.

Another solution for minimizing the cost of evaluating health impact is to add fortification-relevant questions to existing data-collection systems. For instance, for the 2014 Demographic and Health Survey conducted in Cambodia, decision makers added a micronutrient module for the first time (National Institute of Statistics et al. 2015). This allowed for nationally representative information to be available for several biological markers of nutrient status: iron, vitamin A, vitamin D, calcium, folate, vitamin B12 and iodine status. The resources required to add blood and urine sampling to existing surveys, such as this one, are substantially lower than paying for a stand-alone survey to exclusively measure the health impact of food fortification.

6 Health impact results observed from wheat flour fortification studies

Researchers have employed different study designs to assess if wheat flour fortification affects any of the health outcomes described in Table 2: biological markers of nutritional status and functional outcomes (both positive and negative). What follows are the trends observed from these studies (Figs. 5–7).

Nutrients and health outcomes studied:

- Folic acid is the nutrient added to fortified wheat flour that was most studied, followed distantly by iron, vitamin B12 and zinc.
- Numerous health outcomes were studied. For folic acid, these can be grouped into the following categories: folate status (including folate deficiency and folate-deficiency anemia), neural tube defects (including various sub-types), cancer (including breast, colorectal and pediatric

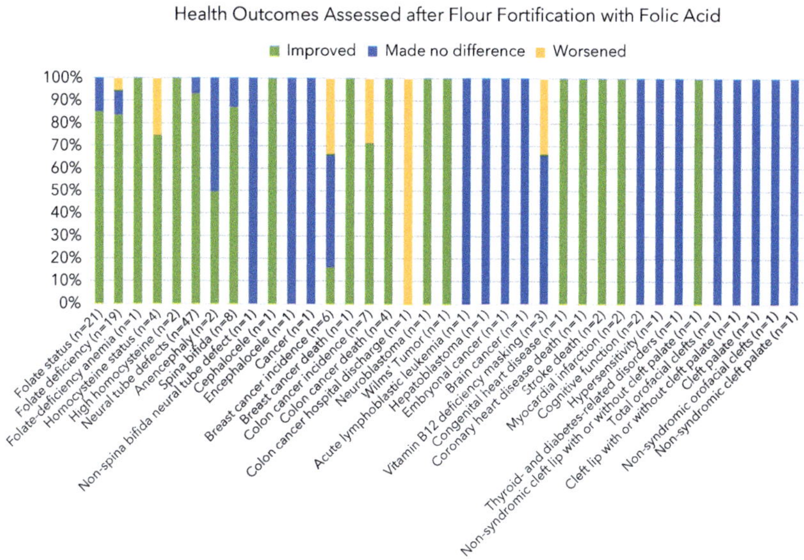

Figure 5 A summary of research on all health outcomes assessed after flour fortification with folic acid: fortification improved, worsened or made no difference in the health outcome. The number in parentheses reflects the number of analyses conducted. If studies reported overall results only, the number reflects the number of studies. If a study reported results by different population groups (e.g. women, children), the number reflects the number of population groups. 'Folate status' as reflected in serum, plasma or red blood cell levels. 'Folate deficiency' as reflected in serum, plasma or red blood cell levels below a cutoff. 'Plasma homocysteine concentration increases when inadequate quantities of folate are available to donate the methyl group that is required to convert homocysteine to methionine' (Institute of Medicine 1998). Serum or plasma homocysteine levels above a cutoff reflect folate deficiency. For 'colon cancer incidence, death and hospital discharge', this may refer to colon cancer alone or to colorectal cancer. Hypersensitivity outcomes include 'asthma, allergy and atopic disease, wheeze, hypersensitivity test, eczema and food allergy' (National Toxicology Program 2015).

cancers), homocysteine status (including high homocysteine), orofacial clefts, heart health (coronary heart disease, stroke, myocardial infarction) and others (vitamin B12 deficiency masking, congenital heart disease, cognitive function, hypersensitivity, and thyroid- and diabetes-related disorders).

- Health outcomes studied for added iron included iron status, iron deficiency and iron-deficiency anemia.
- Health outcomes studies for vitamin B12 and zinc were status and deficiency biomarkers.
- Hemoglobin and anemia, biomarkers potentially linked to multiple nutrients, were also studied.

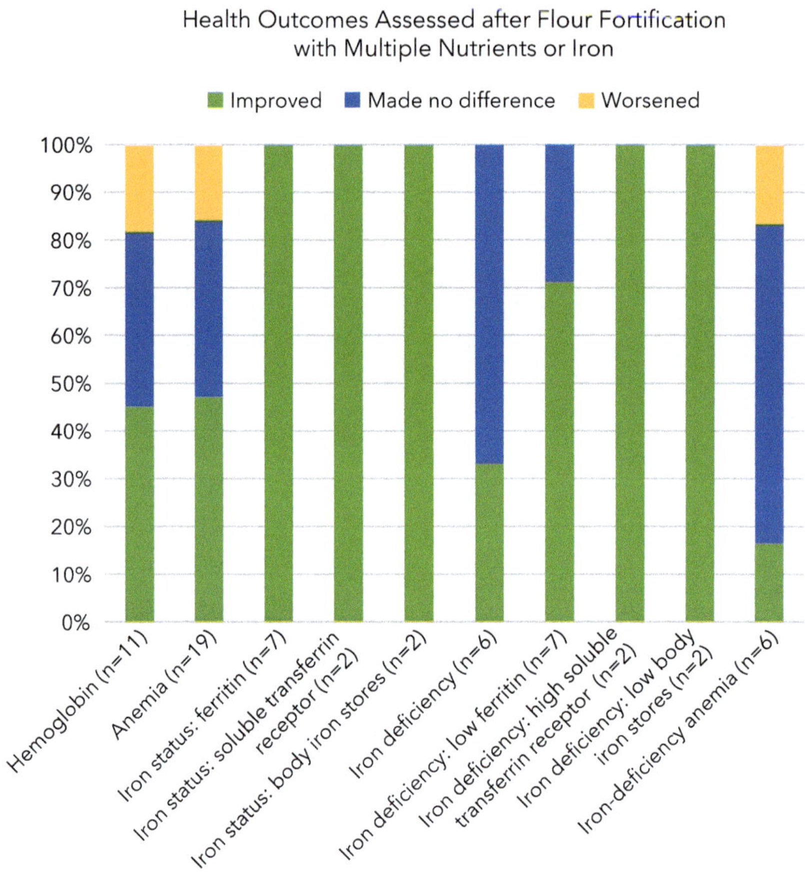

Figure 6 A summary of research on health outcomes assessed after flour fortification with multiple nutrients or iron: fortification improved, worsened or made no difference in the health outcome. The number in parentheses reflects the number of analyses conducted. If studies reported only the overall results, the number reflects the number of studies. If a study reported results by different population groups (e.g. women, children), the number reflects the number of population groups. Hemoglobin level and prevalence of anemia can be affected by multiple nutrients; they can also be affected by non-nutritional factors such as malaria infection. 'Iron deficiency' as defined by the authors.

Results from the most studied outcomes:

- The most studied outcomes were neural tube defects, cancer, folate status, folate deficiency, anemia, iron deficiency, iron status, hemoglobin and iron-deficiency anemia.
- Except for iron-deficiency anemia, most analyses showed improvements in all of these outcomes after fortification. For example, cases of neural tube defects and cancers overwhelmingly decreased and levels of folate and iron in biomarkers increased after fortification.

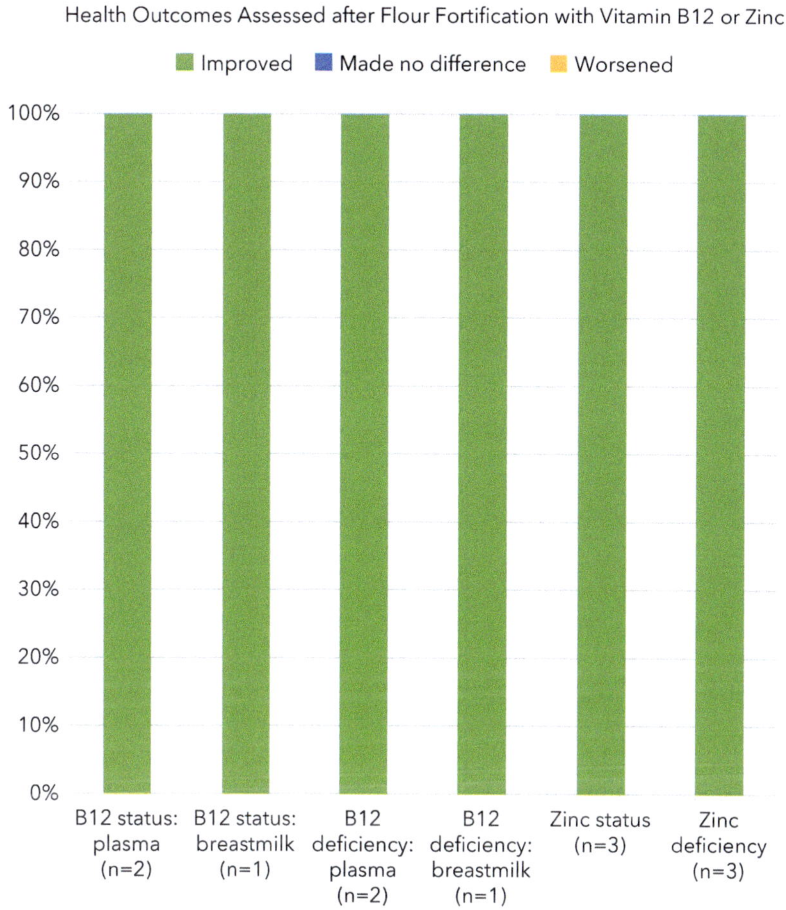

Figure 7 A summary of research on health outcomes assessed after flour fortification with vitamin B12 or zinc: fortification improved, worsened or made no difference in the health outcome. The number in parentheses reflects the number of analyses conducted. If studies reported overall results only, the number reflects the number of studies. If a study reported results by different population groups (e.g. women, children), the number reflects the number of population groups.

- Few analyses showed that health outcomes worsened after fortification. This was the case for cancer, folate deficiency, anemia, hemoglobin and iron-deficiency anemia. That is, the cases of cancer, folate deficiency, anemia and iron-deficiency anemia increased and levels of hemoglobin decreased after fortification.
- Conflicting results for cancer may be explained by the difference in years since fortification was initiated (e.g. increased incidence of colon cancer after fortification is suggested by studies published in the 2000s; the

opposite is observed in studies published in the 2010s) or by the sample size in studies (e.g. increased breast cancer incidence after fortification is observed in studies with sample sizes <2000; studies that observed no difference or a decreased incidence after fortification have sample sizes >2000 and going into the millions).

- Conflicting results for hemoglobin, anemia and iron-deficiency anemia (i.e. some studies show improvements and some show worsening after fortification) may be explained by (1) the existence of non-nutritional causes of anemia which cannot be addressed by fortification, (2) nutritional causes of anemia not addressed by fortification because a limited number of nutrients were added through fortification and (3) levels of nutrients or fortification compounds used in fortification do not follow international guidelines.

Results from other outcomes:

- All remaining outcomes had data from four or fewer analyses: folate-deficiency anemia, homocysteine status, high homocysteine, vitamin B12 deficiency masking, congenital heart disease, coronary heart disease, stroke, myocardial infarction, cognitive function, hypersensitivity, thyroid- and diabetes-related disorders, orofacial clefts, vitamin B12 status, vitamin B12 deficiency, zinc status and zinc deficiency.
- Cautiously, outcomes with two to four analyses suggest the following relationship with fortification.
- Homocysteine status, high homocysteine, stroke death, myocardial infarction, orofacial clefts, vitamin B12 status, vitamin B12 deficiency, zinc status and zinc deficiency trend toward improvement after fortification.
- Vitamin B12 deficiency masking and cognitive function trend towards showing no difference before and after fortification.
- None of the outcomes with two, three or four analyses show a worsening of health after fortification.

6.1 Results from studies in individual countries that assessed health outcomes before and after fortification

'Before and after' studies are those where health outcomes are measured before food fortification is implemented in a country and then after. Here, the before or pre-fortification period is considered the control group for the after or post-fortification period. Health information can be collected on the same individuals, or there can be different individuals in the pre-fortification period and the post-fortification period. What follows are studies conducted in single countries.

6.1.1 Results from studies where the same individuals were measured before and after fortification

Studies where the same individuals were measured before and after fortification in a single country are summarized in Table 4. One example is from Chile with red blood cell folate data from the same women of reproductive age (Hertrampf et al. 2003). Women's blood was taken before initiation of fortification of wheat flour with folic acid and it was taken 12 months after fortification had started. In these women, red blood cell folate levels were 290 ± 102 nmol/L in the pre-fortification period and increased to 707 ± 179 nmol/L in the post-fortification period. Red blood cell folate levels increased within 12 months after fortification started, suggesting that fortification of wheat flour with folic acid improved a biological marker of folate status.

Sometimes, pre- and post-fortification studies may not show clear improvements in nutrient status. For example, South Africa experienced an improvement in the nutritional status of one nutrient added through fortification (i.e. folic acid which was further supported by reductions in neural tube defects (Sayed et al. 2008)) but not in another nutrient added through fortification (i.e. iron) (Modjadji et al. 2007). The results from the Modjadji study suggested that an iron compound more bioavailable than the electrolytic iron specified in the country standard could be warranted (UNICEF and Food Fortification Initiative 2004).

6.1.2 Results from studies where different individuals were measured before and after fortification

Studies where different individuals were measured before and after fortification in a single country are summarized in Table 5 for neural tube defects and Table 6 for other health outcomes.

A study from Iran included neural tube defect data collected from different babies (Abdollahi et al. 2011) (Table 5). The researchers reported neural tube defects before fortification of wheat flour with folic acid (years 2006–2007) and after fortification (2007–2008). There were 31.6 and 21.9 neural tube defects per 10 000 live and still births between the time periods, respectively, pointing to a 31% reduction in neural tube defects after fortification of wheat flour with folic acid. Wheat flour fortification with folic acid improved a functional outcome; this was a consistent finding in all countries which studied neural tube defects.

In Cameroon, women and pre-school children's nutritional status was measured before and after initiation of oil fortification with vitamin A and wheat flour fortification with multiple nutrients: folic acid, iron, vitamin B12 and zinc (Engle-Stone et al. 2017) (Table 6). Plasma vitamin B12 levels were higher in women and children in the post-fortification period than in the pre-fortification period; the same was true for breastmilk vitamin B12 levels in lactating

Table 4 Health outcome results from studies where the same individuals were measured before and after wheat flour fortification[a]

Country	Study	Nutrient[b]	Health outcome[c]	Individuals studied	Results
Chile	Hirsch et al. 2002[d]	Folic acid	Serum folate; folate deficiency	Older adults (women and men)	Higher serum folate levels in the post-fortification than pre-fortification period; prevalence of folate deficiency lower in post-fortification than pre-fortification period
Chile	Hertrampf et al. 2003	Folic acid	Serum folate; red blood cell (RBC) folate; low serum folate, low RBC folate	Women of childbearing age	Higher serum and RBC folate levels in the post-fortification than pre-fortification period; prevalence of low serum and RBC folate lower in post-fortification than pre-fortification period
Iran	Sadighi et al. 2008	Iron[e]	Serum ferritin; hemoglobin; low serum ferritin; anemia; iron-deficiency anemia	Women of childbearing age	No difference between the pre- and post-fortification periods for serum ferritin and hemoglobin; no difference between the pre-and post-fortification periods in the prevalence of low serum ferritin, anemia and iron-deficiency anemia
South Africa[a]	Modjadji et al. 2007	Iron	Serum ferritin; low serum ferritin	Women of childbearing age	No difference in serum ferritin levels between pre- and post-fortification periods; no difference in prevalence of low serum ferritin between pre- and post-fortification periods

South Africa[a]	Modjadji et al. 2007	Folic acid	Serum folate; RBC folate; low serum folate; low RBC folate	Women of childbearing age	Higher serum and RBC folate levels in the post-fortification than pre-fortification period; prevalence of low serum and RBC folate lower in post-fortification than pre-fortification period
South Africa[a]	Modjadji et al. 2007	Multiple[f]	Hemoglobin; low hemoglobin	Women of childbearing age	Higher hemoglobin levels in the post-fortification than pre-fortification period; no difference in prevalence of low hemoglobin between pre- and post-fortification periods
USA[a]	Enquobahrie et al. 2012	Folic acid	Serum folate; serum homocysteine	Adolescents	Higher serum folate and homocysteine[g] levels in the post-fortification than pre-fortification period
USA[a]	Stolzenberg-Solomon et al. 2006; Lin et al. 2008; Stevens et al. 2010; Houghton et al. 2019a,b	Folic acid	Breast cancer	Women 30–55 years (Houghton et al. 2019a), 32–53 years (Houghton et al. 2019b), 45 years or older (Lin et al. 2008), 50–74 years (Stevens et al. 2010), 55–74 years (Stolzenberg-Solomon et al. 2006)	Higher breast cancer incidence in the post-fortification than pre-fortification period (Lin et al. 2008; Houghton et al. 2019b);[h] no difference in breast cancer incidence between pre- and post-fortification periods (Stolzenberg-Solomon et al. 2006; Stevens et al. 2010; Houghton et al. 2019a)

(*Continued*)

Table 4 *(Continued)*

Country	Study	Nutrient[b]	Health outcome[c]	Individuals studied	Results
USA[a]	Gibson et al. 2011	Folate[i]	Colorectal cancer incidence	Adults 50–71 years	Lower colorectal cancer incidence with higher folate intake in the post-fortification period, this trend was not observed in the pre-fortification period

RBC, red blood cell

[a] Wheat flour was the only grain fortified with one or more nutrients in most countries. Wheat and maize flour was fortified with nutrients in South Africa and the USA (in addition to rice); the independent effect of any one of these fortified foods on health outcomes cannot be discerned with this study.

[b] Nutrient added to wheat flour through fortification which is purported to affect the health outcome.

[c] Some of the health outcomes measure the concentration of nutrients or other constituents in the blood or breastmilk and some refer to the prevalence (i.e. percentage of people who have the condition), the incidence (i.e. number of people newly diagnosed with the condition), or deaths (i.e. number of people who die due to the condition).

[d] Even though the title of this article refers to masking of vitamin B12 deficiency, the authors presented no data that could support or refute masking of B12 deficiency by fortification with folic acid (Hirsch et al. 2002).

[e] Wheat flour in Iran is fortified with two nutrients that contribute to hemoglobin synthesis: iron and folic acid (Sadighi et al. 2008).

[f] As reported in the article, South Africa fortified wheat and maize flour with folic acid, iron, vitamin A, thiamin, riboflavin, niacin, pyridoxine and zinc; many of which can contribute to hemoglobin synthesis.

[g] The homocysteine results are the opposite of what is expected. When folate levels are adequate, homocysteine levels are usually low (Institute of Medicine 1998). The authors suggest that in adolescence, there is a trend toward an increase in homocysteine levels which was not affected by fortification (Enquobahrie et al. 2012).

[h] This result is the opposite of what is expected.

[i] Folate from any source: foods naturally rich in folate, supplements containing folic acid and folic acid-fortified food (Gibson et al. 2011).

Table 5 Results for neural tube defects as the health outcome from studies where different individuals were measured before and after wheat flour fortification[a]

Country	Study	Nutrient[b]	Health outcome[c]	Individuals studied	Results
Argentina	López-Camelo et al. 2010; Bidondo et al. 2015; Sargiotto et al. 2015	Folic acid	Neural tube defects	Live births and still births weighing 500 g or more	Prevalence of NTDs lower in post-fortification than pre-fortification period
Australia	Botto et al. 2006; Hilder 2016	Folic acid	Neural tube defects	Live births, still births and terminations	Prevalence of NTDs lower in post-fortification than pre-fortification period[d]
Brazil[a]	Pacheco et al. 2009; López-Camelo et al. 2010; Pacheco Santos et al. 2016	Folic acid	Neural tube defects	Live births (all studies) and still births weighing 500 g or more (López-Camelo et al. 2010; Pacheco Santos et al. 2016)	Prevalence of two types of NTDs lower in post-fortification than pre-fortification period: anencephaly, cephalocele (López-Camelo et al. 2010); no difference in prevalence of one type of NTD between the pre- and post-fortification periods: total spina bifida (López-Camelo et al. 2010); no difference in NTD prevalence between post-fortification and pre-fortification periods (Pacheco et al. 2009); prevalence of NTDs lower in post-fortification than pre-fortification period (Pacheco Santos et al. 2016)
Canada	Ray et al. 2002; Botto et al. 2006	Folic acid	Neural tube defects	Live births, still births and terminations	Prevalence of NTDs lower in post-fortification than pre-fortification period

(*Continued*)

Table 5 (*Continued*)

Country	Study	Nutrient[b]	Health outcome[c]	Individuals studied	Results
Chile	Castilla et al. 2003; López-Camelo et al. 2005, 2010; Corral et al. 2006; Nazer et al. 2007; Cortés et al. 2012; Nazer and Cifuentes 2013	Folic acid	Neural tube defects	Live births (all studies) and still births weighing 500 g or more (Castilla et al. 2003; López-Camelo et al. 2005, 2010; Corral et al. 2006; Nazer et al. 2007; Cortés et al. 2012; Nazer and Cifuentes 2013)	Prevalence of NTDs lower in post-fortification than pre-fortification period
Costa Rica[a]	Tacsan Chen and Ascencio Rivera 2004; Barboza Argüello and Umaña Solís 2011; Barboza Argüello et al. 2015	Folic acid	Neural tube defects	Live births (all studies) and still births weighing 500 g or more (Tacsan Chen and Ascencio Rivera 2004; Barboza Argüello and Umaña Solís 2011; Barboza Argüello et al. 2015)	Prevalence of NTDs lower in post-fortification than pre-fortification period[e]
Iran	Abdollahi et al. 2011	Folic acid	Neural tube defects	Live births and still births weighing 500 g or more with a gestational age of 20 weeks or more	Prevalence of NTDs lower in post-fortification than pre-fortification period

Peru	Ricks et al. 2012; Sanabria Rojas et al. 2013; Tarqui-Mamani et al. 2016	Folic acid	Neural tube defects	Live births and still births (Ricks et al. 2012; Sanabria Rojas et al. 2013); live births (Tarqui-Mamani et al. 2016)	No difference in the prevalence of NTDs between post-fortification and pre-fortification periods (Ricks et al. 2012); prevalence of NTDs lower in post-fortification than pre-fortification period (Sanabria Rojas et al. 2013); prevalence of spina bifida lower in post-fortification than pre-fortification period (Tarqui-Mamani et al. 2016); no difference in the prevalence of anencephaly and encephalocele between post-fortification and pre-fortification periods (Tarqui-Mamani et al. 2016)
South Africa[a]	Sayed et al. 2008	Folic acid	Neural tube defects	Live births and still births	Prevalence of NTDs lower in post-fortification than pre-fortification period
USA[a]	Honein et al. 2001; CDC 2004; Botto et al. 2006	Folic acid	Neural tube defects	Live births (Honein et al. 2001) Live births, still births and terminations (Botto et al. 2006); live births, still births, terminations, and fetal deaths[f] (CDC 2004)	Prevalence of NTDs lower in post-fortification than pre-fortification period

NTDs, neural tube defects

[a] Wheat flour was the only grain fortified with folic acid in most countries. Wheat and maize flour were fortified with folic acid in Brazil, Costa Rica (in addition to rice and milk), South Africa and the USA (in addition to rice); the independent effect of any one of these fortified foods on health outcomes cannot be discerned with this study.

[b] Nutrient added to wheat flour through fortification which is purported to affect the health outcome.

[c] In this table, the health outcome refers to the prevalence (or percentage) of livebirths, still births, fetal deaths and/or terminations affected by a neural tube defect such as spina bifida or anencephaly.

[d] Australia experienced decreases in the prevalence of NTDs when the country went from no fortification to voluntary fortification (Botto et al. 2006) and from voluntary to mandatory fortification (Hilder 2016).

[e] Wheat flour was the first food mandated to be fortified with folic acid in Costa Rica. With the addition of subsequent foods mandated to be fortified with folic acid (maize flour, milk, rice) and with an increase in the folic acid levels in wheat flour (from 1.5 mg/kg to 1.8 mg/kg), NTDs continued to decrease (Barboza Argüello and Umaña Solís 2011).

[f] 'Fetal death refers to the spontaneous intrauterine death of a fetus at any time during pregnancy. Fetal deaths later in pregnancy (at 20 weeks of gestation or more, or 28 weeks or more, for example) are also sometimes referred to as stillbirths.' (CDC 2020).

women. Consistent with these findings, the prevalence of low plasma (women and children) and breastmilk (women) vitamin B12 levels was lower in post-fortification than pre-fortification period. These vitamin B12 results suggest that wheat flour fortification with vitamin B12 improved nutritional outcomes in the country.

Results for the nutritional status of folic acid, zinc and hemoglobin/anemia also suggested that fortification was adequately implemented in Cameroon. For iron, three indicators of nutritional status were measured: plasma ferritin, soluble transferrin receptor and body iron stores. These were used to measure the prevalence of low plasma ferritin and high-soluble transferrin receptors (both markers of iron deficiency); together with hemoglobin, plasma ferritin was used to calculate the prevalence of simultaneous iron deficiency and anemia. Most of these measures pointed toward improvements in iron status for women and children except for the prevalence of low plasma ferritin (women) and iron-deficiency anemia (women and children) which was not different between the pre- and post-fortification periods.

The same study design was used to investigate potential negative health impacts of fortification (Table 6). In one study conducted in the USA, researchers surmised that people with low vitamin B12 deficiency and no anemia who consumed grains (i.e. wheat flour, maize flour and rice) fortified with folic acid could be at risk of developing vitamin B12 deficiency (Qi et al. 2014). In other words, since folic acid is provided through grain fortification in the USA, these individuals will not develop anemia. However, since vitamin B12 is not provided through grain fortification in the USA, they may develop vitamin B12 deficiency. The researchers assessed if the prevalence of older adults with vitamin B12 deficiency and no anemia changed between pre and post folic-acid fortification periods. If folic acid was masking vitamin B12 deficiency, one would expect an increase in the post-fortification period in the prevalence of vitamin B12 deficiency and no anemia in the same adults. There was no change in the prevalence from the pre- to the post-fortification period, suggesting there was no masking of vitamin B12 deficiency by grain fortification with folic acid.

6.2 Results from trend studies in individual countries that assessed health outcomes multiple times after fortification

When information on a health outcome is available for many years after fortification has started, trend studies can be completed (Table 7). For example, Saudi Arabia began voluntary wheat flour fortification with folic acid and other nutrients in 2001 (Safdar et al. 2007). Investigators had information on the number of babies born with neural tube defects for 3 years before fortification started (~15, 30 and 20 per 10000 births, respectively) and for 5 years after fortification started (~15, 12, 10, 10 and 9 per 10000 births, respectively). The

Table 6 Results for health outcomes other than neural tube defects from studies where different individuals were measured before and after wheat flour fortification[a]

Country	Study	Nutrient[b]	Health outcome[c]	Individuals studied	Results
Australia	Brown et al. 2011	Folic acid	Serum folate; red blood cell (RBC) folate; low serum folate; low RBC folate	Blood samples analyzed in a public hospital's laboratory	Higher serum folate and RBC folate levels in the post-fortification than pre-fortification period; prevalence of low serum folate and low RBC folate lower in post-fortification than pre-fortification period in women and children
Australia	Beckett et al. 2017	Folic acid	Plasma homocysteine; high homocysteine; serum folate; low serum folate; RBC folate	Women and men 65 years or older	Higher serum folate and RBC folate levels in the post-fortification than pre-fortification period; lower plasma homocysteine[d] levels in the post-fortification than pre-fortification period; prevalence of high homocysteine levels (hyperhomocysteinemia)[d] and low serum folate lower in post-fortification than pre-fortification period
Brazil[a]	Britto et al. 2014	Folic acid	Serum folate; red blood cell folate	Pregnant women, children, adolescents, adults, elderly	Higher serum folate and red blood cell folate levels in the post-fortification than pre-fortification period in all age groups
Brazil[a]	Assunção and Santos 2007; Assunção et al. 2007; Costa et al. 2009; Fujimori et al. 2011, Assunção et al. 2012	Iron	Hemoglobin (Assunção et al. 2007, Fujimori et al. 2011; Assunção et al. 2012); anemia (all studies)	Children less than 6 years (Assunção et al. 2007, 2012; Costa et al. 2009) Pregnant women (Fujimori et al. 2011)	No difference in hemoglobin levels (Assunção et al. 2007); higher hemoglobin levels in post-fortification than pre-fortification period (Fujimori et al. 2011); lower hemoglobin levels in post-fortification than pre-fortification period (Assunção et al. 2012); prevalence of anemia higher in post-fortification than pre-fortification (Assunção et al. 2007, 2012);[e] prevalence of anemia lower in post-fortification than pre-fortification (Costa et al. 2009; Fujimori et al. 2011)[f]

(*Continued*)

Table 6 *(Continued)*

Country	Study	Nutrient[b]	Health outcome[c]	Individuals studied	Results
Cameroon	Engle-Stone et al. 2017[g]	Folic acid	Plasma folate; low plasma folate	Women of childbearing age, children 12–59 months	Higher plasma folate levels in the post-fortification than pre-fortification period in women and children; prevalence of low plasma folate lower in post-fortification than pre-fortification period in women and children
Cameroon	Engle-Stone et al. 2017[g]	Iron	Plasma ferritin; soluble transferrin receptor; body iron stores; low plasma ferritin; high soluble transferrin receptor; iron-deficiency anemia	Women of childbearing age	Higher plasma ferritin levels and body iron stores in the post-fortification than pre-fortification period; lower soluble transferrin receptor levels in the post-fortification than pre-fortification period;[h] no difference in prevalence of low plasma ferritin and iron-deficiency anemia between post-fortification and pre-fortification period; prevalence of high-soluble transferrin receptor and low body iron stores lower in post-fortification than pre-fortification period
Cameroon	Engle-Stone et al. 2017[g]	Iron	Plasma ferritin; soluble transferrin receptor; body iron stores; low plasma ferritin; high soluble transferrin receptor; iron-deficiency anemia	Children 12–59 months	Higher plasma ferritin levels and body iron stores in the post-fortification than pre-fortification period; lower soluble transferrin receptor levels in the post-fortification than pre-fortification period;[h] no difference in prevalence of iron-deficiency anemia between post-fortification and pre-fortification period; prevalence of low plasma ferritin lower in post-fortification than pre-fortification period in children; prevalence of high soluble transferrin receptor and low body iron stores lower in post-fortification than pre-fortification period

Cameroon	Engle-Stone et al. 2017[g]	Vitamin B12	Plasma vitamin B12; breastmilk vitamin B12 (women only); prevalence low vitamin B12; prevalence low breastmilk vitamin B12 (women only)	Women of childbearing age, children 12–59 months	Higher plasma (women and children) and breastmilk (women) vitamin B12 levels in the post-fortification than pre-fortification period; prevalence of low plasma (women and children) and breastmilk (women) vitamin B12 lower in post-fortification than pre-fortification period
Cameroon	Engle-Stone et al. 2017[g]	Zinc	Plasma zinc; low plasma zinc	Women of childbearing age, children 12–59 months	Higher plasma zinc levels in the post-fortification than pre-fortification period in women and children; prevalence of low plasma zinc lower in post-fortification than pre-fortification period in women and children
Cameroon	Engle-Stone et al. 2017[g]	Multiple	Hemoglobin; anemia	Women of childbearing age, children 12–59 months	In women and children, no difference in hemoglobin levels between the post-fortification and pre-fortification period; prevalence of anemia lower in post-fortification than pre-fortification period in women, no difference in prevalence of anemia in post-fortification and pre-fortification period in children
Canada	Ray et al. 2003; Liu et al. 2004	Folic acid	Serum folate (Ray et al. 2003; Liu et al. 2004); RBC folate (Liu et al. 2004); plasma homocysteine (Liu et al. 2004); folate deficiency (Ray et al. 2003)	Women 65 years and older (Ray et al. 2003), women 19–44 years and women and men 65 years and older (Liu et al. 2004)	In women and older adults: higher serum and RBC folate levels in the post-fortification than pre-fortification period, lower plasma homocysteine levels in the post-fortification than pre-fortification period; in women 65 years and older: prevalence of folate deficiency lower in post-fortification than pre-fortification period
Canada	Mason et al. 2007	Folic acid	Incidence of colorectal cancer	All ages	Incidence of colorectal cancer higher in post-fortification than pre-fortification period[e]

(Continued)

Table 6 *(Continued)*

Country	Study	Nutrient[b]	Health outcome[c]	Individuals studied	Results
Chile	Hirsch et al. 2009	Folic acid	Hospital discharge due to colon cancer	Patients discharged from public or private hospitals	Following the secular trend, higher hospital discharges due to colon cancer in the post-fortification than pre-fortification period[i]
Costa Rica[a]	Tacsan Chen and Ascencio Rivera 2004	Folic acid	Serum folate; folate deficiency	Women	In urban and rural areas: higher serum folate levels in the post-fortification than pre-fortification period; in urban and rural areas: prevalence of folate deficiency lower in the post-fortification than pre-fortification period
Costa Rica[a]	Martorell et al. 2015	Iron	Serum ferritin; hemoglobin; iron deficiency; iron-deficiency anemia; anemia	Children 1–7 years	Higher serum ferritin and hemoglobin in the post-fortification than pre-fortification period; prevalence of iron deficiency, iron-deficiency anemia and anemia lower in the post-fortification than pre-fortification period
Costa Rica[a]	Martorell et al. 2015	Vitamins A, B1, B2, B9, B12, E Minerals iron, zinc[f]	Hemoglobin; anemia	Women 15–45 years	Higher hemoglobin levels in the post-fortification than pre-fortification period; prevalence of anemia lower in the post-fortification than pre-fortification period
Fiji	National Food and Nutrition Centre 2012	Folic acid, iron, niacin, riboflavin, thiamin, zinc	Serum ferritin; hemoglobin; serum folate; serum zinc; low serum ferritin; anemia; low serum folate; low serum zinc	Women of childbearing age	Higher serum ferritin, hemoglobin, serum folate and serum zinc levels in the post-fortification than pre-fortification period; prevalence of low serum ferritin, anemia, low serum folate and low serum zinc lower in post-fortification than pre-fortification period

Iran	Abdollahi et al. 2011	Folic acid	Serum folate; plasma homocysteine; low serum folate; high plasma homocysteine	Women of childbearing age	Higher serum folate levels in the post-fortification than pre-fortification period; lower plasma homocysteine levels in the post-fortification than pre-fortification period;[j] prevalence of low serum folate and high plasma homocysteine lower in post-fortification than pre-fortification period
Iran	Sadighi et al. 2009	Iron[k]	Serum ferritin; hemoglobin; low serum ferritin; anemia; iron-deficiency anemia	Women of childbearing age	Higher serum ferritin levels in the post-fortification than pre-fortification period, lower hemoglobin levels[e] in the post-fortification than pre-fortification period; prevalence of low serum ferritin lower in post-fortification than pre-fortification period; prevalence of anemia[e] and iron-deficiency anemia[e] higher in post-fortification than pre-fortification period
USA[a]	CDC 2003; Dietrich et al. 2005; Pfeiffer et al. 2012, 2019	Folic acid	Serum folate (all studies); RBC folate (all studies); low serum folate (Dietrich et al. 2005; Pfeiffer et al. 2012); low RBC folate (Dietrich et al. 2005; Pfeiffer et al. 2012)	Children (Pfeiffer et al. 2012, 2019) Adults (Dietrich et al. 2005; Pfeiffer et al. 2012, 2019) Adult women (CDC 2003)	Higher serum and RBC folate levels in the post-fortification than pre-fortification period; prevalence of low serum folate and low RBC folate lower in post-fortification than pre-fortification period
USA[a]	Mills et al. 2003; Qi et al. 2014	Folic acid	Masking of vitamin B12 deficiency	Older adults with both low vitamin B12 status (or vitamin B12 deficiency) and no anemia (or macrocytosis[lb])	Prevalence of low vitamin B12 status and no anemia the same in the pre- and post-fortification periods; no evidence of masking;[m] prevalence of vitamin B12 deficiency and no anemia (or macrocytosis) the same (or lower) in post-fortification than pre-fortification period; no evidence of masking[m]

(Continued)

Table 6 *(Continued)*

Country	Study	Nutrient[b]	Health outcome[c]	Individuals studied	Results
USA[a]	Mason et al. 2007	Folic acid	Incidence of colorectal cancer	All ages	Incidence of colorectal cancer higher in post-fortification than pre-fortification period[e]
Venezuela[a]	Layrisse et al. 1996, 2002	Iron[m]	Serum ferritin (both studies); low serum ferritin (both studies); anemia (both studies); iron-deficiency anemia (Layrisse et al. 1996)	Children and adolescents	Higher serum ferritin levels in the post-fortification than pre-fortification period (both studies); prevalence of low serum ferritin and anemia lower in post-fortification than pre-fortification period (Layrisse et al. 1996); no difference in prevalence of low serum ferritin or anemia in post-fortification and pre-fortification period (Layrisse et al. 2002); no difference in prevalence of iron-deficiency anemia in post-fortification and pre-fortification period (Layrisse et al. 1996)

NTDs, neural tube defects; RBC, red blood cell

[a] Wheat flour was the only grain fortified with one or more nutrients in most countries. Wheat and maize flour were fortified with nutrients in Brazil, Costa Rica (in addition to rice and milk), the USA (in addition to rice) and Venezuela; the independent effect of any one of these fortified foods on health outcomes cannot be discerned with this study.

[b] Nutrient added to wheat flour through fortification which is purported to affect the health outcome.

[c] Some of the health outcomes measure the concentration of nutrients or other constituents in the blood or breastmilk and some refer to the prevalence (i.e. percentage of people who have the condition), the incidence (i.e. number of people newly diagnosed with the condition), or deaths (i.e. number of people who die due to the condition).

[d] 'Plasma homocysteine concentration increases when inadequate quantities of folate are available to donate the methyl group that is required to convert homocysteine to methionine' (Institute of Medicine 1998). When folate levels are high, homocysteine levels are usually low and the prevalence of people with high homocysteine levels (e.g. hyperhomocysteinemia) are usually low.

[e] This result is the opposite of what is expected.

[f] The post-fortification prevalence of anemia was compared to the pre-fortification prevalence of anemia for preschool children (1) in the same southeast region of the country as well as (2) other regions of the country. The results were the same.

[g] For this study from Cameroon, adjusted results are presented (not unadjusted results).

[h] A decrease in soluble transferrin receptor levels reflect an improvement in iron status.

[i] The upward trend in Chile mirrors the increase in deaths due to colon cancer evident from 1990 to 2003 (Donoso et al. 2006) - preceding the introduction of fortification with folic acid in the country in 2000.
[j] A decrease in homocysteine levels reflects an improvement in folate status.
[k] Wheat flour in Iran is fortified with two nutrients that contribute to hemoglobin synthesis: iron and folic acid (Sadighi et al. 2008).
[l] Macrocytosis means enlarged red blood cells.
[m] If the prevalence of individuals with vitamin B12 deficiency (or low vitamin B12 status) and no anemia (or no macrocytosis) increased between the pre and post folic-acid fortification periods, that would be evidence of fortification with folic acid masking of vitamin B12 deficiency.
[n] Wheat and maize flour in Venezuela are fortified with other nutrients that contribute to hemoglobin synthesis: iron, vitamin A, thiamin and riboflavin (Layrisse et al. 1996).

60% reduction in neural tube defects from pre- to post fortification periods suggests that wheat flour fortification with folic acid improved a functional outcome.

The same study design was used to investigate potential negative health impacts of fortified food, such as cancer. Vollset et al. (2013) collated the number of colorectal deaths per 100 000 population in the USA from 1950 to 2010. In the country, voluntary fortification of breakfast cereals with folic acid began in 1973 and mandatory fortification of grains with folic acid became effective in 1998. If fortification with folic acid accelerates death from cancer, cancer deaths from 1973 (or 1998) until 2010 should increase. The data show the opposite trend for women and men: during this 60-year period, there was a decline in colorectal cancer deaths. These results suggest that fortification with folic acid does not cause cancer deaths.

6.3 Results from cross-sectional studies in individual countries that assessed health outcomes and fortification exposure simultaneously

Cross-sectional studies are another type of design that can inform the health impact of a fortification program. Cross-sectional means that information was collected at one point of time only. These studies are especially useful in cases where no pre-fortification information is available, so it is not possible to complete a before-and-after study. Three countries have completed such studies (Table 8).

One example is from Oman where a one-time, cross-sectional survey was conducted in 2004 (Grimm et al. 2012). Ferritin and C-reactive protein were assessed in non-pregnant women of childbearing age. This information was used to calculate the percentage of women with iron deficiency, a biomarker of iron status. Families were asked how much wheat flour they consumed in the previous two months and the total number of individuals living or working in the household during this time. Additionally, wheat flour samples were taken from homes and analyzed for the presence of fortificant iron. This information was used to calculate the monthly per capita consumption of fortified wheat flour.

The researchers then completed a dose-response analysis and found that the prevalence of iron deficiency was lowest in women whose households consumed the highest amount of fortified wheat flour: 26.8% compared with 38.8%. These results are in the direction one would expect if flour fortified with iron is being produced and consumed in the country. While the study design does not allow one to conclude that fortification caused a reduction in iron deficiency, the results suggest that fortification is contributing to improving the iron status of women in Oman.

Table 7 Results from trend studies where health outcomes were measured multiple times after wheat flour fortification began[a]

Country	Study	Nutrient[b]	Health outcome[c]	Individuals studied	Results
Canada	Persad et al. 2002; De Wals et al. 2003; Liu et al. 2004; De Wals et al. 2007	Folic acid	Neural tube defects	Live births, still births and terminations	Prevalence of NTDs lower in post-fortification than pre-fortification period
Canada	French et al. 2003; Grupp et al. 2011	Folic acid	Pediatric cancers	Children less than 17 years (French et al. 2003) Children less than 4 years (Grupp et al. 2011)	Prevalence of neuroblastoma and Wilms' tumor lower in post-fortification than pre-fortification period; no difference in the prevalence of acute lymphoblastic leukemia, hepatoblastoma, embryonal cancers or brain cancers between post- and pre-fortification periods
Canada	Ionescu-Ittu et al. 2009	Folic acid	Congenital heart disease	Live births, still births	Prevalence of congenital heart disease[d] lower in post-fortification than pre-fortification period
Canada	Yang et al. 2006	Folic acid	Stroke death	Adults 40 years of age and older	Decrease in deaths from stroke was greater in the post-fortification period (−5.4% annually) than pre-fortification period (−1.0% annually)
Jordan	Amarin and Obeidat 2010	Folic acid	Neural tube defects	Live births	Prevalence of NTDs lower in post-fortification than pre-fortification period
Oman	Alasfoor et al. 2010	Folic acid	Spina bifida[e]; non-spina bifida neural tube defects[f]	Live births	Prevalence of spina bifida lower in post-fortification than pre-fortification period; no difference in the prevalence of other, non-spina-bifida NTDs between post- and pre-fortification periods

(*Continued*)

Table 7 *(Continued)*

Country	Study	Nutrient[b]	Health outcome[c]	Individuals studied	Results
Saudi Arabia	Safdar et al. 2007	Folic acid	Neural tube defects	Live births	Prevalence of NTDs lower in post-fortification than pre-fortification period
Tanzania[a]	Noor et al. 2017	Folic acid	Plasma folate; folate deficiency	Women	Higher plasma folate levels in the post-fortification than pre-fortification period; prevalence of folate deficiency lower in post-fortification than pre-fortification period
USA[a]	Vollset et al. 2013; Keum and Giovannucci 2014; Siegel et al. 2019	Folic acid	Colorectal cancer incidence (Keum and Giovannucci 2014; Siegel et al. 2019) and mortality (Vollset et al. 2013; Keum and Giovannucci 2014; Siegel et al. 2019); breast cancer incidence and mortality (Siegel et al. 2019)	All ages and sexes (Siegel et al. 2019) Adults (Vollset et al. 2013; Keum and Giovannucci 2014)	Incidence of colorectal cancer decreased for women (from 1998 to 2008), men (from 1998 to 2015) (Siegel et al. 2019) and all adults (from 1975 to 2009) (Keum and Giovannucci 2014); incidence of breast cancer decreased for women (from 1999 to 2004) (Siegel et al. 2019); death from colorectal cancer decreased in women and men (from 1950 to 2010) (Vollset et al. 2013); death from colorectal cancer decreased in black and white women and men (from 1975 to 2009) (Keum and Giovannucci 2014); death from colorectal cancer decreased in women (from 1975 to 2016) and men (from 1987 to 2016) (Siegel et al. 2019); death from breast cancer decreased in women (from 1990 to 2016) (Siegel et al. 2019)

USA[a]	Mathews et al. 2002; Williams et al. 2002, 2005, 2015	Folic acid	Neural tube defects	Live births (all studies), still births (Williams et al. 2002, 2015), fetal deaths[g] (Williams et al. 2002, 2005) and terminations (Williams et al. 2002, 2005, 2015)	Prevalence of NTDs lower in post-fortification than pre-fortification period (Mathews et al. 2002; Williams et al. 2002); for Hispanic and non-Hispanic whites: prevalence of NTDs lower in post-fortification than pre-fortification period (Williams et al. 2005); for non-Hispanic blacks: no difference in prevalence of NTDs between post-fortification and pre-fortification periods (Williams et al. 2005)
USA[a]	Yang et al. 2006	Folic acid	Stroke death	Adults 40 years of age and older	Decrease in deaths from stroke was greater in the post-fortification period (−2.9% annually) than pre-fortification period (−0.3% annually)

[a] Wheat flour was the only grain fortified with folic acid in most countries. Wheat and maize flour were fortified with folic acid in Tanzania and the USA (in addition to rice); the independent effect of any one of these fortified foods on health outcomes cannot be discerned with this study.
[b] Nutrient added to wheat flour through fortification which is purported to affect the health outcome.
[c] Some of the health outcomes measure the concentration of nutrients or other constituents in the blood or breastmilk and some refer to the prevalence (i.e. percentage of people who have the condition), the incidence (i.e. number of people newly diagnosed with the condition), or deaths (i.e. number of people who die due to the condition).
[d] Severe congenital heart defects studied were as follows: 'tetralogy of Fallot, endocardial cushion defects, univentricular hearts, truncus arteriosus, or transposition complexes' (Ionescu-Ittu et al. 2009).
[e] Spina bifida is a type of neural tube defect.
[f] The other neural tube defects that were not spina bifida were not specified (Alasfoor et al. 2010).
[g] 'Fetal death refers to the spontaneous intrauterine death of a fetus at any time during pregnancy. Fetal deaths later in pregnancy (at 20 weeks of gestation or more, or 28 weeks or more, for example) are also sometimes referred to as stillbirths.' (CDC 2020).

Table 8 Results from cross-sectional studies in individual countries that assessed health outcomes and fortification exposure simultaneously.

Country	Study	Nutrient[a]	Health outcome[b]	Individuals studied	Results
Oman	Grimm et al. 2012	Iron	Iron deficiency	Women of childbearing age	Prevalence of iron deficiency lower in women whose households consumed the most fortified wheat flour compared with households who purchased the least fortified flour
Colombia	Fothergill et al. 2019	Iron	Low serum ferritin	Women of childbearing age Children 2–4 and 5–12 years	Prevalence of low serum ferritin did not differ by intake of wheat flour-containing foods
Colombia	Fothergill et al. 2019	Folic acid, iron, riboflavin, thiamin	Anemia	Women of childbearing age Children 2–4 and 5–12 years	Prevalence of anemia was lowest in children 2–4 years who were in the highest quartile for intake of wheat flour-containing foods; prevalence of anemia did not differ by intake of wheat flour-containing foods[c] for women and children 5–12 years
Uzbekistan	Hund et al. 2013[d]	Iron	Iron depletion[e]	Women of childbearing age	Prevalence of iron depletion did not differ by household possession of flour that should be fortified[f] and bread that should be fortified
Uzbekistan	Hund et al. 2013[d]	Folic acid	Folate deficiency	Women of childbearing age	Prevalence of folate deficiency was higher in households possessing flour that should be fortified;[f,g] prevalence of folate deficiency was lower in households possessing bread that should be fortified

Uzbekistan	Hund et al. 2013[d]	Folic acid, iron, riboflavin, thiamin, zinc	Anemia	Women of childbearing age	Prevalence of anemia did not differ by household possession of flour that should be fortified[f] and bread that should be fortified

[a] Nutrient added to wheat flour through fortification which is purported to affect the health outcome.

[b] Some of the health outcomes measure the concentration of nutrients or other constituents in the blood or breastmilk and some refer to the prevalence (i.e. percentage of people who have the condition), the incidence (i.e. number of people newly diagnosed with the condition), or deaths (i.e. number of people who die due to the condition).

[c] Wheat-flour containing foods includes foods such as bread, pasta and cookies; the fortification status of the flour used to make these foods could not be confirmed (Fothergill et al. 2019).

[d] The unpublished report by Northrop-Clewes et al. (2013) contains the same information as this published article.

[e] Serum ferritin <12 μg/L; in summarizing the results in this chapter, this health outcome was classified as iron deficiency.

[f] While the fortification status of household wheat flour was measured (41.6% of flour was fortified), the researchers did not compare iron depletion, folate deficiency or anemia prevalence between households with fortified flour and those with non-fortified flour.

[g] This result is the opposite of what is expected.

Published by Burleigh Dodds Science Publishing Limited, 2021.

6.4 Other relevant evidence from individual countries

6.4.1 Result from cost-effectiveness studies in individual countries

Economists compare the costs of operating programs, such as fortification, with the effectiveness of such programs. They do this at two time points: before a program has started using hypothesized costs and outcomes (e.g. Dalziel et al. 2009), and after a program has operated using actual costs and outcomes. The latter studies are described here.

After fortification initiation, three countries compared the costs of adding folic acid to flour, the costs of treating people with spina bifida, a type of neural tube defect, and the effectiveness of fortification in reducing neural tube defects (Table 9). Each study showed significant annual net savings in healthcare expenses when spina bifida is prevented through fortification: 2.0–2.6 million international dollars in Chile (Llanos et al. 2007), 40.6 million Rand in South Africa (Sayed et al. 2008) and 88–603 million US dollars in the USA (Grosse et al. 2005, 2016). Since these are annual figures, every year of fortification leads to these net savings.

6.4.2 Results from cross-sectional studies in individual countries that assessed health outcomes only in the post-fortification period

As noted earlier, some researchers publish only post-fortification results. These are rarely informative without a comparison to pre-fortification values. Table 10 lists studies highlighting an outcome for which there are no pre- and post-fortification results: folate deficiency for Canada and folate-deficiency anemia for the USA. In both cases, the post-fortification prevalence of these outcomes is <1% suggesting that fortification with folic is contributing to keeping these values low.

6.4.3 Results from modeling the health impact of wheat flour fortification in individual countries

With information from singles countries, it is possible to statistically model the health impact that fortification is having (Table 11). Tice et al. (2001) modeled the impact of mandatory fortification of wheat flour, maize flour and rice with folic acid on myocardial infarctions (heart attacks) and death from coronary heart disease (CHD) in the USA. Using conservative assumptions of how much fortification would reduce homocysteine levels (i.e. by 5 μmol/L) and what those reductions would be due to the risk of coronary heart disease (i.e. decrease by 9%), they estimated that up to 1% of heart attacks and deaths from CHD could

Table 9 Results from cost-effectiveness studies in individual countries that assessed costs and health outcomes in the post-fortification period[a]

Country	Study	Nutrient[b]	Health outcome[c]	Individuals studied	Results
Chile	Llanos et al. 2007	Folic acid	Spina bifida[d]	Live births and fetal deaths[e] weighing 500 g or more	Annual net savings in healthcare expenses when spina bifida is prevented through fortification: 2.0-2.6 million international dollars
South Africa[a]	Sayed et al. 2008	Folic acid	Spina bifida[d]	Live births and still births	Annual net savings in healthcare expenses when spina bifida is prevented through fortification: 40.6 million Rand
USA[a]	Grosse et al. 2005, 2016	Folic acid	Spina bifida[d]	Live births	Annual net savings in healthcare expenses when spina bifida is prevented through fortification: 88–145 million US dollars (Grosse et al. 2005) and 299-603 million US dollars (Grosse et al. 2016)
USA[a]	Bentley et al. 2009	Folic acid	Neural tube defects (NTDs); myocardial infarctions (MIs); colon cancer incidence; vitamin B12 masking	Adults 15 years or older	Annual net savings through fortification at current levels (140 μg/100 g) when NTDs, MIs and colon cancer are averted and when masking of B12 deficiency occurs: 780.5 million US dollars

[a] Wheat flour was the only grain fortified with folic acid in Chile. Wheat and maize flour were fortified with folic acid in South Africa and the USA (in addition to rice); the independent effect of any one of these fortified foods on health outcomes cannot be discerned with this study.
[b] Nutrient added to wheat flour through fortification which is purported to affect the health outcome.
[c] Some of the health outcomes measure the concentration of nutrients or other constituents in the blood or breastmilk and some refer to the prevalence (i.e. percentage of people who have the condition), the incidence (i.e. number of people newly diagnosed with the condition), or deaths (i.e. number of people who die due to the condition).
[d] Spina bifida is a type of neural tube defect.
[e] 'Fetal death refers to the spontaneous intrauterine death of a fetus at any time during pregnancy. Fetal deaths later in pregnancy (at 20 weeks of gestation or more, or 28 weeks or more, for example) are also sometimes referred to as stillbirths' (CDC 2020).

Table 10 Results from cross-sectional studies in individual countries that assessed health outcomes in the post-fortification period only[a]

Country	Study	Nutrient[a]	Health outcome[b]	Individuals studied	Results
Canada	Colapinto et al. 2011	Folic acid	Folate deficiency	Females and males 6–79 years	Prevalence of folate deficiency in the post-fortification period was <1%
USA[a]	Odewole et al. 2013	Folic acid	Folate deficiency, folate-deficiency anemia	Adults 50 years and older	Prevalence of folate deficiency in the post-fortification period was 0.1%, prevalence of folate-deficiency anemia in the post-fortification period was 0.1%

[a] Wheat flour was the only grain fortified with folic acid in Canada. Wheat and maize flour and rice were fortified with folic acid in the USA; the independent effect of any one of these fortified foods on health outcomes cannot be discerned with this study.
[b] Nutrient added to wheat flour through fortification which is purported to affect the health outcome.
[c] Some of the health outcomes measure the concentration of nutrients or other constituents in the blood or breastmilk and some refer to the prevalence (i.e. percentage of people who have the condition), the incidence (i.e. number of people newly diagnosed with the condition), or deaths (i.e. number of people who die due to the condition).

be prevented over a 10-year period. Using less-conservative assumptions (i.e. homocysteine levels reduced by 11 μmol/L and CHD reduced by 29%), they estimated that up to 13% of heart attacks and deaths from CHD could be prevented over a 10-year period.

Table 11 Results from modeling the health impact of wheat flour fortification for individual countries[a]

Study	Country	Nutrient[b]	Health outcome[c]	Individuals studied	Results
Tice et al. 2001	USA[a]	Folic acid	Myocardial infarction incidence (MIs); coronary heart disease (CHD) deaths	Adults	Cereal grain fortification could reduce MI incidence and CHD deaths by 1–13% over a 10-year period

[a] In the United States, cereal grains that must be fortified with folic acid include wheat flour, maize flour and rice (Global Fortification Data Exchange 2020d). The independent effect of any one of these fortified foods on health outcomes cannot be discerned with this study.
[b] Nutrient added to wheat flour through fortification which is purported to affect the health outcome.
[c] Some of the health outcomes measure the concentration of nutrients or other constituents in the blood or breastmilk and some refer to the prevalence (i.e. percentage of people who have the condition), the incidence (i.e. number of people newly diagnosed with the condition), or deaths (i.e. number of people who die due to the condition).

6.5 Results from systematic reviews of multiple studies from multiple countries

Systematic reviews of the literature compare and contrast the health outcomes reported from multiple studies which can come from the same country but also often have information from different countries. Systematic reviews focused on wheat flour fortification or that include wheat flour fortification are summarized in Table 12.

One systematic review assessed the impact of flour fortification with folic acid on neural tube defects (Castillo-Lancellotti et al. 2013). Twenty seven studies were obtained from nine countries, of which most only fortified wheat flour with folic acid (Chile, Argentina, Canada, Iran, Jordan) and some fortified multiple foods with folic acid (Brazil and South Africa - wheat and maize flour; Costa Rica - wheat and maize flour, rice, milk; USA - wheat flour, maize flour, rice). The authors concluded that 'Fortification of flour with folic acid has had a major impact on [neural tube defects] in all countries where this has been reported.'

The review by van Gool et al. (2018) was difficult to interpret. The document reviewed evidence linking folic acid (from any source, including fortified food) to positive outcomes such as decreasing neural tube defects and to negative outcomes such as masking vitamin B12 deficiency. There was no succinct summary of each of the outcomes. Instead the authors concluded 'the risks carried by a high daily intake of folate equivalents do not outweigh the benefits of folic acid fortification of staple foods, as long as concentrations of serum un-metabolized folic acid, RBC folate, and serum vitamin B12 can be monitored periodically'.

6.6 Results from meta-analyses of multiple studies from multiple countries

Meta-analyses go one step further from systematic reviews and take numeric results from multiple studies and re-analyze them, to come up with a new estimate of what the relationship is between fortification and the health outcome. As with systematic reviews, meta-analyses are completed with data from multiple studies and they can be from the same country, or, more often than not, from different countries (Table 13).

Keats et al. (2019) published a meta-analysis of 17 studies of which 16 evaluated a national fortification program. In all cases, wheat and maize flour were fortified with folic acid and the fortification took place for 1–11 years in study countries. Several health outcomes were analyzed: serum folate, folate deficiency and neural tube defects. In eight studies with 6765 women, serum folate increased by 11.94 nmol/L from the pre- to the post-fortification period. In

Table 12 Results from systematic reviews of multiple studies from multiple countries that evaluated the health impact of wheat flour fortification

Study	Countries (n)	Nutrient[a]	Health outcome[b]	Individuals studied	Results
Assunção and Santos 2007	6	Iron	Anemia	Children	Prevalence of anemia lower in post-fortification than pre-fortification period
Castillo-Lancellotti et al. 2013	9	Folic acid	Neural tube defects (NTDs)	Differed depending on article reviewed	Prevalence of NTDs lower in post-fortification than pre-fortification period
Pacheco Santos and Zanon Pereira 2007	5	Folic acid	Neural tube defects	Live births	Prevalence of NTDs lower in post-fortification than pre-fortification period
Pachón et al. 2015	13	Multiple	Iron deficiency, anemia	Children and women	Prevalence of low ferritin lower in post-fortification than pre-fortification period for women; no difference in the prevalence of low ferritin in post-fortification and pre-fortification period in children; no difference in the prevalence of anemia in post-fortification and pre-fortification period in women and children
Rosenthal et al. 2014	15	Folic acid	Neural tube defects	Live births	Prevalence of NTDs lower in post-fortification than pre-fortification period
van Gool et al. 2018	Not specified	Folic acid	Multiple	Multiple	Not clear
van Gool et al. 2020	Not specified	Folic acid	Cognitive function	Not specified	'The hypothesis that cognitive impairment in "subclinical" cyanocobalamin deficiency is folate-mediated is untenable.'

[a] Nutrient added to wheat flour through fortification which is purported to affect the health outcome.
[b] Some of the health outcomes measure the concentration of nutrients or other constituents in the blood or breastmilk and some refer to the prevalence (i.e. percentage of people who have the condition), the incidence (i.e. number of people newly diagnosed with the condition), or deaths (i.e. number of people who die due to the condition).

four studies with 4645 women, the prevalence of women with folate deficiency decreased by 80% from the pre- to the post-fortification period. From eight studies with 19 million data points, the odds of a baby being born with neural tube defects declined by 41% between the pre- and the post-fortification period.

The National Toxicology Program (2015) reviewed research assessing adverse health outcomes after consuming high levels of folate (whether from food sources, supplements containing folic acid or foods fortified with folic acid). Four health outcomes were considered high priority (cancer, cognition, hypersensitivity, and thyroid- and diabetes-related disorders) and thus meta-analyses were conducted. In summary, 96% of 27 million data points showed no relationship between high folate and cancer, and for cognition, hypersensitivity (such as asthma) and thyroid- and diabetes-related disorders, results were 'not supportive' or 'inconclusive' of a relationship between these outcomes and high folate (Table 13). The effect of wheat flour fortification with folic acid cannot be isolated from these studies; however, these results suggest that high folate levels (independent of the source) are not associated with negative health outcomes.

For other health outcomes, that were deemed lower priority by the National Toxicology Program, research results were briefly summarized in their 2015 report. From these summaries, it is not clear what the folate source(s) were. Nevertheless, they consistently found no relationship between folate intake from any source and adverse health outcomes, as follows:

- Cardiovascular outcomes. 'None of the 39 identified meta-analyses reported any adverse effects associated with folic acid intake.'
- Twinning and multiple births. 'While it may be biologically plausible that periconceptional vitamin use plays a role in the incidence of multiple births, the available evidence has been well explored – the most recent human study identified was published in 2006 – so this was not determined to be a high priority topic for this review.'
- Autism. 'Due to weaknesses in the design of studies reporting adverse effects, the currently available literature did not support consideration of autism as a high priority outcome for this review.'
- Other neurological outcomes. 'None of the 10 identified meta-analyses reported adverse effects of folic acid.'
- Other immunological outcomes. 'The majority of other immunological outcomes which were not considered hypersensitive-related, such as autoimmune diseases, did not suggest any adverse effects of folic acid and were not considered a high priority category.'
- Other endocrine and metabolic disease outcomes. 'Body weight, body composition, and BMI constituted the largest group of studies (~50),

Table 13 Results from meta-analyses of multiple before and after studies from multiple countries that evaluated the health impact of wheat flour fortification

Study	Countries or studies (n)	Nutrient[a]	Health outcome[b]	Individuals studied	Results
Atta et al. 2016	52 countries	Folic acid	Spina bifida[c]	Live births; live births and still births; live births, still births and terminations	Prevalence of spina bifida lower in countries with mandatory fortification than voluntary fortification
Blencowe et al. 2010	6 countries	Folic acid	Neural tube defects	Not specified	Prevalence of NTDs lower in post-fortification than pre-fortification period
Das et al. 2013	10 countries	Folic acid	Serum folate; neural tube defects	Women of reproductive age	No difference in serum folate levels between pre- and post-fortification periods; prevalence of NTDs lower in post-fortification than pre-fortification period
Keats et al. 2019	12[d] countries	Folic acid	Serum folate; folate deficiency; neural tube defects	Live births; still births	Higher serum folate levels in the post-fortification than pre-fortification period; prevalence of folate deficiency and NTDs lower in post-fortification than pre-fortification period
Millacura et al. 2017	9 countries	Folic acid	Orofacial clefts	Total births	Prevalence of non-syndromic cleft lip with or without cleft palate is lower in post-fortification than pre-fortification period; no difference in total orofacial clefts, cleft lip with or without cleft palate, cleft palate, non-syndromic orofacial clefts, and non-syndromic cleft palate between pre- and post-fortification periods

National Toxicology Program 2015	43 pooled and meta-analysis studies	Folate[e]	Cancer	Not specified	From 27 million data points, 96% showed no relationship or a protective relationship between high folate (intake or blood levels) and cancer and 4% showed a harmful relationship between high folate (intake or blood levels) and cancer
National Toxicology Program 2015	28 studies and 2 meta-analyses[f]	Folate[e]	'Cognition in conjunction with vitamin B12 deficiency'	Adults	Results 'not supportive' of a relationship between high folate and cognitive impairment in the presence of vitamin B12 deficiency
National Toxicology Program 2015	42 studies and 1 meta-analysis[f]	Folate[e]	Hypersensitivity (e.g. asthma, eczema)	All age groups	Results 'not supportive' or 'inconclusive' of a relationship between high folate and hypersensitivity
National Toxicology Program 2015	72 studies and 1 meta-analysis[f]	Folate[e]	Thyroid- and diabetes-related disorders (e.g. insulin resistance, metabolic syndrome)		Results "not supportive" of a relationship between high folate and thyroid or diabetes-related disorders

[a] Nutrient added to wheat flour through fortification which is purported to affect the health outcome.
[b] Some of the health outcomes measure the concentration of nutrients or other constituents in the blood or breastmilk and some refer to the prevalence (i.e. percentage of people who have the condition), the incidence (i.e. number of people newly diagnosed with the condition), or deaths (i.e. number of people who die due to the condition).
[c] Spina bifida is a type of neural tube defect.
[d] Studies from these 12 countries did not all report on the three outcomes: serum folate, folate deficiency and neural tube defects.
[e] Researchers assessed high levels of folate from any source: non-fortified food, supplements containing folic acid or foods fortified with folic acid; the effect of wheat flour fortification with folic acid cannot be isolated from this study (National Toxicology Program 2015).
[f] Unable to download the meta-analysis results from the study website (https://hawcproject.org/assessment/67/downloads/; https://hawcproject.org/assessment/48/downloads/) (National Toxicology Program 2015).

with only 2 studies reporting any significant relationship between higher folate intake or level and increased body weight. No studies of folate and polycystic ovary syndrome or pancreatitis reported any adverse associations.'

- Other reproductive outcomes. 'None of the 9 meta-analyses reported an adverse effect of folic acid, so reproductive effects were not considered a high priority category.'
- Mortality. '18 meta-analyses have been conducted for several mortality outcomes with a sufficient number of available studies (e.g. all-cause, cardiovascular, cancer, perinatal) and none report any statistically significant adverse meta-estimates.'

6.7 Results from modeling the health impact of wheat flour fortification for multiple countries

With information from multiple countries, it is possible to model fortification's health impact (Table 14). For example, researchers estimated how much of the neural tube defects that can be prevented with folic acid is being prevented through fortification of wheat and/or maize flour with folic acid (Kancherla et al. 2018). The investigators modeled the impact of fortification on three groups of countries: those with high prevention potential, because they have fortification programs in place; those with no prevention potential because they have no fortification programs; and those with modest prevention potential because their fortification programs do not have high coverage or high-enough levels of folic acid to prevent neural tube defects. They estimated 50 270 birth defects were prevented in 2017 where flour was fortified with folic acid.

6.8 Studies from multiple countries that did not assess the independent contribution of mandatorily fortified wheat flour

Additional systematic reviews and meta-analyses were completed that analyzed the health impact of many fortified foods simultaneously, including wheat flour. However, they were not included in this chapter for one of two reasons (Table 15). One, the wheat flour was not fortified as part of the country's mandatory fortification program, but rather for the explicit purposes of conducting the research project. Two, the results were presented combined, for all foods together, and it was not possible to isolate the contribution of fortified wheat flour.

For example, Best and colleagues (2011) reviewed the impact of foods fortified with multiple micronutrients on many health outcomes including nutritional status, growth and cognitive development. Of 12 studies included

Table 14 Results from modeling the health impact of wheat flour fortification, alone or in combination with maize flour, for multiple countries

Study	Countries (n)	Nutrient[a]	Health outcome[b]	Individuals studied	Results
Youngblood et al. 2013	65	Folic acid	Neural tube defects	Live births	In 2012, flour fortification with folic acid prevented 25% of neural tube defects[c]
Arth et al. 2016	58	Folic acid	Neural tube defects	Live births	In 2015, flour fortification with folic acid prevented 13.2% of neural tube defects[c]
Kancherla et al. 2018	59	Folic acid	Neural tube defects	Live births	In 2017, flour fortification with folic acid prevented 18% of neural tube defects[c]

[a] Nutrient added to wheat flour through fortification which is purported to affect the health outcome.

[b] Some of the health outcomes measure the concentration of nutrients or other constituents in the blood or breastmilk and some refer to the prevalence (i.e. percentage of people who have the condition), the incidence (i.e. number of people newly diagnosed with the condition), or deaths (i.e. number of people who die due to the condition).

[c] These are the neural tube defects that can be prevented by women having optimum blood folate levels around the time of conception, known as folic acid preventable spina bifida and anencephaly.

Table 15 Systematic reviews or meta-analyses that evaluated wheat flour in the assessment of fortification's health impact and reason for exclusion from this chapter

Study	Nutrient[a]	Health outcome[b]	Reason for exclusion[c]
Athe et al. 2014	Iron	Hemoglobin	Effect of wheat flour could not be isolated
Best et al. 2011	Multiple	Multiple	Flour not fortified per mandatory fortification program
Black et al. 2012	Vitamin D	Vitamin D status	Flour not fortified per mandatory fortification program
Castillo-Lancellotti et al. 2012	Folic acid	Breast cancer risk	Effect of wheat flour could not be isolated
Chen et al. 2014	Folic acid	Breast cancer risk	Effect of wheat flour could not be isolated
Centeno Tablante et al. 2019	Folic acid	Multiple	For most studies, flour not fortified per mandatory fortification program
Eichler et al. 2019	Multiple	Multiple	For most studies, flour not fortified per mandatory fortification program
Gera et al. 2012	Iron	Multiple	Effect of wheat flour could not be isolated
Ghanchi et al. 2019	Iron	Diarrhea	Flour not fortified per mandatory fortification program
Hess et al. 2016	Multiple	Multiple	Flour not fortified per mandatory fortification program
Hombali et al. 2019	Vitamin A	Multiple	Effect of wheat flour could not be isolated
Kennedy et al. 2011	Folic acid	Colorectal cancer	Effect of wheat flour could not be isolated
Mendu et al. 2019	Vitamin A	Vitamin A status	Effect of wheat flour could not be isolated
O'Donnell et al. 2008	Vitamin D	Vitamin D status	Effect of wheat flour could not be isolated
Petry et al. 2016	Iron, Zinc	Multiple	Effect of wheat flour could not be isolated
Salam et al. 2019	Multiple	Multiple	Effect of wheat flour could not be isolated
Shah et al. 2016	Zinc	Multiple	Flour not fortified per mandatory fortification program
Tam et al. 2020	Multiple	Multiple	Effect of wheat flour could not be isolated

Study	Nutrient[a]	Health outcome[b]	Reason for exclusion[c]
Tio et al. 2014	Folic acid	Breast cancer risk	Effect of wheat flour could not be isolated
Yang et al. 2016	Folic acid	Multiple	Effect of wheat flour could not be isolated

[a] Nutrient added to foods through fortification which is purported to affect the health outcome.
[b] Some of the health outcomes measure the concentration of nutrients or other constituents in the blood or breastmilk and some refer to the prevalence (i.e. percentage of people who have the condition), the incidence (i.e. number of people newly diagnosed with the condition), or deaths (i.e. number of people who die due to the condition).
[c] Reason why the research was excluded from this chapter: (1) the wheat flour was not fortified under the rubric of the country's mandatory fortification program and/or (2) the results were presented combined for all fortified foods and the independent contribution of fortified wheat flour could not be isolated.

in their review, two assessed the impact of fortified biscuits. Both of these studies used flour that was fortified to meet the researchers' scientific interests; the flour was not fortified according to the country's mandatory fortification program (or the country did not have such a national program at that time).

7 Summary

- The health impact of wheat flour fortification after large-scale implementation in countries has been studied employing different study designs and in a diversity of health outcomes that span all age groups and many systems of the human body.
- Folic acid was the most studied nutrient. To a lesser extent, fortification with iron, vitamin B12 and zinc was also examined.
- The most studied outcomes were neural tube defects, cancer, folate status, folate deficiency, anemia, iron deficiency, iron status, hemoglobin and iron-deficiency anemia.
- For all of these outcomes except iron-deficiency anemia, the majority of studies showed improvements after fortification. That is, these studies suggest that fortification reduces neural tube defects, cancer, folate deficiency, anemia, and iron deficiency, and that fortification improves folate status, iron status, and hemoglobin levels.
- For some of the outcomes (cancer, anemia, hemoglobin, folate deficiency), there were studies that indicated health outcomes worsened after fortification. These analyses suggest that cancer, anemia and folate deficiency increased and hemoglobin levels decreased after fortification.
- Discrepant values for cancer may be due to the difference in years since fortification was initiated (e.g. increased incidence of colon cancer after fortification is suggested by studies published in the 2000s; the opposite

is observed in studies published in the 2010s) or by the sample size in studies (e.g. increased breast cancer incidence after fortification is observed in studies with sample sizes <2000; studies that observed no difference or a decreased incidence after fortification have sample sizes >2000 and going into the millions).

- Discrepant values for anemia and hemoglobin may be due to the (1) existence of non-nutritional causes of anemia which cannot be addressed by fortification, (2) nutritional causes of anemia not addressed by fortification because a limited number of nutrients were added through fortification and (3) levels of nutrients or fortification compounds used in fortification do not follow international guidelines.
- Many more outcomes were studied which only have results for four or fewer analyses: folate-deficiency anemia, homocysteine status, high homocysteine, vitamin B12 deficiency masking, congenital heart disease, coronary heart disease, stroke, myocardial infarction, cognitive function, hypersensitivity, thyroid- and diabetes-related disorders, orofacial clefts, vitamin B12 status, vitamin B12 deficiency, zinc status and zinc deficiency.
- Homocysteine status, high homocysteine, stroke death, myocardial infarction, orofacial clefts, vitamin B12 status, vitamin B12 deficiency, zinc status and zinc deficiency trend toward improvement after fortification.
- Vitamin B12 deficiency masking and cognitive function trend toward showing no difference before and after fortification.
- None of the outcomes with two, three or four analyses shows a worsening of health after fortification.
- While none of the study designs employed can be used to confirm a causal relationship between fortification and health outcomes, the preponderance of the evidence suggests that wheat flour fortification improves many health outcomes.
- Because fortification may also be associated with negative health outcomes such as cancer, health monitoring should continue to assess these outcomes in countries with fortification.

8 Future trends in research

Program decision makers are urged to consider several actions that can facilitate the health impact evaluation of their flour fortification programs:

- Plan for health impact evaluations while planning for the implementation of fortification.
- Use existing data or data-collection systems to evaluate the health impact of fortification at lower cost than planning a stand-alone fortification evaluation.

- Regularly review all kinds of information on the fortification program to determine if it is likely to be having health impact, such as intake, coverage and compliance. And, review this information before proceeding with an impact evaluation.
- Craft program impact pathways to document the plausibility that fortification contributed to the health outcome observed.
- Make fortification information publicly available for scrutiny by interested parties.
- Ensure that multiple programs (such as fortification, supplementation, micronutrient powders, biofortification) are not providing an excess of nutrients to the population. If they are, review and adjust programs accordingly.
- Continue to assess for potential negative health outcomes of fortification. This is especially important for outcomes that may take years or decades to manifest, such as cancer.

9 Where to look for further information

9.1 World Health Organization guidelines

The World Health Organization (WHO) offers several guidelines related to food fortification generally and wheat flour fortification specifically.

WHO's website on wheat flour fortification.

WHO. e-Library of Evidence for Nutrition Actions (eLENA): fortification of wheat flour. https://www.who.int/elena/titles/wheat-flour-fortification/en/.

WHO and FAO's book with basic principles of food fortification.

WHO and FAO. (2006). Guidelines on food fortification with micronutrients. https://www.who.int/nutrition/publications/micronutrients/9241594012/en/.

WHO's recommendations for wheat and maize flour fortification.

World Health Organization, Food and Agriculture Organization of the United Nations, the United Nations Children's Fund, Global Alliance for Food Fortification, Micronutrient Initiative, and the Food Fortification Initiative. (2009). Recommendations on wheat and maize flour fortification. Meeting report: interim consensus statement. https://www.who.int/nutrition/publications/micronutrients/wheat_maize_fortification/en/.

Special issue of the Food and Nutrition Bulletin journal that summarizes the evidence that generated WHO's 2009 recommendations.

(2010). The opportunity of flour fortification: building on the evidence to move forward. https://journals.sagepub.com/toc/fnba/31/1_suppl1.

9.2 Flour Millers toolkit

The Food Fortification Initiative provides basic specifications for fortifying flour at the wheat mill.

Food Fortification Initiative. Flour Millers Toolkit for implementing wheat and maize flour fortification. http://www.ffinetwork.org/tools_training/flour_millers_toolkit.html.

9.3 Best practices for foundational fortification documents

A review of best practices for fortification legislation and standard documents and monitoring protocols and an assessment of how closely countries with mandatory fortification of wheat flour, maize flour and rice follow these best practices.

Marks, K. J. et al. (2018), Review of grain fortification legislation, standards, and monitoring documents. https://www.ghspjournal.org/content/6/2/356.

9.4 Government monitoring of fortification

Monitoring by governments is essential to ensure flour is adequately and consistently fortified. This document provides guidance on the minimum elements that should be included in a country's monitoring plan.

Global Alliance for Improved Nutrition and Project Healthy Children. (2018). Regulatory monitoring of national food fortification programs: A policy guidance document. https://fortificationdata.org/resources/.

9.5 Country statistics on wheat flour fortification

These can be found at two websites:

Global Fortification Data Exchange. https://fortificationdata.org.
Food Fortification Initiative. http://www.ffinetwork.org.

9.6 Book on food fortification

A recently published book on fortification offers over 40 chapters on different aspects of food fortification, including health impact evaluations.

Mannar, M. G. V. and Hurrell, R. F. (2018), Food fortification in a globalized world. https://www.elsevier.com/books/food-fortification-in-a-globalized-world/mannar/978-0-12-802861-2.

10 References

Abdollahi, Z., Elmadfa, I., Djazayery, A., Golalipour, M. J., Sadighi, J., Salehi, F. and Sharif, S. S. (2011). Efficacy of flour fortification with folic acid in women of childbearing age in Iran, *Ann. Nutr. Metab.* 58(3), 188–96.

Alasfoor, D., Elsayed, M. K. and Mohammed, A. J. (2010). Spina bifida and birth outcome before and after fortification of flour with iron and folic acid in Oman, *East. Mediterr. Health J.* 16(5), 533–8.

Allen, L. H., Miller, J. W., de Groot, L., Rosenberg, I. H., Smith, A. D., Refsum, H. and Raiten, D. J. (2018). Biomarkers of Nutrition for Development (BOND): vitamin B-12 review, *J. Nutr.* 148(suppl_4), 1995S–2027S.

Amarin, Z. O. and Obeidat, A. Z. (2010). Effect of folic acid fortification on the incidence of neural tube defects, *Paediatr. Perinat. Epidemiol.* 24(4), 349–51.

Arth, A., Kancherla, V., Pachón, H., Zimmerman, S., Johnson, Q. and Oakley, G. P. Jr. (2016). A 2015 global update on folic acid-preventable spina bifida and anencephaly, *Birth Defects Res. A* 106(7), 520–9.

Assunção, M. C. F. and Santos, I. S. (2007). Effect of food fortification with iron on childhood anemia: a review study [Efeito da fortificação de alimentos com ferro sobre anemia em crianças: um estudo de revisão], *Cad. Saúde Pública* 23(2), 269–81.

Assunção, M. C., Santos, I. S., Barros, A. J., Gigante, D. P. and Victora, C. G. (2007). Effect of iron fortification of flour on anemia in preschool children in Pelotas, Brazil, *Rev. Saude Publ.* 41(4), 539–48.

Assunção, M. C., Santos, I. S., Barros, A. J., Gigante, D. P. and Victora, C. G. (2012). Flour fortification with iron has no impact on anaemia in urban Brazilian children, *Public Health Nutr.* 15(10), 1796–801. Erratum in: *Public Health Nutr.* (2013) 16(1), 188.

Athe, R., Rao, M. V. and Nair, K. M. (2014). Impact of iron-fortified foods on Hb concentration in children (<10 years): a systematic review and meta-analysis of randomized controlled trials, *Public Health Nutr.* 17(3), 579–86.

Atta, C. A., Fiest, K. M., Frolkis, A. D., Jette, N., Pringsheim, T., St Germaine-Smith, C., Rajapakse, T., Kaplan, G. G. and Metcalfe, A. (2016). Global birth prevalence of spina bifida by folic acid fortification status: a systematic review and meta-analysis, *Am. J. Public Health* 106(1), e24–34.

Avagliano, L., Massa, V., George, T. M., Qureshy, S., Bulfamante, G. P. and Finnell, R. H. (2018). Overview on neural tube defects: from development to physical characteristics, *Birth Defects Res.* 111(19), 1455–67.

Barboza Argüello, M. P. and Umaña Solís, L. M. (2011). 'Impacto de la fortificación de alimentos con ácido fólico en los defectos del tubo neural en Costa Rica', *Rev. Panam. Salud Publica..*, 30(1), 1–6.

Barboza Argüello, M. L., Umaña-Solís, L. M., Azofeifa, A., Valencia, D., Flores, A. L., Rodríguez-Aguilar, S., Alfaro-Calvo, T. and Mulinare, J. (2015). Neural tube defects in Costa Rica, 1987-2012: origins and development of birth defect surveillance and folic acid fortification, *Matern. Child Health J.* 19(3), 583-90.

Bauernfeind, J. C. and DeRitter, E. (1991). Foods considered for nutrient addition: cereal grain products. In: Bauernfeind, J. C. and Lachance, P. A. (Eds) *Nutrient Additions to Food: Nutritional, Technological and Regulatory Aspects*. Food & Nutrition Press, Inc, Trumbull, pp. 143-209.

Beckett, E. L., Martin, C., Boyd, L., Porter, T., King, K., Niblett, S., Yates, Z., Veysey, M. and Lucock, M. (2017). Reduced plasma homocysteine levels in elderly Australians following mandatory folic acid fortification–a comparison of two cross-sectional cohorts, *J. Nutr. Intermed. Metab.* 8, 14-20.

Bentley, T. G., Weinstein, M. C., Willett, W. C. and Kuntz, K. M. (2009). A cost-effectiveness analysis of folic acid fortification policy in the United States, *Public Health Nutr.* 12(4), 455-67.

Berry, R. J. (2019). Lack of historical evidence to support folic acid exacerbation of the neuropathy caused by vitamin B12 deficiency, *Am. J. Clin. Nutr.* 110(3), 554-61.

Best, C., Neufingerl, N., Del Rosso, J. M., Transler, C., van den Briel, T. and Osendarp, S. (2011). Can multi-micronutrient food fortification improve the micronutrient status, growth, health, and cognition of schoolchildren? a systematic review, *Nutr. Rev.* 69(4), 186-204.

Bidondo, M. P., Liascovich, R., Barbero, P. and Groisman, B. (2015). Prevalencia de defectos del tubo neural y estimación de casos evitados posfortificación en Argentina, *Arch. Argent. Pediatr.* 113(6), 498-501.

Black, L. J., Seamans, K. M., Cashman, K. D. and Kiely, M. (2012). An updated systematic review and meta-analysis of the efficacy of vitamin D food fortification, *J. Nutr.* 142(6), 1102-8.

Blencowe, H., Cousens, S., Modell, B. and Lawn, J. (2010). Folic acid to reduce neonatal mortality from neural tube disorders, *Int. J. Epidemiol.* 39 (Suppl. 1), i110-21.

Blencowe, H., Kancherla, V., Moorthie, S., Darlison, M. W. and Modell, B. (2018). Estimates of global and regional prevalence of neural tube defects for 2015: a systematic analysis, *Ann. N.Y. Acad. Sci.* 1414(1), 31-46.

Botto, L. D., Lisi, A., Bower, C., Canfield, M. A., Dattani, N., De Vigan, C., De Walle, H., Erickson, D. J., Halliday, J., Irgens, L. M., Lowry, R. B., McDonnell, R., Metneki, J., Poetzsch, S., Ritvanen, A., Robert-Gnansia, E., Siffel, C., Stoll, C. and Mastroiacovo, P. (2006). Trends of selected malformations in relation to folic acid recommendations and fortification: an international assessment, *Birth Defects Res. A* 76(10), 693-705.

Britto, J. C., Cançado, R. and Guerra-Shinohara, E. M. (2014). Concentrations of blood folate in Brazilian studies prior to and after fortification of wheat and cornmeal (maize flour) with folic acid: a review, *Rev. Bras. Hematol. Hemoter.* 36(4), 275-86.

Brown, R. D., Langshaw, M. R., Uhr, E. J., Gibson, J. N. and Joshua, D. E. (2011). The impact of mandatory fortification of flour with folic acid on the blood folate levels of an Australian population, *Med. J. Aust.* 194(2), 65-7.

Castilla, E. E., Orioli, I. M., Lopez-Camelo, J. S., da Graca Dutra, M. and Nazer-Herrera, J. and for the ECLAMC: Latin American Collaborative Study of Congenital Malformations. (2003). Preliminary data on changes in neural tube defect prevalence rates after folic acid fortification in South America, *Am. J. Med. Genet. A* 123A(2), 123-8.

Castillo-Lancellotti, C., Tur, J. A. and Uauy, R. (2012). Folatos y riesgo de cáncer de mama: revisión sistemática [Folate and breast cancer risk: a systematic review], *Rev. Med. Chil.* 140(2), 251–60.

Castillo-Lancellotti, C., Tur, J. A. and Uauy, R. (2013). Impact of folic acid fortification of flour on neural tube defects: a systematic review, *Public Health Nutr.* 16(5), 901–11. Erratum in: *Public Health Nutr.* (2013) 16(8), 1527.

Centeno Tablante, E., Pachón, H., Guetterman, H. M. and Finkelstein, J. L. (2019). Fortification of wheat and maize flour with folic acid for population health outcomes, *Cochrane Database Syst. Rev.* 7(7), CD012150.

Centers for Disease Control and Prevention. (2003). Folate status in women of childbearing age, by race/ethnicity-United states, 1999–2000, *M.M.W.R. Morb. Mortal. Wkly. Rep.* 51, 808–10.

Centers for Disease Control and Prevention. (2004). Spina bifida and anencephaly before and after folic acid mandate- United States, 1995–1996 and 1999–2000, *M.M.W.R. Morb. Mortal. Wkly. Rep.* 53(17), 362–5.

Centers for Disease Control and Prevention. (2020). Fetal deaths. Available at: https://www.cdc.gov/nchs/nvss/fetal_death.htm; Accessed 10 May 2020.

Chaparro, C. M. and Suchdev, P. S. (2019). Anemia epidemiology, pathophysiology, and etiology in low- and middle-income countries, *Ann. N.Y. Acad. Sci.* 1450(1), 15–31.

Chen, P., Li, C., Li, X., Li, J., Chu, R. and Wang, H. (2014). Higher dietary folate intake reduces the breast cancer risk: a systematic review and meta-analysis, *Br. J. Cancer* 110(9), 2327–38.

Colapinto, C. K., O'Conner, D. L. and Tremblay, M. S. (2011). Folate status of the population in the Canadian Health Measures Survey, *C.M.A.J.* 183(2), E100–6.

Corral, S. E., Moreno, S. R., Pérez, G. G., Ojeda, B. M. E, Valenzuela, G. H., Reascos, M. M. and Sepúlveda L. W. (2006). Defectos congénitos cráneo-encefálicos: variedades y respuesta a la fortificación de la harina con ácido fólico, *Rev. Méd. Chile* 134(9), 1129–34. https://scielo.conicyt.cl/scielo.php?script=sci_arttext&pid=S0034-98872006000900007

Cortés, F., Mellado, C., Pardo, R. A., Villarroel, L. A. and Hertrampf, E. (2012). Wheat flour fortification with folic acid: changes in neural tube defects rates in Chile, *Am. J. Med. Genet. A* 158A(8), 1885–90.

Costa, C. A., Machado, E. H., Colli, C., Latorre, W. C. and Szarfarc, S. C. (2009). [Anemia in pre-school children attending day care centers of São Paulo: perspectives of the wheat and maize flour fortification], *Nutrire* 34(1), 59–74.

Czeizel, A. E. and Dudás, I. (1992). Prevention of the first occurrence of neural-tube defects by periconceptional vitamin supplementation, *N. Engl. J. Med.* 327(26), 1832–5.

Dalziel, K., Segal, L. and Katz, R. (2009). Cost-effectiveness of mandatory folate fortification v. other options for the prevention of neural tube defects: results from Australia and New Zealand, *Public Health Nutr.* 13(4), 566–78.

Das, J. K., Salam, R. A., Kumar, R. and Bhutta, Z. A. (2013). Micronutrient fortification of food and its impact on woman and child health: a systematic review, *Syst. Rev.* 2, 67.

De Wals, P., Rusen, I. D., Lee, N. S., Morin, P. and Niyonsenga, T. (2003). Trend in prevalence of neural tube defects in Quebec, *Birth Defects Res. A* 67(11), 919–23.

De Wals, P., Tairou, F., Van Allen, M. I., Uh, S. H., Lowry, R. B., Sibbald, B., Evans, J. A., Van den Hof, M. C., Zimmer, P., Crowley, M., Fernandez, B., Lee, N. S. and Niyonsenga, T. (2007). Reduction in neural-tube defects after folic acid fortification in Canada, *N. Engl. J. Med.* 357(2), 135–42.

Dietrich, M., Brown, C. J. P. and Block, G. (2005). The effect of folate fortification of cereal-grain products on blood folate status, dietary folate intake, and dietary folate sources among adult non-supplement users in the United States, *J. Am. Coll. Nutr.* 24(4), 266–74.

Donoso D, A., Villarroel del, P. L. and Pinedo, M. G. (2006). Aumento de la mortalidad por cáncer de colon en Chile, 1990–2003, *Rev. Méd. Chile* 134(2), 152–8.

Eichler, K., Hess, S., Twerenbold, C., Sabatier, M., Meier, F. and Wieser, S. (2019). Health effects of micronutrient fortified dairy products and cereal food for children and adolescents: a systematic review, *PLoS ONE* 14(1), e0210899.

Engle-Stone, R., Nankap, M., Ndjebayi, A. O., Allen, L. H., Shahab-Ferdows, S., Hampel, D., Killilea, D. W., Gimou, M. M., Houghton, L. A., Friedman, A., Tarini, A., Stamm, R. A. and Brown, K. H. (2017), Iron, zinc, folate, and vitamin B-12 status increased among women and children in Yaoundé and Douala, Cameroon, 1 year after introducing fortified wheat flour, *J. Nutr.* 147, 1–11.

Enquobahrie, D. A., Feldman, H. A., Hoelscher, D. H., Steffen, L. M., Webber, L. S., Zive, M. M., Rimm, E. B., Stampfer, M. J. and Osganian, S. K. (2012),. 'Serum homocysteine and folate concentrations among a US cohort of adolescents before and after folic acid fortification', *Public Health Nutr.*, 15(10), 1818--26.

Food Fortification Initiative. (2020). *2019 Annual Report.* Food Fortification Initiative, Atlanta.

Fothergill, A., Fonseca Centeno, Z. Y., Ocampo Téllez, P. R. and Pachón, H. (2019). Consumption of fortified wheat flour and associations with anemia and low serum ferritin in Colombia, *Perspect. Nutr. Hum.* 21(2), 159–71.

French, A. E., Grant, R., Weitzman, S., Ray, J. G., Vermeulen, M. J., Sung, L., Greenberg, M. and Koren, G. (2003). Folic acid food fortification is associated with a decline in neuroblastoma, *Clin. Pharmacol. Ther.* 74(3), 288–94.

Fujimori, E., Sato, A. P., Szarfarc, S. C., Veiga, G. V., Oliveira, V. A., Colli C, Moreira-Araújo, R. S., Arruda, I. K., Uchimura, T. T., Brunken, G. S., Yuyama, L. K., Muniz, P. T., Priore, S. E., Tsunechiro, M. A., Frazão, A. d., Passoni, C. R. and Araújo, C. R. (2011). Anemia in Brazilian pregnant women before and after flour fortification with iron, *Rev. Saude Publ.* 45(6), 1027–35.

Gera, T., Sachdev, H. S. and Boy, E. (2012). Effect of iron-fortified foods on hematologic and biological outcomes: systematic review of randomized controlled trials, *Am. J. Clin. Nutr.* 96(2), 309–24.

Ghanchi, A., James, P. T. and Cerami, C. (2019). Guts, germs, and iron: a systematic review on iron supplementation, iron fortification, and diarrhea in children aged 4–59 months, *Curr. Dev. Nutr.* 3(3), nzz005.

Gibson, R. S. (1990). *Principles of Nutritional Assessment.* Oxford University Press, New York.

Gibson, T. M., Weinstein, S. J., Pfeiffer, R. M., Hollenbeck, A. R., Subar, A. F., Schatzkin, A., Mayne, S. T. and Stolzenberg-Solomon, R. (2011). Pre- and postfortification intake of folate and risk of colorectal cancer in a large prospective cohort study in the United States, *Am. J. Clin. Nutr.* 94(4), 1053–62.

Global Fortification Data Exchange. (2020a). Interactive map: fortification legislation. Available at: https://fortificationdata.org/interactive-map-fortification-legislation/#; Accessed 10 February 2020.

Global Fortification Data Exchange. (2020b). Interactive map: nutrient levels in fortification standards (mid-range or average). Available at: https://fortificationdata.org/map-nutrient-levels-in-fortification-standards/; Accessed 10 February 2020.

Global Fortification Data Exchange. (2020c). Brazil fortification dashboard. Available at: https://fortificationdata.org/country-fortification-dashboard/?alpha3_code=BRA&lang=en; Accessed 12 April 2020.

Global Fortification Data Exchange. (2020d). United States of America fortification dashboard. Available at: https://fortificationdata.org/country-fortification-dashboard/?alpha3_code=USA&lang=en; Accessed 12 April 2020.

Grimm, K. A., Sullivan, K. M., Alasfoor, D., Parvanta, I., Mohammad Suleiman, A. J., Kaur, M., Al-Hatmi, F. O. and Ruth, L. J. (2012). Iron-fortified wheat flour and iron deficiency among women, *Food Nutr. Bull.* 33(3), 180-5.

Grosse, S. D., Berry, R. J., Tilford, J. M., Kucik, J. E. and Waitzman, N. J. (2016). Retrospective assessment of cost savings from prevention: folic acid and spina bifida in the U.S, *Am. J. Prev. Med.* 50 (Suppl. 1), S74-80.

Grosse, S. D., Waitzman, N. J., Romano, P. S. and Mulinare, J. (2005). Reevaluating the benefits of folic acid fortification in the United States: economic analysis, regulation, and public health, *Am. J. Public Health* 95(11), 1917-22.

Grupp, S. G., Greenberg, M. L., Ray, J. G., Busto, U., Lanctôt, K. L., Nulman, I. and Koren, G. (2011). Pediatric cancer rates after universal folic acid flour fortification in Ontario, *J. Clin. Pharmacol.* 51(1), 60-5.

Hertrampf, E., Cortés, F., Erickson, J. D., Cayazzo, M., Freire, W., Bailey, L. B., Howson, C., Kauwell, G. P. and Pfeiffer, C. (2003). Consumption of folic acid-fortified bread improves folate status in women of reproductive age in Chile, *J. Nutr.* 133(10), 3166-9.

Hess, S., Tecklenburg, L. and Eichler, K. (2016). Micronutrient fortified condiments and noodles to reduce anemia in children and adults--a literature review and meta-analysis, *Nutrients* 8(2), 88.

Hilder, L. (2016). Neural tube defects in Australia, 2007-2011: before and after implementation of the mandatory folic acid fortification standard. Technical Report 11488, National Perinatal Epidemiology and Statistics Unit, University of New South Wales, Wales.

Hirsch, S., de la Maza, P., Barrera, G., Gattás, V., Petermann, M. and Bunout, D. (2002). The Chilean flour folic acid fortification program reduces serum homocysteine levels and masks vitamin B-12 deficiency in elderly people, *J. Nutr.* 132(2), 289-91.

Hirsch, S., Sanchez, H., Albala, C., de la Maza, M. P., Barrera, G., Leiva, L. and Bunout, D. (2009). Colon cancer in Chile before and after the start of the flour fortification program with folic acid, *Eur. J. Gastroenterol. Hepatol.* 21(4), 436-9.

Hombali, A. S., Solon, J. A., Venkatesh, B. T., Nair, N. S. and Peña-Rosas, J. P. (2019). Fortification of staple foods with vitamin A for vitamin A deficiency, *Cochrane Database Syst. Rev.* 5(5), CD010068.

Honein, M. A., Paulozzi, L. J., Mathews, T. J., Erickson, J. D. and Wong, L. Y. (2001). Impact of folic acid fortification of the US food supply on the occurrence of neural tube defects, *J.A.M.A.* 285(23), 2981-6.

Houghton, S. C., Eliassen, A. H., Zhang, S. M., Selhub, J., Rosner, B. A., Willett, W. C. and Hankinson, S. E. (2019a). Plasma B-vitamin and one-carbon metabolites and risk of breast cancer before and after folic acid fortification in the United States, *Int. J. Cancer* 144(8), 1929–40.

Houghton, S. C., Eliassen, A. H., Zhang, S. M., Selhub, J., Rosner, B. A., Willett, W. C. and Hankinson, S. E. (2019b). Plasma B-vitamins and one-carbon metabolites and the risk of breast cancer in younger women, *Breast Cancer Res. Treat.* 176(1), 191-203.

Hund, L., Northrop-Clewes, C. A., Nazario, R., Suleymanova, D., Mirzoyan, L., Irisova, M., Pagano, M. and Valadez, J. J. (2013). A novel approach to evaluating the iron and folate status of women of reproductive age in Uzbekistan after 3 years of flour fortification with micronutrients, *PLoS ONE* 8(11), e79726.

Hurrell, R. F., Ranum, P., de Pee, S., Biebinger, R., Hulthen, L., Johnson, Q. and Lynch, S. (2010). Revised recommendations for iron fortification of wheat flour and an evaluation of the expected impact of current national wheat flour fortification programs, *Food Nutr. Bull.* 31(1), S7–21.

Institute of Medicine (US) Standing Committee on the Scientific Evaluation of Dietary Reference Intakes and its Panel on Folate, Other B Vitamins, and Choline. (1998). Dietary Reference Intakes for Thiamin, Riboflavin, Niacin, Vitamin B6, Folate, Vitamin B12, Pantothenic Acid, Biotin, and Choline. National Academies Press (US), Washington, DC.

Ionescu-Ittu, R., Marelli, A. J., Mackie, A. S. and Pilote, L. (2009). Prevalence of severe congenital heart disease after folic acid fortification of grain products: time trend analysis in Quebec, Canada, *B.M.J.* 338, b1673.

Kancherla, V., Wagh, K., Johnson, Q. and Oakley, G. P. Jr. (2018). A 2017 global update on folic acid-preventable spina bifida and anencephaly, *Birth Defects Res.* 110(14), 1139–47.

Kassebaum, N. J., Jasrasaria, R., Naghavi, M., Wulf, S. K., Johns, N., Lozano, R., Regan, M., Weatherall, D., Chou, D. P., Eisele, T. P., Flaxman, S. R., Pullan, R. L., Brooker, S. J. and Murray, C. J. (2014). A systematic analysis of global anemia burden from 1990 to 2010, *Blood* 123(5), 615–24.

Keats, E. C., Neufeld, L. M., Garrett, G. S., Mbuya, M. N. N. and Bhutta, Z. A. (2019). Improved micronutrient status and health outcomes in low- and middle-income countries following large-scale fortification: evidence from a systematic review and meta-analysis, *Am. J. Clin. Nutr.* 109(6), 1696–708.

Kennedy, D. A., Stern, S. J., Moretti, M., Matok, I., Sarkar, M., Nickel, C. and Koren, G. (2011). Folate intake and the risk of colorectal cancer: a systematic review and meta-analysis, *Cancer Epidemiol.* 35(1), 2–10.

Keum, N. and Giovannucci, E. L. (2014). Folic acid fortification and colorectal cancer risk, *Am. J. Prev. Med.* 46(3) (Suppl. 1), S65–72.

Kraemer, K. and Zimmermann, M. B. (2007). *Nutritional Anaemia*. Sight and Life Press, Basel, Switzerland.

Layrisse, M., Chaves, J. F., Mendez-Castellano, H., Bosch, V., Tropper, E., Bastardo, B. and González, E. (1996). Early response to the effect of iron fortification in the Venezuelan population, *Am. J. Clin. Nutr.* 64(6), 903–7.

Layrisse, M., García-Casal, M. N., Méndez-Castellano, H., Jiménez, M., Henry, O., Chávez, J. E. and González, E. (2002). Impact of fortification of flours with iron to reduce the prevalence of anemia and iron deficiency among schoolchildren in Caracas, Venezuela: a follow-up, *Food Nutr. Bull.* 23(4), 384–9.

Lin, J., Lee, I. M., Cook, N. R., Selhub, J., Manson, J. E., Buring, J. E. and Zhang, S. M. (2008). Plasma folate, vitamin B-6, vitamin B-12, and risk of breast cancer in women, *Am. J. Clin. Nutr.* 87(3), 734–43.

Liu, S., West, R., Randell, E., Longerich, L., Steel O'Connor, K. S., Scott, H., Crowley, M., Lam, A., Prabhakaran, V. and McCourt, C. (2004). A comprehensive evaluation of food fortification with folic acid for the primary prevention of neural tube defects, *BMC Preg. Childbirth* 4(1), 20.

Llanos, A., Hertrampf, E., Cortes, F., Pardo, A., Grosse, S. D. and Uauy, R. (2007). Cost-effectiveness of a folic acid fortification program in Chile, *Health Policy* 83(2–3), 295–303.

López-Camelo, J. S., Castilla, E. E., Orioli, I. M., INAGEMP (Instituto Nacional de Genética Médica Populacional) and ECLAMC (Estudio Colaborativo Latino Americano de Malformaciones Congénitas). (2010). Folic acid flour fortification: impact on the frequencies of 52 congenital anomaly types in three South American countries, *Am. J. Med. Genet. A* 152A(10), 2444–58.

López-Camelo, J. S., Orioli, I. M., da Graça Dutra, M., Nazer-Herrera, J., Rivera, N., Ojeda, M. E., Canessa, A., Wettig, E., Fontannaz, A. M., Mellado, C. and Castilla, E. E. (2005). Reduction of birth prevalence rates of neural tube defects after folic acid fortification in Chile, *Am. J. Med. Genet. A* 135(2), 120–5.

Marks, K. J., Luthringer, C. L., Ruth, L. J., Rowe, L. A., Khan, N. A., De-Regil, L. M., López, X. and Pachón, H. (2018). Review of grain fortification legislation, standards, and monitoring documents, *Glob. Health Sci. Pract.* 6(2), 356–71.

Martinez, H., Weakland, A. P., Bailey, L. B., Botto, L. D., De-Regil, L. M. and Brown, K. H. (2018). Improving maternal folate status to prevent infant neural tube defects: working group conclusions and a framework for action, *Ann. N.Y. Acad. Sci.* 1414(1), 5–19.

Martorell, R., Ascencio, M., Tacsan, L., Alfaro, T., Young, M. F., Addo, O. Y., Dary, O. and Flores-Ayala, R. (2015). Effectiveness evaluation of the food fortification program of Costa Rica: impact on anemia prevalence and hemoglobin concentrations in women and children, *Am. J. Clin. Nutr.* 101(1), 210–7.

Mason, J. B., Dickstein, A., Jacques, P. F., Haggarty, P., Selhub, J., Dallal, G. and Rosenberg, I. H. (2007). A temporal association between folic acid fortification and an increase in colorectal cancer rates may be illuminating important biological principles: a hypothesis, *Cancer Epidemiol. Biomarkers Prev.* 16(7), 1325–9.

Mathews, T. J., Honein, M. A. and Erickson, J. D. (2002). Spina bifida and anencephaly prevalence–United States, 1991-2001, *M.M.W.R. Recomm. Rep.* 51(RR-13), 9–11.

Mendu, V. V. R., Nair, K. P. M. and Athe, R. (2019). Systematic review and meta-analysis approach on vitamin A fortified foods and its effect on retinol concentration in under 10 year children, *Clin. Nutr. E.S.P.E.N.* 30, 126–30.

Millacura, N., Pardo, R., Cifuentes, L. and Suazo, J. (2017). Effects of folic acid fortification on orofacial clefts prevalence: a meta-analysis, *Public Health Nutr.* 20(12), 2260–8.

Mills, J. L., Von Kohorn, I., Conley, M. R., Zeller, J. A., Cox, C., Williamson, R. E. and Dufour, D. R. (2003). Low vitamin B-12 concentrations in patients without anemia: the effect of folic acid fortification of grain, *Am. J. Clin. Nutr.* 77(6), 1474–7.

Modjadji, S. E. P., Alberts, M. and Mamabolo, R. L. (2007). Folate and iron status of South African non-pregnant rural women of childbearing age, before and after fortification of foods, S.A.J.C.N. 20(3), 89–93.

MRC Vitamin Study Research Group. (1991). Prevention of neural tube defects: results of the Medical Research Council Vitamin Study. *Lancet*, 338(8760), 131–7.

Muthayya, S., Thankachan, P., Hirve, S., Amalrajan, V., Thomas, T., Lubree, H., Agarwal, D., Srinivasan, K., Hurrell, R. F., Yajnik, C. S. and Kurpad, A. V. (2012). Iron fortification of whole wheat flour reduces iron deficiency and iron deficiency anemia and increases body iron stores in Indian school-aged children, *J. Nutr.* 142(11), 1997–2003.

National Food and Nutrition Centre. (2012). *Impact of Iron Fortified Flour in Child Bearing Age (CBA) Women in Fiji, 2010 Report*. National Food and Nutrition Centre, Suva.

National Institute of Statistics, Directorate General for Health and ICF International. (2015). *Cambodia Demographic and Health Survey 2014*. National Institute of Statistics, Directorate General for Health, and ICF International, Phnom Penh, Cambodia and Rockville, MD.

National Toxicology Program (2015). *NTP Monograph: Identifying Research Needs for Assessing Safe Use of High Intakes of Folic Acid*. United States Department of Health and Human Services, Washington DC.

Nazer, H. J. and Cifuentes, O. L. (2013). Effects of wheat flour fortification with folic acid on the prevalence of neural tube defects in Chile, *Rev. Med. Chile* 141(6), 751–7.

Nazer, H. J., Cifuentes, O. L., Aguila, R. A., Juárez, H. M. E., Cid, R. M. P., Godoy, V. M. L., García, A. K. and Melibosky, R. F. (2007). Effects of folic acid fortification in the rates of malformations at birth in Chile, *Rev. Méd. Chile* 135(2), 198–204.

Noor, R. A., Abioye, A. I., Ulenga, N., Msham, S., Kaishozi, G., Gunaratna, N. S., Mwiru, R., Smith, E., Dhillon, C. N., Spiegelman, D. and Fawzi, W. (2017). Large-scale wheat flour folic acid fortification program increases plasma folate levels among women of reproductive age in urban Tanzania, *PLoS ONE* 12(8), e0182099.

Northrop-Clewes, C., Hund, L., Valadez, J., Mirzoyan, L. and Irisova, M. (2013). LC-LQAS survey report. Global Alliance for Improved Nutrition, Geneva.

Oakley, G. P. Jr. (2009). The scientific basis for eliminating folic acid-preventable spina bifida: a modern miracle from epidemiology, *Ann. Epidemiol.* 19(4), 226–30.

Odewole, O. A., Williamson, R. S., Zakai, N. A., Berry, R. J., Judd, S. E., Qi, Y. P., Adedinsewo, D. A. and Oakley, G. P. (2013). Near-elimination of folate-deficiency anemia by mandatory folic acid fortification in older US adults: reasons for geographic and racial differences in stroke study 2003–2007, *Am. J. Clin. Nutr.* 98(4), 1042–7.

O'Donnell, S., Cranney, A., Horsley, T., Weiler, H. A., Atkinson, S. A., Hanley, D. A., Ooi, D. S., Ward, L., Barrowman, N., Fang, M., Sampson, M., Tsertsvadze, A. and Yazdi, F. (2008). Efficacy of food fortification on serum 25-hydroxyvitamin D concentrations: systematic review, *Am. J. Clin. Nutr.* 88(6), 1528–34.

Pacheco Santos, L. M., Reyes Lecca, R. C., Cortez-Escalante, J. J., Sanchez, M. N. and Rodrigues, H. G. (2016). Prevention of neural tube defects by the fortification of flour with folic acid: a population-based retrospective study in Brazil, *Bull. World Health Organ.*, 94(1), 22–9.

Pacheco, S. S., Braga, C., Souza, A. I. and Figueiroa, J. N. (2009). Effects of folic acid fortification on the prevalence of neural tube defects, *Rev. Saúde Publ.* 43(4), 565–71.

Pacheco Santos, L. M. and Zanon Pereira, M. (2007). Efeito da fortificação com ácido fólico na redução dos defeitos do tubo neural [The effect of folic acid fortification on the reduction of neural tube defects], *Cad. Saúde Publ.* 23(1), 17–24.

Pachón, H., Spohrer, R., Mei, Z. and Serdula, M. K. (2015). Evidence of the effectiveness of flour fortification programs on iron status and anemia: a systematic review, *Nutr. Rev.* 73(11), 780–95.

Persad, V. L., Van den Hof, M. C., Dubé, J. M. and Zimmer, P. (2002). Incidence of open neural tube defects in Nova Scotia after folic acid fortification, *C.M.A.J.* 167(3), 241–5.

Petry, N., Olofin, I., Boy, E., Donahue Angel, M. and Rohner, F. (2016). The effect of low dose iron and zinc intake on child micronutrient status and development during the first 1000 days of life: a systematic review and meta-analysis, *Nutrients* 8(12), pii: E773.

Pfeiffer, C. M., Hughes, J. P., Lacher, D. A., Bailey, R. L., Berry, R. J., Zhang, M., Yetley, E. A., Rader, J. I., Sempos, C. T. and Johnson, C. L. (2012). Estimation of trends in serum and RBC folate in the U.S. population from pre- to postfortification using assay-adjusted data from the NHANES 1988–2010, *J. Nutr.* 142(5), 886–93.

Pfeiffer, C. M., Sternberg, M. R., Zhang, M., Fazili, Z., Storandt, R. J., Crider, K. S., Yamini, S., Gahche, J. J., Juan, W., Wang, C. Y., Potischman, N., Williams, J. and LaVoie, D. J. (2019). Folate status in the US population 20 y after the introduction of folic acid fortification, *Am. J. Clin. Nutr.* 110(5), 1088–97.

Qi, Y. P., Do, A. N., Hamner, H. C., Pfeiffer, C. M. and Berry, R. J. (2014). The prevalence of low serum vitamin B-12 status in the absence of anemia or macrocytosis did not increase among older U.S. adults after mandatory folic acid fortification, *J. Nutr.* 144(2), 170–6.

Ray, J. G., Meier, C., Vermeulen, M. J., Boss, S., Wyatt, P. R. and Cole, D. E. (2002). Association of neural tube defects and folic acid food fortification in Canada, *Lancet* 360(9350), 2047–8.

Ray, J. G., Vermeulen, M. J., Langman, L. J., Boss, S. C. and Cole, D. E. (2003). Persistence of vitamin B12 insufficiency among elderly women after folic acid food fortification, *Clin. Biochem.* 36(5), 387–91.

Ricks, D. J., Rees, C. A., Osborn, K. A., Crookston, B. T., Leaver, K., Merrill, S. B., Velásquez, C. and Ricks, J. H. (2012). Peru's national folic acid fortification program and its effect on neural tube defects in Lima, *Rev. Panam. Salud Publ.* 32(6), 391–8.

Rosenthal, J., Casas, J., Taren, D., Alverson, C. J., Flores, A. and Frias, J. (2014). Neural tube defects in Latin America and the impact of fortification: a literature review, *Public Health Nutr.* 17(3), 537–50.

Sadighi, J., Mohammad, K., Sheikholeslam, R., Amirkhani, M. A., Torabi, P., Salehi, F. and Abdolahi, Z. (2009). Anaemia control: lessons from the flour fortification programme, *Public Health* 123(12), 794–9.

Sadighi, J., Sheikholeslam, R., Mohammad, K., Pouraram, H., Abdollahi, Z., Samadpour, K., Kolahdooz, F. and Naghavi, M. (2008). Flour fortification with iron: a mid-term evaluation, *Public Health* 122(3), 313–21.

Safdar, O. Y., Al-Dabbagh, A. A., Abuelieneen, W. A. and Kari, J. A. (2007). Decline in the incidence of neural tube defects after the national fortification of flour (1997–2005), *Saudi Med. J.* 28(8), 1227–9.

Salam, R. A., Das, J. K., Ahmed, W., Irfan, O., Sheikh, S. S. and Bhutta, Z. A. (2019). Effects of preventive nutrition interventions among adolescents on health and nutritional status in low- and middle-income countries: a systematic review and meta-analysis, *Nutrients* 12(1), pii: E49.

Samet, J. M., Wipfli, H., Platz, E. A. and Bhavsar, N. (2008). In: Porta, M. (Eds). *A Dictionary of Epidemiology* (5th edn.). Oxford University Press, New York.

Sanabria Rojas, H. A., Tarqui-Mamani, C. B., Arias Pachas, J. and Lam Figueroa, N. M. (2013). Impacto de la fortificación de la harina de trigo con ácido fólico en los defectos del tubo neural, en Lima, Perú, *An. Fac. Med.* 74(3), 175–80.

Sargiotto, C., Bidondo, M. P., Liascovich, R., Barbero, P. and Groisman, B. (2015). Descriptive study on neural tube defects in Argentina, *Birth Defects Res. A* 103(6), 509–16.

Sayed, A.-R., Bourne, D., Pattinson, R., Nixon, J. and Henderson, B. (2008). Decline in the prevalence of neural tube defects following folic acid fortification and its cost-benefit in South Africa, *Birth Defects Res. A* 82(4), 211-6.

Shah, D., Sachdev, H. S., Gera, T., De-Regil, L. M. and Peña-Rosas, J. P. (2016). Fortification of staple foods with zinc for improving zinc status and other health outcomes in the general population, *Cochrane Database Syst. Rev.* 6(6), CD010697.

Siegel, R. L., Miller, K. D. and Jemal, A. (2019). Cancer statistics, 2019, *C.A. Cancer J. Clin.* 69(1), 7-34.

Smarter Futures. (n.d.). *FORTIMAS: An Approach for Tracking the Population Coverage and Impact of a Flour Fortification Program*. International Federation for Spina Bifida and Hydrocephalus, Brussels.

Smith, A. D., Kim, Y. I. and Refsum, H. (2008). Is folic acid good for everyone?, *Am. J. Clin. Nutr.* 87(3), 517–33.

Solomons, N. W. and Allen, L. H. (1983). The functional assessment of nutritional status: principles, practice and potential, *Nutr. Rev.* 41(2), 33–50.

Stevens, G. A., Finucane, M. M., De-Regil, L. M., Paciorek, C. J., Flaxman, S. R., Branca, F., Peña-Rosas, J. P., Bhutta, Z. A. and Ezzati, M. (2013). Global, regional, and national trends in haemoglobin concentration and prevalence of total and severe anaemia in children and pregnant and non-pregnant women for 1995–2011: a systematic analysis of population-representative data, *Lancet Glob. Health* 1(1), e16-25.

Stevens, V. L., McCullough, M. L., Sun, J. and Gapstur, S. M. (2010). Folate and other one-carbon metabolism-related nutrients and risk of postmenopausal breast cancer in the cancer prevention study II nutrition cohort, *Am. J. Clin. Nutr.* 91(6), 1708–15.

Stolzenberg-Solomon, R. Z., Chang, S. C., Leitzmann, M. F., Johnson, K. A., Johnson, C., Buys, S. S., Hoover, R. N. and Ziegler, R. G. (2006). Folate intake, alcohol use, and postmenopausal breast cancer risk in the Prostate, Lung, Colorectal, and Ovarian Cancer Screening Trial, *Am. J. Clin. Nutr.* 83(4), 895–904.

Tacsan Chen, L. and Ascencio Rivera, M. (2004). The Costa Rican experience: reduction of neural tube defects following food fortification programs, *Nutr. Rev.* 62(6 Pt 2), S40–3.

Tam, E., Keats, E. C., Rind, F., Das, J. K. and Bhutta, A. Z. A. (2020). Micronutrient supplementation and fortification interventions on health and development outcomes among children under-five in low- and middle-income countries: a systematic review and meta-analysis, *Nutrients* 12(2), pii: E289.

Tarqui-Mamani, C. B. (2013). Sobre el estudio de evaluación de los defectos de tubo neural en Lima, *Rev. Panam. Salud Publ.* 34(3), 210-.

Tarqui-Mamani, C. B., Sanabria-Rojas, H., Rossi de Chiarella, G., Arana-Panduro, M., Altamirano, H., et al. (2016). Impact of wheat flour folic acid fortification on neural tube defects in three cities in Peru, *J. Glob. Health*. https://pubmed.ncbi.nlm.nih.gov/24233115/.

Tice, J. A., Ross, E., Coxson, P. G., Rosenberg, I., Weinstein, M. C., Hunink, M. G. M., Goldman, P. A., Williams, L. and Goldman, L. (2001). Cost-effectiveness of vitamin

Published by Burleigh Dodds Science Publishing Limited, 2021.

therapy to lower plasma homocysteine levels for the prevention of coronary heart disease: effect of grain fortification and beyond, *J.A.M.A.* 286(8), 936–43.
Tio, M., Andrici, J. and Eslick, G. D. (2014). Folate intake and the risk of breast cancer: a systematic review and meta-analysis, *Breast Cancer Res. Treat.* 145(2), 513–24.
UNICEF and the Food Fortification Initiative. (2004). Monitoring of flour fortification: the case of South Africa. Available at: http://www.ffinetwork.org/monitor/Documents/SouthAfricaCS.pdf; Accessed 9 March 2020.
United Nations Statistics Division. (2020). Standard country or area codes for statistical use (vol. M49). Available at: https://unstats.un.org/unsd/methodology/m49/; Accessed 9 March 2020.
United States Department of Agriculture. (2020). FoodData Central: wheat flour, white, all-purpose, unenriched. Available at: https://fdc.nal.usda.gov/fdc-app.html#/food-details/169761/nutrients; Accessed 10 February 2020.
van Gool, J. D., Hirche, H., Laz, H. and De Schaepdrijver, L. (2018). Folic acid and primary prevention of neural tube defects: a review, *Reprod. Toxicol.* 80, 73–84.
van Gool, J. D., Hirche, H., Lax, H. and Schaepdrijver, L. (2020). Fallacies of clinical studies on folic acid hazards in subjects with a low vitamin B12 status, *Crit. Rev. Toxicol.* 31, 1–11.
Victora, C. G., Habicht, J. P. and Bryce, J. (2004). Evidence-based public health: moving beyond randomized trials, *Am. J. Public Health* 94(3), 400–5.
Vollset, S. E., Clarke, R., Lewington, S., Ebbing, M., Halsey, J., Lonn, E., Armitage, J., Manson, J. E., Hankey, G. J., Spence, J. D., Galan, P., Bønaa, K. H., Jamison, R., Gaziano, J. M., Guarino, P., Baron, J. A., Logan, R. F., Giovannucci, E. L., den Heijer, M., Ueland, P. M., Bennett, D., Collins, R. and Peto, R. (2013). Effects of folic acid supplementation on overall and site-specific cancer incidence during the randomised trials: meta-analyses of data on 50 000 individuals, *Lancet* 381(9871), 1029–36.
Williams, J., Mai, C. T., Mulinare, J., Isenburg, J., Flood, T. J., Ethen, M., Frohnert, B., Kirby, R. S. and Centers for Disease Control and Prevention (2015). Updated estimates of neural tube defects prevented by mandatory folic acid fortification–United States, 1995–2011, *M.M.W.R. Morb. Mortal. Wkly. Rep.* 64(1), 1–5.
Williams, L. J., Mai, C. T., Edmonds, L. D., Shaw, G. M., Kirby, R. S., Hobbs, C. A., Sever, L. E., Miller, L. A., Meaney, F. J. and Levitt, M. (2002). Prevalence of spina bifida and anencephaly during the transition to mandatory folic acid fortification in the United States, *Teratology* 66(1), 33–9.
Williams, L. J., Rasmussen, S. A., Flores, A., Kirby, R. S. and Edmonds, L. D. (2005). Decline in the prevalence of spina bifida and anencephaly by race/ethnicity: 1995–2002, *Pediatrics* 116(3), 580–86.
World Bank. (2020). How does the World Bank classify countries?. Available at: https://datahelpdesk.worldbank.org/knowledgebase/articles/378834-how-does-the-world-bank-classify-countries; Accessed 9 March 2020.
World Health Organization. (2020). Anaemia. Available at: https://www.who.int/health-topics/anaemia#tab=tab_1; Accessed 27 April 2020.
World Health Organization and Food and Agriculture Organization of the United Nations. (2006). *Guidelines on Food Fortification with Micronutrients*. World Health Organization, Geneva.
World Health Organization, Food and Agriculture Organization of the United Nations, the United Nations Children's Fund is a United Nations, Global Alliance for Food Fortification, Micronutrient Initiative, and the Food Fortification Initiative. (2009).

Recommendations on wheat and maize flour fortification. Meeting report: interim consensus statement. World Health Organization, Geneva.

Yang, Q., Botto, L. D., Erickson, J. D., Berry, R. J., Sambell, C., Johansen, H. and Friedman, J. M. (2006). Improvement in stroke mortality in Canada and the United States, 1990 to 2002, *Circulation* 113(10), 1335–43.

Yang, X., Chen, H., Du, Y., Wang, S. and Wang, Z. (2016). Periconceptional folic acid fortification for the risk of gestational hypertension and pre-eclampsia: a meta-analysis of prospective studies, *Matern. Child Nutr.* 12(4), 669–79.

Youngblood, M. E., Williamson, R., Bell, K. N., Johnson, Q., Kancherla, V. and Oakley, G. P. (2013). 2012 Update on global prevention of folic acid-preventable spina bifida and anencephaly, *Birth Defects Res. A* 97(10), 658–63.

Chapter 9

Developing hulled wheat-based cereal products with enhanced nutritional properties: emmer, einkorn and spelt

Dagmar Janovská and Petra Hlásná Čepková, Crop Research Institute, Czech Republic; Stefano D'Amico, Austrian Agency for Health and Food Safety, Austria; Andrea Brandolini, Council for Agricultural Research and Economics, Italy; and Heinrich Grausgruber, University of Natural Resources and Life Sciences Vienna, Austria

1 Introduction

Hulled wheats can be distinguished from free-threshing (syn. naked) wheats by a distinct feature. If a spike is threshed, it breaks to single spikelet, each consisting of one or more grains enclosed by tough glumes which are attached to a rachis segment. In contrast, in naked wheats the grain falls free from the glumes and rachis. The hulled character is typical of wild relatives of wheat but is also present in domesticated forms of wheat (Nesbitt and Samuel, 1996):

- einkorn (*Triticum monococcum* L.; 2n = 2x = 14; A^mA^m);
- emmer (*T. dicoccum* Schrank ex Schübler; 2n = 4x = 28; BBAA);
- Colchis emmer (*T. paleocolchicum* Menabde; 2n = 4x = 28; BBAA);
- Esfahanian emmer (*T. ispahanicum* Heslot; 2n = 4x = 28; BBAA);
- Zanduri wheat (*T. timopheevii* (Zhuk.) Zhuk.; 2n = 4x = 28; GGAA);

http://dx.doi.org/10.19103/AS.2021.0087.20

- spelt (*T. spelta* L.; 2n = 6x = 42; BBAADD);
- Georgian spelt (*T. macha* Dekapr. & Menabde; 2n = 6x = 42; BBAADD);
- Vavilov's wheat (*T. vavilovii* (Tumanian) Jakubz.; 2n = 6x = 42; BBAADD); and
- Zhukovsky's wheat (*T. zhukovskyi* Menabde & Ericzjan; 2n = 6x = 42; $BBAAA^mA^m$).

While einkorn and emmer are linked to the emergence of agriculture in the Fertile Crescent during the Neolithic Revolution (Nesbitt and Samuel, 1996; Lev-Yadun et al., 2000), the other hulled wheats arose regionally through natural hybridization between species and following adaptation to specific environments (Perrino et al., 1996; Khoshbakht, 2009; Dvorak et al., 2012; Jorjadze et al., 2014). Over millennia the cultivation of hulled wheats was given up in favour of the higher yielding naked species durum (*T. durum*) and common wheat (*T. aestivum*). By the middle of the twentieth century, hulled wheats almost disappeared from farmers' fields, surviving only in remote and mountainous areas (Nesbitt and Samuel, 1996; Perrino et al., 1996). The cultivation of spelt in Central Europe began to grow again in the early 1970s and has experienced a boom in recent years. Interest in einkorn and emmer also grew and these species are now often marketed as 'ancient wheats'.

There are various reasons for the increased interest in hulled wheats. Consumers connect 'ancient wheats' with higher nutritional value, a more natural method of production and no perceived negative impacts from modern plant breeding and are thus willing to pay higher prices for products. This attracts growers and processors. The growth in organic farming has also stimulated cultivation of hulled wheat. Under high-input production hulled wheats demonstrate a significant yield gap compared to free-threshing wheats. However, they adapt better to low-input conditions resulting in a higher marginal income for the farmers due to higher prices. While almost all einkorn and emmer germplasm used for cultivation is derived from old landraces, spelt has undergone more intense breeding activity in the last five decades. Today old landraces and selections and/or crosses between them are cultivated alongside modern short-strawed varieties selected from crosses with common wheat (Grausgruber, 2018).

2 Protein: content, immunogenic and allergenic proteins

2.1 Protein content and composition

Hulled wheats contain generally higher amounts of protein compared to common wheat and durum (Table 1). Under organic conditions in the UK, emmer showed a slightly higher protein content than einkorn (Costanzo et al., 2019). Protein composition showed a high variability among wheat species, predominantly affected by the ploidy level. Hexaploids contain higher amounts

Table 1 Variation in crude protein content (%) in hulled and naked wheat species in selected studies (number of investigated genotypes in brackets)

	Hulled wheat			Naked wheat	
Reference	Einkorn	Emmer	Spelt	Durum	Common
Brandolini et al. (2008)	15.5–22.8 (65)	14.0–16.0 (2)	17.1–18.7 (2)	14.0–18.3 (5)	13.7–15.7 (5)
Call et al. (2020)	15.5–19.8 (7)	14.9–18.3 (3)	14.9–18.3 (7)	12.7–16.4 (4)	11.9–16.8 (18)
Konvalina et al. (2014)	14.8–16.1 (4)	14.1–16.9 (8)	15.1–16.0 (7)	-	11.8–13.8 (7)
Geisslitz et al. (2018b)	11.6–13.9 (8)	11.2–12.4 (8)	10.8–16.1 (8)	13.0–15.5 (8)	9.3–13.3 (8)
Mohan Kumar et al. (2017)	-	13.7–15.9 (3)	-	11.3–14.6 (6)	9.0–14.7 (25)
Longin et al. (2016)	11.4–15.1 (15)	11.8–14.1 (15)	11.1–14.6 (15)	12.2–15.8 (15)	10.0–12.7 (15)

of glutenins, especially high molecular weight subunits (HMWs). Generally, diploid and tetraploid species contain more gliadins and less glutelins, especially HMWs. Albumins and globulins were present at a similar level among all wheat species (Geisslitz et al., 2018b; Call et al., 2020). Differences in protein fractions significantly affected the gliadin to glutenin ratio (Gli/Glu). Einkorn exhibited the highest ratios (4.2-12.0; $\bar{x}$ = 8.0), followed by emmer (3.6–6.7; $\bar{x}$ = 5.2) and durum (2.2–5.3; $\bar{x}$ = 3.6). Common wheat (2.0–3.2; $\bar{x}$ = 2.6) and spelt (2.8–4.0; $\bar{x}$ = 3.5) displayed the lowest ratios (Geisslitz et al., 2018b). Similar patterns between wheat species were found by Frakolaki et al. (2018), Grausgruber et al. (2019) and Call et al. (2020) but with distinctly lower Gli/Glu ratios. Differences between wheat species with respect to loaf volume and baking performance are probably attributed to their glutenin content and Gli/Glu ratio (Frakolaki et al., 2018; Geisslitz et al., 2018b; Callejo et al., 2019).

A group of tryptophan-rich proteins, called puroindolines, are responsible for grain hardness and thus milling performance. They are controlled by the genes *Pina* and *Pinb* (Morris, 2002). Einkorn has a very soft texture and common wheat a medium soft/hard texture, whereas durum and emmer are characterized by very hard kernels (Bonafede et al., 2007; Szymanski and Morris, 2015). Taddei et al. (2009) showed that puroindolines can be found in high concentrations in einkorn compared to common wheat, whereas these proteins are completely absent in durum.

2.2 Immunogenic and allergenic proteins

The higher protein content and greater abundance of gliadins suggest a stronger immunogenic potential of hulled and other 'ancient wheats' in relation to celiac disease (CD). However, several studies have reported a lower immunogenic response of einkorn and emmer as measured by ELISA, whereas the reaction to spelt was similar to common wheat or even more toxic (Gell et al., 2015; Gélinas and McKinnon, 2016; Escarnot et al., 2018; Grausgruber et al., 2019). The lower toxicity of diploid and tetraploid wheats towards CD was confirmed by Schopf and Scherf (2018). However, the monoclonal antibodies G12 and R5 were found to be at very low levels to detect for einkorn and emmer, whereas polyclonal antibodies were more active against gliadins from einkorn and emmer (Schopf and Scherf 2018). Quantitation of the immunodominant 33-mer peptide by LC-MS/MS revealed that it was missing in diploid and tetraploid wheat species (Schalk et al., 2017), but that spelt contained similar amounts to common wheat. Molberg et al. (2005) showed that the D genome, which is absent in diploid einkorn and tetraploid emmer, encodes most of the harmful gluten proteins. This probably explains the low response of einkorn and emmer to the monoclonal antibody G12. However, sequencing of einkorn cDNA clones related to storage protein genes revealed the presence of 4 toxic and 13 immunogenic peptides, demonstrating that einkorn has full potential to induce CD (Vaccino et al., 2009). Ozuna et al. (2015) and Ruiz-Carnicer et al. (2019) investigated the α-gliadin epitopes of various wheat species and wheat wild relatives by next generation sequencing (NGS) and Sanger sequencing and found that the total abundance of canonical CD epitopes was highest in einkorn and *Aegilops tauschii*, classifying einkorn as one of the most immunogenic species, whereas tetraploid wheat species showed the lowest abundance of canonical epitopes. Studying more than 120 spelt accessions, Dubois et al. (2018) observed great variability regarding CD-related epitope expression, but no influence of environmental factors and/or the breeding period on the amount of epitope transcripts.

IgA reactivity and IgE reactivity of serum from CD patients with gliadins from emmer, durum and common wheat provide an ambiguous picture: IgE reactivity was lower in tetraploid wheats, whereas no differences were observed for IgA reactivity (Mohan Kumar et al. 2017). Sievers et al. (2020) examined two varieties each of einkorn, emmer and spelt with respect to wheat-induced food allergy by immunoblotting. IgE bound to a great number of proteins in all fractions, both in hulled and naked wheat grains. Only one protein band corresponding to an amylase-trypsin inhibitor (ATI) was not found in einkorn. This suggests that einkorn might have a reduced potential for people sensitized to ATI.

It has also been suggested that reduced immune-toxicity of einkorn in vivo depends on digestion, as gliadins from einkorn showed an increased breakdown and lower immune stimulatory activity after in vitro gastrointestinal digestion compared to common wheat (Gianfrani et al., 2015) and durum (Di Stasio et al., 2020). Beside differences between techniques applied for quantification of gluten toxicity (ELISA, mass spectroscopy or immunostaining), variation in protein digestibility obviously plays a crucial role in the different immunogenic potential of wheat species.

ATIs are potential culprits in triggering the immune reaction associated with non-celiac wheat sensitivity (NCWS). A high variation of single ATIs in wheat genotypes was detected by targeted LC-MS/MS. Total ATI content in emmer and spelt was higher than in their free-threshing relatives, that is, durum and common wheat, whereas einkorn had a very low content (Geisslitz et al., 2018a; Call et al., 2020). Other studies have confirmed the absence of several ATIs in einkorn (D'Amico et al., 2018; Sievers et al., 2020).

Although the data shows discrepancies with respect to allergenic and immunogenic potential of hulled wheats compared to durum and/or common wheat, they cannot be regarded as safe for people with adverse reactions to wheat proteins. In fact, an in vivo research study by Zanini et al. (2015a,b) clearly showed that einkorn, apparently well tolerated by most CD patients, was toxic based on histological and serological results. A probable lower negative impact of einkorn for NCWS patients could be explained by a better digestibility of gliadins and the absence or very low abundance of several ATIs. This assumption was confirmed by Zevallos et al. (2017), who observed a significantly lower activation of the toll-like receptor 4 (TLR4) by ATIs from einkorn and emmer, compared to hexaploid wheat.

3 Starch

Starch is a primary component of cereals and usually represents 60–75% of the grain and 70–80% of the flour. Wheat starch is one of the major sources of energy in the human diet and is also used in many industrial applications (Shevkani et al., 2017). Starch is mainly found in the endosperm, is basically a polymer of glucose and consists structurally of two main components, that is, amylose and amylopectin (Van Hung et al., 2006). Common wheat starch is composed of 20–30% amylose and 70–80% amylopectin. Starch properties have a significant influence on the texture and quality of wheat-based food products (Chibbar and Chakraborty, 2005).

Among cereals, common wheat contains a similar amount of starch as oats, but less than barley and rice (Marcotuli et al., 2020). Hulled wheats did not show significantly different amounts of starch compared to durum and/or common wheat. According to Abdel-Aal et al. (1995) and Abdel-Aal and Hucl (2005), starch content of spelt varied between 61% and 66%, compared to 61-62% in common wheat, 65% in durum and 62% in einkorn. Lower contents were reported for spelt and emmer varieties by Galterio et al. (2003). However, this was probably due to a (different methodology) used for starch determination compared to other studies. According to Brandolini et al. (2008) and Hidalgo and Brandolini (2019) einkorn yielded on average 66% starch, ranging from 61% to 71%. Einkorn possesses smaller starch granules to common wheat, while the size distribution of the small granules is much wider than in other wheat species (Frégeau-Reid and Abdel-Aal, 2005).

After baking, the nutritional starch fractions and digestion indices of spelt products change and the values obtained are close to common wheat. Milling and fermentation have been shown to have no significant influence on the starch content of spelt (Abdel-Aal and Rabalski, 2008; Starzyńska-Janiszewska et al. 2019). In the case of einkorn, the efficiency of starch extraction is the lowest compared to durum and common wheat (Frégeau-Reid and Abdel-Aal, 2005). Compared to modern wheats, Marconi and Cubadda (2005) attributed the low starch contents in hulled wheats to their lower capacity for starch accumulation compared to modern wheats.

The content of resistant starch (RS) has an impact on processing and nutritional quality of wheat products as RS is not absorbed in the small intestine of humans and has, therefore, a positive effect on digestion. Abdel-Aal and

Rabalski (2008) revealed that spelt contains about 8–10-fold more RS (3.7%) than common wheat (0.4%). In the case of einkorn, the values of RS can significantly vary (Hidalgo and Brandolini, 2019). The amylose content in einkorn ranged up to 30% (Mohammadkhani et al., 1998; Frégeau-Reid and Abdel-Aal, 2005; Brandolini et al., 2011) and is comparable with other hulled wheat species. Amylose content in einkorn accessions was lower than in emmer (Mohammadkhani et al., 1998).

The last decade has seen the development of varieties with a waxy type of starch. Waxy wheat starch with ≤1% amylose shows different physicochemical properties important for specific end uses (Van Hung et al., 2006). Waxy wheats are characterized by the presence of the waxy genes *Wx-A1*, *Wx-B1* and *Wx-D1*. Three alleles have been identified for *Wx-B1* in emmer (Guzmán et al., 2011). A new waxy allele (*Wx-D1g*) has also been detected in spelt. Waxy varieties have enlarged the gene pool in common wheat for differing end uses (Guzmán et al., 2010). In hulled wheats, however, waxy mutants have not so far been exploited for breeding for different end uses.

4 Dietary fibre

The EU regulation 1169/2011 defines fibre as 'carbohydrate polymers with three or more monomeric units, which are neither digested nor absorbed in the human small intestine'. The recommended daily intake in the EU for adults is 25–30 g per day, but these recommendations have not been achieved in practice, except for males in Norway, Germany and partially in Hungary. Since cereal products are the most important source of fibre in the western world and account for about 50% of the total daily intake, it is important to include enough wholegrain products in the diet. Dietary fibres have several health benefits, including reduction in risk of diseases such as cardiovascular diseases, diabetes, cancer as well as reduced risk of obesity (JRC, 2020). Because of these benefits dietary fibre can be the basis of a health claim in food labelling according to EU regulation (EC) No 1924/2006.

The major constituent of dietary fibre is cell wall components or hemicelluloses in the outer layers of kernels. The most abundant carbohydrates in cereals are arabinoxylans, followed by fructans, cellulose, β-glucan and Klason lignin (Shewry and Hey, 2015). Total dietary fibre (TDF) content varies among wheat species. Emmer and einkorn generally contain the lowest fibre content of all samples investigated. Both common wheat and durum wheat contain broadly similar levels of dietary fibre, with intermediate content observed for spelt (Løje et al., 2003; Grausgruber et al., 2004; Gebruers et al., 2008; Shewry and Hey, 2015; Fares et al., 2019). Fibre is further classified according to its solubility as soluble and insoluble fibres. The insoluble fibre content of grain is commonly 2.5–4 times higher than the soluble fibre content (Fares et al., 2008).

Quantification of pyrolysis fragments from lignin and polysaccharides revealed distinct quality differences between spring and winter emmers compared to durum and einkorn (D'Antuono et al., 1998).

Arabinoxylans are also important from the processing and technological point of view in that they increase, for example, water-holding capacity, which is positively correlated with loaf volume, dough stability time and viscosity, rheological properties, and in stabilizing protein foams (Gebruers et al., 2008). The range of arabinoxylan content in refined flours of hulled wheat is generally narrower compared to common and durum wheat. The total arabinoxylan content in einkorn varied from 1.45% to 2.35%, in emmer from 1.40% to 1.95% and in spelt from 1.60% to 2.15% (Gebruers et al., 2008). The range for common wheat was found to be from 1.35% to 2.75% (spring and winter wheat) and for durum wheat from 1.70% to 2.35%. Similar ranges for refined flours were reported in the overview by Shewry and Hey (2015).

β-Glucans consist of a linear chain of β-D-glucosyl residues linked by both (1–3) and (1–4) linkages (Escarnot et al., 2015). The content of β-glucans is lower in wheat in comparison with barley or oats (Gebruers et al., 2008). There are also differences within wheat species. The lowest mean content was determined in einkorn, emmer and durum wheat. β-Glucan contents in common wheat and spelt were both similar in the range of 0.54% to 0.8% (Løje et al., 2003; Grausgruber et al., 2004; Gebruers et al., 2008; Escarnot et al., 2015). The higher amounts in hexaploid wheats were confirmed by Shewry and Hey (2015).

Fructans belonging to the group of FODMAPs (Fermentable Oligo-, Di-, Monosaccharides and Polyols) have attracted more attention recently due to their negative impact for people with intolerances like irritable bowel syndrome (IBS) and NCWS. Only a few studies have analysed fructans in wheat species, showing higher concentrations of fructans in einkorn and spelt compared to common wheat and emmer. However, fructans are strongly degraded during fermentation by yeast enzymes and the final content in bakery products is much lower, especially for long sourdough fermentations (Ziegler et al., 2016b; Longin et al., 2020).

It can be concluded that the nutritional value of hulled wheats with respect to fibre is lower than for dehulled wheats. β-glucan is also more abundant in hexaploid wheats. The consumption of modern common wheat is highly recommended for people without adverse reactions to wheat if consumed as a whole grain product.

5 Antioxidants: polyphenols, carotenoids and tocols

5.1 Polyphenols

Polyphenols are secondary plant metabolites with a potential health benefit due to their antioxidative behaviour. They can be classified according to the number of phenol rings and the structural elements that link these rings. Phenolic acids (PAs) are the major group of cereal polyphenols which are involved in the plant's defence mechanism against biotic and abiotic stresses. PAs can provide significant protection in humans against the development of several chronic diseases (Pandey and Rizvi, 2009). The predominant PA in cereals is ferulic acid (FA). *p*-Hydroxybenzoic, vanillic, syringic, *p*-coumaric and sinapic acids can also be found. The majority of PAs are linked to cell wall polysaccharides, mainly arabinoxylans, and are therefore insoluble. They occur primarily in the outer layers of the wheat kernel (Barański et al., 2020). Flavanols are less abundant in wheat (Şahin et al., 2017), while flavone glucosides are the major flavonoids (Geng et al., 2016). Isoflavones like isovitexin, vitexin and daidzein have only been found in traces (Montevecchi et al., 2019). Beside PAs, alk(en)ylresorcinols as phenolic lipids play a distinct role in wheat for human nutrition (Ross et al., 2003). Anthocyanins seldom occur in the wheat grain, and are restricted to a few durum and common wheat genotypes with a purple pericarp or blue aleurone. The latter trait is also characteristic for wild einkorn, *T. boeoticum* (Grausgruber et al., 2014; Liu et al., 2020).

Various studies have investigated total phenolic content (TPC) in wheat species, but the results are contradictory and hard to compare due to differences in:

i the methodology used for extraction and determination as well as the compounds used for calibration;
ii the amount and genetics of the germplasm used; and
iii the environmental conditions in which the germplasm was grown.

As an example, TPC was significantly higher in einkorn compared to other wheat species in the studies of Zrcková et al. (2019) and Barański et al. (2020), whereas the HEALTHGRAIN study revealed the lowest concentrations for einkorn and spelt (Shewry and Hey, 2015). On the other hand, a meta-analysis across 28 studies revealed the highest weighted means for total phenolic acids of spelt, followed by emmer, durum, einkorn and common wheat (Shewry and Hey, 2015).

FA, the most abundant PA, has been shown to account for >90% of the insoluble fractions in wheat (Brandolini et al., 2013; Zrcková et al., 2019). Einkorn is similar to common wheat with respect to the total content of free polyphenols (Brandolini et al., 2015; Lachman et al., 2011) and PAs (Li et al., 2008; Serpen et al., 2008). However, einkorn has more conjugated PAs along with less-bound phenolic acids (Brandolini et al., 2013). Overall, the total phenolic acid content of einkorn was in the range 449–816 mg/kg dm (Li et al., 2008; Brandolini et al., 2013). Wheat species affect the composition of free phenolics. FA accounted for 57% of PAs in einkorn and 70% in common wheat, followed by vanillic and syringic acid, revealing a higher diversity of PAs compared to the bound fraction (Brandolini et al., 2013).

Antioxidative capacity is strongly affected by the composition of compounds, depending on the assay used. Although emmer and einkorn contained similar amounts of free and bound phenolics, their antioxidative potential as measured by the DPPH assay differed distinctly, with einkorn revealing at least a doubled radical scavenging activity (Bencze et al., 2020). Fogarasi et al. (2015) confirmed the superior antioxidative properties of einkorn, which was at least three times higher than that of common wheat. In contrast, antioxidant capacity measured by the ORAC assay revealed no differences between einkorn, emmer and common wheat (Engert and Honermeier, 2011). Skrajda-Brdak et al. (2019) observed significantly higher ORAC values in spelt compared to common wheat, but DPPH values were similar. Kerienė et al. (2015) found similar ranges of the main PAs and flavonoids in spelt and common wheat, but antioxidative activity as measured by DPPH and ABTS varied significantly. Yilmaz et al. (2015) demonstrated that also the extraction solvents used affected the results of several spectrophotometric assays.

Alkylresorcinols (ARs) are phenolic lipids composed of a long aliphatic chain and a resorcinol-type phenolic compound. ARs represent a significant proportion of the phytochemicals in wheat, where they are concentrated in the bran. The highest concentrations were found in einkorn and emmer. Durum and spelt contained marginally higher levels than common wheat (Ziegler et al., 2016a). In the HEALTHGRAIN study, hulled wheats had generally higher levels of ARs (Shewry and Hey, 2015). The composition of ARs differs among wheat species. Although AR-21 is predominant in all species (>50% of all ARs), AR-19 is more abundant in hexaploid wheats, whereas AR-23 was found in increased proportions in einkorn and emmer (Andersson et al., 2008; Hammann et al., 2019).

Phenolic compounds in the wheat grain are concentrated in the germ and bran. Milling or pearling therefore has a major impact on antioxidative content of the flour. Further processing steps such as fermentation or cooking and/or baking can reduce or improve the antioxidative properties of wheat products (for a review see Ragaee et al., 2014). Generally, fermentation and baking favour cell wall degradation processes which then increase free phenolics. However, Skrajda-Brdak et al. (2019) have shown that the impact of fermentation type on the content of free phenolic acids varies due to the interaction between different factors.

5.2 Carotenoids

Carotenoids are isoprenoid pigments that cause yellow, orange or red colouring of various plant organs. Besides yellow and orange vegetables and fruits, carotenoids are also abundant in many dark green vegetables (Humphries and Khachik, 2003; Fraser and Bramley, 2004). The two main groups of carotenoids differ in their structure: carotenes (e.g. lycopene, α- and β-carotene) are hydrocarbons composed of a carbon skeleton with different hydrogenation levels, and xanthophylls (e.g. lutein, zeaxanthin, violaxanthin) also possess oxygen-containing functional groups (Cazzonelli, 2011; Maoka, 2020).

Carotenoids can act as free-radical scavengers and are very effective in quenching singlet oxygen (Ramel et al., 2012). They therefore play important roles in photosynthesis and photoprotection, as well as in nutrition and human health (Fraser and Bramley, 2004; Cazzonelli, 2011). α-Carotene, β-carotene and β-cryptoxanthin have provitamin A activity (Tang and Russell, 2009), while zeaxanthin and lutein are the major carotenoids in the macular region of the retina in humans (Bernstein et al., 2001).

Durum wheat is specifically selected for high yellow endosperm pigmentation in order to provide the yellow colour of pasta, whereas the yellow colouring is not desirable for common wheat flour used for baking. In hulled wheat species, considerable carotenoid concentrations are found in einkorn (Table 2), significantly surpassing those of durum wheat (Hidalgo et al., 2006; Ziegler et al., 2015). Lutein is by far the most important carotenoid in wheat grains, followed by traces of zeaxanthin, β-carotene and β-cryptoxanthin (Hidalgo et al., 2006; Abdel-Aal et al., 2007; Lachman et al., 2013; Hussain et al., 2015). In einkorn, the highest lutein concentration is in the germ (32.1 mg/kg), followed by the endosperm (6.3 mg/kg) and the bran (4.3 mg/kg). Since the endosperm represents about 74% of the total kernel, refined flour still contains most of the lutein (Hidalgo and Brandolini, 2008).

A significant proportion of carotenoids is lost during milling or pasta making due to oxidative degradation by lipoxygenase. A deletion in the *Lpx-1*

Table 2 Lutein concentration (mg/kg) in whole meals of hulled wheat species (number of investigated genotypes in brackets)

Reference	Einkorn	Emmer	Spelt
Abdel-Aal et al. (2002)	1.78-8.69 (15)	3.68 (1)	1.03-2.58 (4)
Panfili et al. (2004)	-	1.78 (3)	1.46 (3)
Hidalgo et al. (2006)	3.92-12.64 (54)	1.89-1.90 (2)	2.45-2.71 (2)
Abdel-Aal et al. (2007)	6.37-8.46 (26)	3.21-4.69 (13)	0.93-2.01 (4)
Serpen et al. (2008)	3.22-10.04 (6)	3.26-4.14 (12)	-
Lachman et al. (2013)	3.03-8.74 (5)	0.29-1.31 (5)	-
Hussain et al. (2015)	-	1.06-1.61 (2)	1.06-2.22 (11)
Ziegler et al. (2015)	4.03-8.95 (15)	0.64-2.27 (15)	0.62-2.42 (15)

locus and a positive *Lpx-3* allele improved semolina and pasta colour of durum wheat significantly (Carrera et al., 2007). Across wheat species, lipoxygenase activity was shown to be lowest in einkorn (Leenhardt et al., 2006a; Hidalgo and Brandolini, 2012), highly variable in emmer and high in spelt (Hidalgo and Brandolini, 2012).

During breadmaking carotenoids are degraded. Lutein is degraded by approximately 40% and zeaxanthin by about 25% compared to the concentration in flour (Antognoni et al., 2017). Besides the baking process itself, major losses occur during the kneading stage (approximately 15% on average), when the incorporation of oxygen in the dough fosters lipoxygenase-catalysed oxidation of carotenoids (Hidalgo et al., 2010). In contrast, fermentation has little effect on carotenoid content, possibly due to the consumption of oxygen by yeast, which prevents lipoxygenase-mediated carotenoid degradation (Leenhardt et al., 2006b; Hidalgo et al., 2010). Baking has a marked influence on carotenoid loss in bread crust (29%) but not in bread crumb (3%) (Hidalgo et al., 2010). Sourdough fermentation does not significantly affect carotenoid content, suggesting higher degradation due to higher oxygen incorporation. It seems that the acidic environment and/or the microbial metabolism do not favour the activity of carotenoid-degrading enzymes (Antognoni et al., 2017).

Bioaccessibility of lutein and zeaxanthin in einkorn breads has been found to be higher than 60% in some samples (Antognoni et al., 2017), whereas significant lower values have been reported for einkorn flatbreads, muffins and cookies (Read et al., 2015). Improved carotenoid bioaccessibility has been shown to occur with higher fat content (Read et al., 2015). The type of food matrix or the presence of fibres can also influence carotenoid bioavailability. Sourdough breads have lower carotenoid bioaccessibility due to increased dietary fibre content via greater solubilization of arabinoxylans (O'Connell et al., 2008). Hidalgo et al. (2018) studied the bioaccessibility of carotenoids, tocols and phenolic acids of einkorn biscuits during in vitro gastric and intestinal digestion and observed that only hydrophilic molecules were solubilised by gastric digestion. After intestinal digestion, 77% and 67% of carotenoids and tocols, respectively, were released. Soluble-conjugated phenolic acids increased (30%) and insoluble-bound forms decreased (17%), suggesting partial conversion from bound to conjugated form.

5.3 Tocols

The total tocols content of einkorn is higher than that of bread and durum wheats (Hidalgo et al., 2006; Lampi et al., 2008; Lachman et al., 2013). The average concentration in einkorn has been found to be 78 mg/kg dry matter (dm) and the maximum value 116 mg/kg dm, compared to 63 and 53 mg/kg dm in common and durum wheat, respectively (Hidalgo et al., 2006). The most abundant compound in einkorn is β-tocotrienol (62%), followed by α-tocotrienol and α-tocopherol (each 16%), and β-tocopherol (6%) (Hidalgo et al., 2006).

6 Micronutrients: vitamins and minerals

6.1 Vitamins

Wholegrain wheat is generally a rich source of B vitamins, thiamine (B1), riboflavin (B2), niacin (B3), pyridoxine (B6) and folates (B9). Piironen et al. (2009) found a very high diversity of vitamin levels in bread wheat. However, little research has been carried out on vitamins in hulled wheats. Abdel-Aal et al. (1995) found similar B1 concentrations in einkorn, spelt and common wheat, with B2 lowest in spelt. B3 was highest in spelt but showed high variability. B6 levels were higher in einkorn, with lower levels found in hexaploid wheats. Another study compared vitamin contents of spelt and common wheat. B1 and B2 were higher in bread wheat, whereas B3 levels were superior in spelt (Ranhotra et al., 1996).

Folic acid has been related in ancient wheat and modern wheat species within the HEALTHGRAIN project. Results showed marginally higher concentrations in durum wheat (0.74 mg/g) and emmer (0.69 mg/g) compared to the other species (0.56–0.58 mg/g). However, a low number of accessions (except for common wheat) from only one harvest year limits the significance of these results (Shewry and Hey, 2015). However, the impact of genotype on vitamin concentrations in wheat breads is small compared to the major effect of processing, especially milling and fermentation times (Batifoulier et al., 2005).

6.2 Minerals

Erba et al. (2011) tested five einkorns and one common wheat in four different environments over two cropping seasons and found that einkorn had higher zinc (72 mg/kg vs. 35 mg/kg), iron (52 mg/kg vs. 36 mg/kg), manganese (46 mg/kg vs. 30 mg/kg), copper (9 mg/kg vs. 6 mg/kg), magnesium (1512 mg/kg vs. 1125 mg/kg) and phosphorus (5411 mg/kg vs. 3099 mg/kg) concentrations compared to common wheat. Potassium and calcium contents did not differ between species.

Suchowilska et al. (2012) found that, in comparison with common wheat, einkorn has superior concentrations of P (5.20 mg/kg vs. 4.70 g/kg for einkorn and common wheat, respectively), Mg (1.63 mg/kg vs. 1.44 mg/kg), Zn (53 mg/kg vs. 35 mg/kg), Fe (49 mg/kg vs. 37.5 mg/kg), Mn (28 mg/kg vs. 26 mg/kg), Sr (5.4 mg/kg vs. 3.0 mg/kg) and Mo (1.2 mg/kg vs. 0.65 mg/kg). Lachman et al. (2011) reported higher selenium concentration in einkorn (52.3 μg/kg) compared to common wheat (34.8 μg/kg). However, whole meal flour contains phytic acid, a strong chelator of mineral cations (e.g. calcium, iron and zinc).

The resulting salts are excreted which limits micronutrient absorption. To offset micronutrient deficiency, breeders have crossed high Zn progenitors such as emmer, spelt and landraces. Velu et al. (2018) found that a Zn-motif transcription factor and metal-ion binding by phosphatase play a key role in Zn biofortification in wheat. Sequences of SNP markers have been allocated to publicly available databases to support breeding programmes.

7 Case studies: yellow pigments in einkorn

7.1 Degradation of einkorn yellow pigments by baking

Grains of einkorn accession, 'Vorarlberger', grown under organic conditions in St. Bernhard (48°41′ N, 15°35′41″ E) were milled with a stone mill (König Super M1, Grieskirchen, Austria). Bread was prepared from wholemeal flour (500 g), compressed yeast (25 g), salt (9 g), wheat malt flour (5 g) and shortening (5 g). After dough preparation (3 min mixing) the loaves were formed by hand and put in a baking pan. The dough was allowed to rest for 80 min at 27°C in a fermentation chamber (Wiesheu GS8 Mini R, Wolfen, Germany) before it was baked at 220°C for 25 min (Wiesheu Minimat Eldo). Steam was provided at the beginning of baking and again after 15 min. Samples of dough after mixing and fermentation, as well as bread crumb and crust samples were freeze dried before extraction of the yellow pigments.

Extraction of yellow pigments followed the method described by Hentschel et al. (2002). A few modifications were applied: three cycles of extraction with methanol/tetrahydrofuran (1:1, v/v) were carried out at 50°C in a water bath; homogenization was conducted using a vortex; the extract was rotary evaporated at 30°C to dryness, and subsequently reconstituted in hexane. Subsequently the samples were measured spectrophotometrically at a wavelength of 440 nm and the absorption values were transformed into yellow pigment values using a calibration with β-carotene.

From Table 3 it is obvious that a certain loss of yellow pigments is associated with lipoxygenase activity and baking temperature. The first distinct loss of yellow pigments was due to oxidative degradation during flour storage for 3 days and mixing. During fermentation losses were low, that is, ≈6%. Losses due to baking remain low for the bread crumb; however, they are significantly higher (≈40% compared to the original whole meal) for the bread crust, which can be explained by the great difference between the temperatures prevalent during baking in the inner (≈95°C) and outer part (≈200°C) of the dough.

7.2 Degradation of einkorn yellow pigments by cooking

Pasta and pearled einkorn grains were cooked for 10 min and 20 min, respectively. Pasta samples were taken after 10 min of cooking, samples of pearled grains

Table 3 Yellow pigment concentration (YPC; mg β-carotene equivalents/kg) in einkorn flour and products before and after processing (means ± standard error)

Product	Processing step	YPC	Change (%)[a]
Flour	Whole meal	11.5 ± 0.17	-
	Refined flour	12.7 ± 0.26	+10.6
	Bran	9.29 ±0.09	−19.1
Bread	Dough after mixing	9.75 ± 0.43	−15.1
	Dough after fermentation	9.05 ± 0.20	−21.2
	Crumb after baking	8.60 ± 0.29	−25.1
	Crust after baking	6.88 ± 0.40	−40.1
Pasta	Wholemeal; uncooked	10.9 ± 0.31	−5.3
	Wholemeal; cooked 10 min	6.42 ± 0.36	−44.1
	Refined flour; uncooked	9.04 ± 0.17	−28.8
	Refined flour; cooked 10 min	5.89 ± 0.29	−53.6
Pearled grain	Pearls	12.8 ± 0.17	-
	Pearls; cooked 5 min	8.09 ± 0.21	−36.5
	Pearls; cooked 10 min	7.39 ± 0.21	−42.0
	Pearls; cooked 15 min	6.06 ± 0.35	−52.5
	Pearls; cooked 20 min	5.19 ± 0.01	−59.3
	Pearls; steam cooked 20 min	10.3 ± 0.08	−19.4

[a] Changes (%) refer to the original source, that is, whole meal or refined flour for bread and pasta, and uncooked pearled grains.

were taken every 5 min. The samples were freeze dried and further analysed as mentioned above. The reductions in yellow pigments depending on time and type of cooking are demonstrated in Table 3. If noodles or pearled einkorn grains were cooked for 5 min losses in yellow pigments were ≈35%, followed by an additional 6–10% at any 5 min interval. Finally, 20 min-cooked pearled einkorn shows losses of nearly 60%. In contrast, stewing of pearled einkorn for 20 min resulted in a loss of only ≈20%. Since HPLC analysis of the einkorn samples revealed that the major carotenoid was lutein, it can be supposed, in agreement with other studies, that lutein is sensitive to heat.

8 Summary and future trends

Hulled wheats have some distinct features compared to common wheat which allow for processing of healthy and nutritious food. Einkorn is of particular interest due to higher concentrations of some minerals and carotenoids. However, processing such as baking and cooking can significantly degrade phytochemicals. Hence, processing techniques must be adapted, for example, high temperatures avoided, in order to maintain phytochemical levels with their potential health benefits.

Consumers are becoming more aware of the quality of food and its health benefits. There has been significant research into the quality and health benefits of hulled wheats. However, further research is needed to focus on the nutritional quality of spelt, einkorn and emmer varieties/accessions in comparison to other wheat species and other cereals. In addition, it is also necessary to do further research on the bioavailability of phytochemicals in final processed products made from hulled wheats.

9 Where to look for further information

Good introductions to the subject for non-specialists:

- 'Natural variation in grain composition of wheat and related cereals' by Peter Shewry and co-workers (https://pubs.acs.org/doi/pdf/10.1021/jf3054092).
- 'Emmer in Turkey' published by Bioversity International (https://www.bioversityinternational.org/fileadmin/_migrated/uploads/tx_news/Emmer_in_Turkey_1272.pdf).
- 'Ethnobotany of einkorn and emmer in Romania and Slovakia: towards interpretation of archaeological evidence' by Mária Hajnalová and Dagmar Dreslerová (https://www.researchgate.net/publication/281547751).

Key articles and conferences papers:

- 'Comparative study of hulled (einkorn, emmer, and spelt) and naked wheats (durum and bread wheat): agronomic performance and quality traits' by Friedrich Longin and co-workers (https://doi.org/10.2135/cropsci2015.04.0242).
- 'Hulled wheats: a review of nutritional properties and processing methods' by Ademola Hammed and Senay Simsek (https://doi.org/10.1094/CCHEM-09-13-0179-RW).
- 'Hulled wheats', Proceedings from the 1st International Workshop on Hulled Wheats, edited by Stefano Padulosi et al. (https://www.bioversityinternational.org/fileadmin/_migrated/uploads/tx_news/Hulled_wheat_54.pdf).
- 'Rheological and technological quality of minor wheat species and common wheat' by Petr Konvalina et al. (https://doi.org/10.5772/67229).
- 'Current trends in ancient grains-based foodstuffs: insights into nutritional aspects and technological applications' by Fatma Boukid et al. (https://doi.org/10.1111/1541-4337.12315).

Major international and national research projects:

- Growing Pennsylvania Grains Project - value-added small grains; Pennsylvania State University (https://plantscience.psu.edu/research/projects/grains/heritage-grains).
- An integrated approach to diversify the genetic base, improve stress resistance, agronomic management and nutritional/processing quality of minor cereal crops for human nutrition in Europe (http://healthyminorcereals.eu/; https://cordis.europa.eu/project/id/613609/reporting).
- HEALTHGRAIN Forum (https://healthgrain.org/) with available scientific results published within the EU HEALTHGRAIN project (https://www.sciencedirect.com/journal/trends-in-food-science-and-technology/vol/25/issue/2).
- Alkobeer project (http://alkobeerprojekt.hu/); Hungarian project on einkorn breeding and beer brewing (product website at: http://alakorsor.com/).
- Verein für alpine Kulturpflanzen - Schriftenreihe Kulturpflanzen der Schweiz/Geschichte der Kulturpflanzen (http://www.berggetreide.ch/Geschichte_Kulturpflanzen.html); series of brochures (in German and French) by Per Schilperoord on the history and cultivation of crops (e.g. spelt, einkorn, emmer and Rivet wheat) in Alpine regions.

Breeding and production initiatives

- Farro della Garfagnana (http://www.farrodellagarfagnana.it/); webpage (in Italian) of the protected geographical indication (PGI) production of emmer in the Garfagnana valley, Tuscany; (English description at: https://www.qualigeo.eu/en/prodotto-qualigeo/farro-della-garfagnana-igp/).
- Le petit épeautre de Haute Provence (https://www.petitepeautre.com/petit-epeautre/presentation/); webpage (in French) of the PGI production of einkorn in southeast France.
- Urdinkel - das wertvollste Getreide (https://www.urdinkel.ch/); webpage (in German, French, Italian and English) of the Swiss syndicate of growers of traditional spelt varieties.
- Cultivari (https://www.cultivari.de/en/projects/einkorn/einkorn-breeding/); organic einkorn breeding programme.

10 References

Abdel-Aal, E. M. and Rabalski, I. (2008). 'Effect of baking on nutritional properties of starch in organic spelt whole grain product', *Food Chem.*, 11, 150–156.

Abdel-Aal, E.-S. M. and Hucl, P. (2005) 'Spelt: a specialty wheat for emerging food uses', in E.-S. M. Abdel-Aal and P. Wood (Eds.), *Specialty Grains for Food and Feed* (pp. 109–142), St. Paul: American Association of Cereal Chemists, Inc.

Abdel-Aal, E.-S. M., Hucl, P. and Sosulski, F. W. (1995), 'Composition and nutritional characteristics of spring einkorn and spelt wheats', *Cereal Chem.*, 72, 621-624.

Abdel-Aal, E.-S. M., Young, J. C., Rabalski, I., Hucl, P. and Frégeau-Reid, J. (2007), 'Identification and quantification of seed carotenoids in selected wheat species', *J. Agric. Food Chem.*, 55, 787-794.

Abdel-Aal, E.-S. M., Young, J. C., Wood, P. J., Rabalski, I., Hucl, P., Falk, D. and Frégeau-Reid, J. (2002), 'Einkorn: a potential candidate for developing high lutein wheat', *Cereal Chem.*, 79, 455-457.

Andersson, A. A. M., Kamal-Eldin, A., Fraś, A., Boros, D. and Åman, P. (2008), 'Alkylresorcinols in wheat varieties in the healthgrain diversity screen', *J. Agric. Food Chem.*, 56, 9722-9725.

Antognoni, F., Mandrioli, R., Bordoni, A., Di Nunzio, M., Viadel, B., Gallego, E., Villalba, M. P., Tomás-Cobos, L., Taneyo Saa, D. L. and Gianotti, D. L. (2017), 'Integrated evaluation of the potential health benefits of einkorn-based breads', *Nutrients*, 9, 1232.

Barański, M., Lacko-Bartošová, M., Rembiałkowska, E. and Lacko-Bartošová, L. (2020), 'The effect of species and cultivation year on phenolic acids content in ancient wheat', *Agronomy*, 10, 673.

Batifoulier, F., Verny, M.-A., Chanliaud, E., Rémésy, C. and Demigné, C. (2005), 'Effect of different breadmaking methods on thiamine, riboflavin and pyridoxine contents of wheat bread', *J. Cereal Sci.*, 42, 101-108.

Bencze, S., Makádi, M., Aranyos, T., Földi, M., Hertelendy, P., Mikó, P., Bosi, S., Negri, L. and Drexler, D. (2020), 'Re-introduction of ancient wheat cultivars into organic agriculture–Emmer and einkorn cultivation experiences under marginal conditions', *Sustainability*, 12, 1584.

Bernstein, P. S., Khachik, F., Carvalho, L. S., Muir, G. J., Zhao, D.-Y. and Katz, N. B. (2001), 'Identification and quantitation of carotenoids and their metabolites in the tissues of the human eye', *Exp. Eye Res.*, 72, 215-223.

Bonafede, M., Kong, L., Tranquilli, G., Ohm, H. and Dubcovsky, J. (2007), 'Reduction of a *Triticum monococcum* chromosome segment carrying the softness genes *Pina* and *Pinb* translocated to bread wheat', *Crop Sci.*, 47, 821-828.

Brandolini, A., Castoldi, P., Plizzari, L. and Hidalgo, A. (2013), 'Phenolic acids composition, total polyphenols content and antioxidant activity of *Triticum monococcum*, *Triticum turgidum* and *Triticum aestivum*: a two-years evaluation', *J. Cereal Sci.*, 58, 123-131.

Brandolini, A., Hidalgo, A., Gabriele, S. and Heun, M. (2015), 'Chemical composition of wild and feral diploid wheats and their bearing on domesticated wheats', *J. Cereal Sci.*, 63, 122-127.

Brandolini, A., Hidalgo, A. and Moscaritolo, S. (2008), 'Chemical composition and pasting properties of einkorn (*Triticum monococcum* L. subsp. *monococcum*) whole meal flour', *J. Cereal Sci.*, 47, 599-609.

Brandolini, A., Hidalgo, A., Plizzari, L. and Erba, D. (2011), 'Impact of genetic and environmental factors on einkorn wheat (*Triticum monococcum* L. subsp. *monococcum*) polysaccharides', *J Cereal Sci.*, 53, 65-72.

Call, L., Kapeller, M., Grausgruber, H., Reiter, E., Schoenlechner, R. and D'Amico, S. (2020), 'Effects of species and breeding on wheat protein composition', *J. Cereal Sci.*, 93, 102974.

Callejo, M. J., Vargas-Kostiuk, M. E., Ribeiro, M. and Rodríguez-Quijano, M. (2019), '*Triticum aestivum* ssp. *vulgare* and ssp. *spelta* cultivars: 2. Bread-making optimisation', *Eur. Food Res. Technol.*, 245, 1399-1408.

Carrera, A., Echenique, V., Zhang, W., Helguera, M., Manthey, F., Schrager, A., Picca, A., Cervigni, G. and Dubcovsky, J. (2007), 'A deletion at the *Lpx-B1* locus is associated with low lipoxygenase activity and improved pasta color in durum wheat (*Triticum turgidum* ssp. *durum*)', *J. Cereal Sci.*, 45, 67–77.

Cazzonelli, C. I. (2011), 'Carotenoids in nature: insights from plants and beyond', *Funct. Plant Biol.*, 38, 833–847.

Chibbar, R. N. and Chakraborty, M. (2005), 'Waxy wheat', in E.-S. M. Abdel-Aal and P. Wood (Eds.), *Specialty Grains for Food and Feed* (pp. 143–166), St. Paul: American Association of Cereal Chemists, Inc.

Costanzo, A., Amos, D. C., Dinelli, G., Sferrazza, R. E., Accorsi, G., Negri, L. and Bosi, S. (2019), 'Performance and nutritional properties of einkorn, emmer and Rivet wheat in response to different rotational position and soil tillage', *Sustainability*, 11, 6304.

D'Amico, S., Call, L., Edlmayr, M., Marzak, S., Jäger, H., Domig, K., Reiter, E. and Grausgruber, H. (2018), 'Characterization of non-gluten proteins in Austrian wheat varieties from 19th century until today', *Cereal Technology/Getreidetechnologie*, 72, 80–88. doi: 10.23789/1869-2303-2018-2-80.

D'Antuono, L. F., Galletti, G. C. and Bocchini, P. (1998), 'Fiber quality of emmer (*Triticum dicoccum* Schubler) and einkorn wheat (*T. monococcum* L.) landraces as determined by analytical pyrolysis', *J. Sci. Food Agric.*, 78, 213–219.

Di Stasio, L., Picascia, S., Auricchio, R., Vitale, S., Gazza, L., Picariello, G., Gianfrani, C. and Mamone, G. (2020), 'Comparative analysis of *in vitro* digestibility and immunogenicity of gliadin proteins from durum and einkorn wheat', *Front. Nutr.*, 7, 56.

Dubois, B., Bertin, P., Hautier, L., Muhovski, Y., Escarnot, E. and Mingeot, D. (2018), 'Genetic and environmental factors affecting the expression of α-gliadin canonical epitopes involved in celiac disease in a wide collection of spelt (*Triticum aestivum* ssp. *spelta*) cultivars and landraces', *BMC Plant Biol.*, 18, 262.

Dvorak, J., Deal, K. R., Luo, M.-C., You, F. M., von Borstel, K. and Dehghani, H. (2012), 'The origin of spelt and free-threshing hexaploid wheat', *J. Hered.*, 103, 426–441.

Engert, N. and Honermeier, B. (2011), 'Characterization of grain quality and phenolic acids in ancient wheat species (*Triticum* sp.)', *J. Appl. Bot. Food Qual.*, 84, 33–39.

Erba, D., Hidalgo, A., Bresciani, J. and Brandolini, A. (2011), 'Environmental and genotypic influence on trace element and mineral concentrations in whole meal flour of einkorn (*Triticum monococcum* L. subsp. *monococcum*)', *J. Cereal Sci.*, 54, 250–254.

Escarnot, E., Dornez, E., Verspreet, J., Agneessens, R. and Courtin, C. M. (2015), 'Quantification and visualization of dietary fibre components in spelt and wheat kernels', *J. Cereal Sci.*, 62, 124–133.

Escarnot, E., Gofflot, S., Sinnaeve, G., Dubois, B., Bertin, P. and Mingeot, D. (2018), 'Reactivity of gluten proteins from spelt and common wheat accessions towards A1 and G12 antibodies in the framework of celiac disease', *Food Chem.*, 268, 522–532.

Fares, C., Codianni, P., Nigro, F., Platani, C., Scazzina, F. and Pellegrini, N. (2008), 'Processing and cooking effects on chemical, nutritional and functional properties of pasta obtained from selected emmer genotypes', *J. Sci. Food Agric.*, 88, 2435–2444.

Fares, C., Menga, V., Codianni, P., Russo, M., Perrone, D., Suriano, S., Savino, M. and Rascio, A. (2019), 'Phenolic acids variability and grain quality of organically and conventionally fertilised old wheats under a warm climate', *J. Sci. Food Agric.*, 99, 4615–4623.

Fogarasi, A.-L., Kun, S., Tankó, G., Stefanovits-Bányai, É. and Hegyesné-Vecseri, B. (2015), 'A comparative assessment of antioxidant properties, total phenolic content of einkorn, wheat, barley and their malts', *Food Chem.*, 167, 1-6.

Frakolaki, G., Giannou, V., Topakas, E. and Tzia, C. (2018), 'Chemical characterization and breadmaking potential of spelt versus wheat flour', *J. Cereal Sci.*, 79, 50-56.

Fraser, P. D. and Bramley, P. M. (2004), 'The biosynthesis and nutritional uses of carotenoids', *Prog. Lipid Res.*, 43, 228-265.

Frégeau-Reid, J. and Abdel-Aal, E.-S. M. (2005), 'Einkorn: a potential wheat and genetic resources', in E.-S. M. Abdel-Aal and P. Wood (Eds.), *Specialty Grains for Food and Feed* (pp. 37-62), St. Paul: American Association of Cereal Chemists, Inc.

Galterio, G., Codianni, P., Giusti, A. M., Pezzarossa, B. and Cannella, C. (2003), 'Assessment of the agronomic and technological characteristics of *Triticum turgidum* ssp. *dicoccum* Schrank and *T. spelta* L.', *Nahrung/Food*, 47, 51-59.

Gebruers, K., Dornez, E., Boros, D., Fras, A., Dynkowska, W., Bedö, Z., Rakszegi, M., Delcour, J. A. and Courtin, C. M. (2008), 'Variation in the content of dietary fiber and components thereof in wheats in the healthgrain diversity screen', *J. Agric. Food Chem.*, 56, 9740-9749.

Geisslitz, S., Ludwig, C., Scherf, K. A. and Koehler, P. (2018a), 'Targeted LC−MS/MS reveals similar contents of αamylase/trypsin-inhibitors as putative triggers of nonceliac gluten sensitivity in all wheat species except einkorn', *J. Agric. Food Chem.*, 66, 12395-12403.

Geisslitz, S., Wieser, H., Scherf, K. A. and Koehler, P. (2018b), 'Gluten protein composition and aggregation properties as predictors for bread volume of common wheat, spelt, durum wheat, emmer and einkorn', *J. Cereal Sci.*, 83, 204-212.

Gélinas, P. and McKinnon, C. (2016), 'Gluten weight in ancient and modern wheat and the reactivity of epitopes towards R5 and G12 monoclonal antibodies', *Int. J. Food Sci. Technol.*, 51, 1801-1810.

Gell, G. M., Kovács, K., Molnár, I., Bugyi, Z., Tömösközi, S. and Juhász, A. (2015), 'Celiac disease-specific prolamin peptide content of wheat', *Cereal Res. Commun.* 43, 133-143.

Geng, P., Sun, J., Zhang, M., Li, X., Harnly, J. M. and Chen, P. (2016), 'Comprehensive characterization of C-glycosyl flavones in wheat (*Triticum aestivum* L.) germ using UPLC-PDA-ESI/HRMSn and mass defect filtering', *J. Mass Spectrom.*, 51, 914-930.

Gianfrani, C., Camarca, A., Mazzarella, G., Di Stasio, L., Giardullo, N., Ferranti, P., Picariello, G., Rotondi Aufiero, V., Picascia, S., Troncone, R., Pogna, N., Auricchio, S. and Mamone, G. (2015), 'Extensive *in vitro* gastrointestinal digestion markedly reduces the immune-toxicity of *Triticum monococcum* wheat: Implication for celiac disease', *Mol. Nutr. Food Res.*, 59, 1844-1854.

Grausgruber, H. (2018), ''Sorting the wheat from the chaff'–comments on the spelt [wheat] discussion', in Saatgut Austria (Ed.), *Proceedings of the 68th Breeders' Conference, 20-22 Nov. 2017, Raumberg-Gumpenstein* (pp. 69-74), Vienna: University of Natural Resources and Life Sciences, Vienna.

Grausgruber, H., D'Amico, S., Dong, R., Atzgersdorfer, K., Musilova, M. and Böhmdorfer, S. (2014), 'Anthocyanins in cereal grains', in N. Motohashi (Ed.), *Occurrences, Structure, Biosynthesis, and Health Benefits Based on Their Evidences of Medicinal Phytochemicals in Vegetables and Fruits* (Vol. 2, pp. 44-66), New York: Nova Science Publishers, Inc.

Grausgruber, H., Scheiblauer, J., Schonlechner, R., Ruckenbauer, P. and Berghofer, E. (2004), 'Variability in chemical composition and biologically active constituents of cereals', in J. Vollmann, H. Grausgruber and P. Ruckenbauer (Eds.), *Genetic Variation for Plant Breeding* (pp. 23–26), Vienna: University of Natural Resources and Life Sciences, Vienna.

Grausgruber, H., Štěrbová, L., Thrackl, K., Bradová, J., Baumgartner, S. and Dvořáček, V. (2019), 'Content of the immunodominant 33-mer peptide from α2-gliadin in common and ancient wheat flours determined by the G12 sandwich ELISA', in C. Guzman (Ed.), *Proceedings of the 13th International Gluten Workshop* (pp. 116–118), Córdoba: UCO Press.

Guzmán, C., Caballero, L. and Alvarez, J. B. (2011), 'Molecular characterisation of the *Wx-B1* allelic variants identified in cultivated emmer wheat and comparison with those of durum wheat', *Mol. Breed.*, 28, 403–411.

Guzmán, C., Caballero, L. and Moral, A. (2010), 'Genetic variation for waxy proteins and amylose content in Spanish spelt wheat (*Triticum spelta* L.)', *Genet. Resour. Crop Evol.*, 57, 721–725.

Hammann, S., Korf, A., Bull, I. D., Hayen, H. and Cramp, L. J. E. (2019), 'Lipid profiling and analytical discrimination of seven cereals using high temperature gas chromatography coupled to high resolution quadrupole time-of-flight mass spectrometry', *Food Chem.*, 282, 27–35.

Hentschel, V., Kranl, K., Hollmann, J., Lindhauer, M. G., Böhm, V. and Bitsch, R. (2002), 'Spectrophotometric determination of yellow pigment content and evaluation of carotenoids by high-performance liquid chromatography in durum wheat grain', *J. Agric. Food Chem.*, 50, 6663–6668.

Hidalgo, A. and Brandolini, A. (2008), 'Protein, ash, lutein and tocols distribution in einkorn (*Triticum monococcum* spp. *monococcum*) seed fractions', *Food Chem.*, 107, 444–448.

Hidalgo, A. and Brandolini, A. (2012), 'Lipoxygenase activity in wholemeal flours from *Triticum monococcum*, *Triticum turgidum* and *Triticum aestivum*', *Food Chem.*, 131, 1499–1503.

Hidalgo, A. and Brandolini, A. (2019), 'Nutritional, technological, and health aspects of einkorn flour and bread', in V. R. Preedy and R. R. Watson (Eds.), *Flour and Breads and Their Fortification in Health and Disease Prevention* (2nd ed., pp. 99–110), London: Academic Press.

Hidalgo, A., Brandolini, A. and Pompei, C. (2010) 'Carotenoids evolution during pasta, bread and water biscuit preparation from wheat flours', *Food Chem.*, 121, 746–751.

Hidalgo, A., Brandolini, A., Pompei, C. and Piscozzi, R. (2006), 'Carotenoids and tocols of einkorn wheat (*Triticum monococcum* ssp. *monococcum* L.)', *J. Cereal Sci.*, 44, 182–193.

Hidalgo, A., Ferraretto, A., De Noni, I., Bottani, M., Cattaneo, S., Galli, S. and Brandolini, A. (2018), 'Bioactive compounds and antioxidant properties of pseudocereals-enriched water biscuits and their *in vitro* digestates', *Food Chem.*, 240, 799–807.

Humphries, J. M. and Khachik, F. (2003), 'Distribution of lutein, zeaxanthin, and related geometrical isomers in fruit, vegetables, wheat, and pasta products', *J. Agric. Food Chem.*, 51, 1322–1327.

Hussain, A., Larsson, H., Kuktaite, R., Olsson, M. E. and Johansson, E. (2015), 'Carotenoid content in organically produced wheat: relevance for human nutritional health on consumption', *Int. J. Environ. Res. Public Health*, 12, 14068–14083.

Jorjadze, M., Berishvili, T. and Shatberashvili, E. (2014), 'The ancient wheats of Georgia and their traditional use in the southern part of the country', *Emir. J. Food Agric.*, 26, 192–202.

JRC. (2020), 'Dietary fibre', Health Promotion and Disease Prevention Knowledge Gateway, Joint Research Centre, European Commission. https://ec.europa.eu/jrc/en/health-knowledge-gateway/promotion-prevention/nutrition/fibre.

Kerienė, I., Mankevičienė, A., Bliznikas, S., Jablonskytė-Raščė, D., Maikštėnienė, S. and Česnulevičienė, R. (2015), 'Biologically active phenolic compounds in buckwheat, oats and winter spelt wheat', *Zemdirbyste-Agriculture*, 102, 289–296.

Khoshbakht, K. (2009), 'Esfahanian emmer (*Triticum ispahanicum* Heslot)–a case of an extinct on-farm crop', in A. Buerkert and J. Gebauer (Eds.), *Agrobiodiversity and Genetic Erosion, Contributions in Honor of Prof. Dr. Karl Hammer* (pp. 189–195), Kassel: Kassel University Press.

Konvalina, P., Stehno, Z., Capouchová, I. Zechner, E., Berger, S., Grausgruber, H., Janovská, D. and Moudrý Sr., J. (2014), 'Differences in grain/straw ratio, protein content and yield in landraces and modern varieties of different wheat species under organic farming', *Euphytica*, 199, 31–40.

Lachman, J., Hejtmánková, K. and Kotíková, Z. (2013), 'Tocols and carotenoids of einkorn, emmer and spring wheat varieties: selection for breeding and production', *J. Cereal Sci.*, 57, 207–214.

Lachman, J., Miholová, D., Pivec, V., Jírů, K. and Janovská, D. (2011), 'Content of phenolic antioxidants and selenium in grain of einkorn (*Triticum monococcum*), emmer (*Triticum dicoccum*) and spring wheat (*Triticum aestivum*) varieties', *Plant Soil Environ.*, 57, 2235–2243.

Lampi, A.-M., Nurmi, T., Ollilainen, V. and Piironen, V. (2008), 'Tocopherols and tocotrienols in wheat genotypes in the HEALTHGRAIN diversity screen', *J. Agric. Food Chem.*, 56, 9716–9721.

Leenhardt, F., Lyan, B., Rock, E., Boussard, A., Potus, J., Chanliaud, E. and Remesy, C. (2006a), 'Genetic variability of carotenoid concentration, and lipoxygenase and peroxidase activities among cultivated wheat species and common wheat varieties', *Eur. J. Agron.*, 25, 170–176.

Leenhardt, F., Lyan, B., Rock, E., Boussard, A., Potus, J., Chanliaud, E. and Remesy, C. (2006b), 'Wheat lipoxygenase activity induces greater loss of carotenoids than vitamin E during breadmaking', *J. Agric. Food. Chem.*, 54, 1710–1715.

Lev-Yadun, S., Gopher, A. and Abbo, S. (2000), 'The cradle of agriculture', *Science*, 288, 1602–1603.

Li, L., Shewry, P. R. and Ward, J. L. (2008), 'Phenolic acids in wheat varieties in the healthgrain diversity screen', *J. Agric. Food Chem.*, 56, 9732–9739.

Liu, X., Feng, Z., Liang, D., Zhang, M., Liu, X., Hao, M., Liu, D., Ning, S., Yuan, Z., Jiang, B., Chen, X., Chen, X. and Zhang, L. (2020), 'Development, identification, and characterization of blue-grained wheat-*Triticum boeoticum* substitution lines', *J. Appl. Genet.*, 61, 169–177.

Løje, H., Møller, B., Laustsen, A. M. and Hansen, Å. (2003), 'Chemical composition, functional properties and sensory profiling of einkorn (*Triticum monococcum* L.)', *J. Cereal Sci.*, 37, 231–240.

Longin, C. F. H., Beck, H., Gütler, A., Gütler, H., Heilig, W., Zimmermann, J., Bischoff, S. C. and Würschum, T. (2020), 'Influence of wheat variety and dough preparation on FODMAP content in yeast-leavened wheat breads', *J. Cereal Sci.*, 95, 103021.

Longin, C. F. H., Ziegler, J., Schweiggert, R., Koehler, P., Carle, R. and Würschum, T. (2016), 'Comparative study of hulled (einkorn, emmer, and spelt) and naked wheats (durum and bread wheat): agronomic performance and quality traits', *Crop Sci.*, 56, 302–311.

Maoka, T. (2020), 'Carotenoids as natural functional pigments', *J. Nat. Med.*, 74, 1-16.

Marconi, E. and Cubadda, R. (2005), 'Emmer wheat', in E.-S. M. Abdel-Aal and P. Wood (Eds.), *Specialty Grains for Food and Feed* (pp. 63-108), St. Paul: American Association of Cereal Chemists, Inc.

Marcotuli, I., Colauonno, P., Hsieh, Y. S. Y., Fincher, G. B. and Gadaleta, A. (2020), 'No-starch polysaccharides in durum wheat: a review', *Int. J. Mol. Sci.*, 21, 2933.

Mohammadkhani, A., Stoddard, F. L. and Marshall, D. R. (1998), 'Survey of amylose content in *Secale cereale*, *Triticum monococcum*, *T. turgidum* and *T. tauschii*', *J. Cereal Sci.*, 28, 273-280.

Mohan Kumar, B. V., Prasada Rao, U. J. S. and Prabhansankar, P. (2017), 'Immunogenicity characterization of hexaploid and tetraploid wheat varieties related to celiac disease and wheat allergy', *Food Agric. Immunol.*, 28, 888-903.

Molberg, Ø., Ulhen, A. K., Jensen, T., Solheim Flæte, N., Fleckenstein, B., Arentz-Hansen, H., Raki, M., Lundin, K. E. A. and Sollid, L. M. (2005), 'Mapping of gluten T-cell epitopes in the common wheat ancestors: implications for celiac disease', *Gastroenterol.*, 128, 393-401.

Montevecchi, G., Setti, L., Olmi, L., Buti, M., Laviano, L., Antonelli, A. and Sgarbi, E. (2019), 'Determination of free soluble phenolic compounds in grains of ancient wheat varieties (*Triticum* sp. pl.) by liquid chromatography-tandem mass spectrometry', *J. Agric. Food Chem.*, 67, 201-212.

Morris, C. F. (2002), 'Puroindolines: the molecular genetic basis of wheat grain hardness', *Plant Mol. Biol.*, 481, 633-647.

Nesbitt, M. and Samuel, D. (1996), 'From staple crop to extinction? The archaeology and history of the hulled wheats', in S. Padulosi, K. Hammer and J. Heller (Eds.), *Hulled Wheat* (pp. 40-99), Rome: International Plant Genetic Resources Institute.

O'Connell, O., Ryan, L., O'Sullivan, L., Aherne-Bruce, S. A. and O'Brien, N. M. (2008), 'Carotenoid micellarization varies greatly between individual and mixed vegetables with or without the addition of fat or fiber', *Int. J. Vitam. Nutr. Res.*, 78, 238-246.

Ozuna, C. V., Iehisa, J. C. M., Giménez, M. J., Alvarez, J. B., Sousa, C. and Barro, F. (2015), 'Diversification of the celiac disease α-gliadin complex in wheat: a 33-mer peptide with six overlapping epitopes, evolved following polyploidization', *Plant J.*, 82, 794-805.

Pandey, K. B. and Rizvi, S. I. (2009), 'Plant polyphenols as dietary antioxidants in human health and disease', *Oxid. Med. Cell Longev.*, 2, 270-278.

Panfili, G., Fratianni, A. and Irano, M. (2004), 'Improved normal-phase high-performance liquid chromatography procedure for the determination of carotenoids in cereals', *J. Agric. Food Chem.*, 52, 6373-6377.

Perrino, P., Laghetti, G., D'Antuono, L. F., Al Ajlouni, M., Kanbertay, M., Szabó, A. T. and Hammer, K. (1996), 'Ecogeographical distribution of hulled wheat species', in S. Padulosi, K. Hammer and J. Heller (Eds.), *Hulled Wheat* (pp. 100-118), Rome: International Plant Genetic Resources Institute.

Piironen, V., Lampi, A. M., Ekholm, P., Salmenkallio-Marttila, M. and Liukkonen, K. H, (2009), 'Micronutrients and phytochemicals in wheat grain', in K. Khan and P. R. Shewry (Eds.), *Wheat: Chemistry and Technology*, (4th Ed., pp. 179-222), St. Paul: AACC International.

Ragaee, S., Seetharaman, K. and Abdel-Aal, E.-S. M. (2014), 'The impact of milling and thermal processing on phenolic compounds in cereal grains', *Crit. Rev. Food Sci. Nutr.*, 54, 837-849.

Ramel, F., Birtic, S., Cuiné, S., Triantaphylidès, C., Ravanat, J.-L. and Havaux, M. (2012), 'Chemical quenching of singlet oxygen by carotenoids in plants', *Plant Physiol.*, 158, 1267–1278.

Ranhotra, G. S., Gelroth, J. A., Glaser, B. K. and Lorenz, K. J. (1996), 'Nutrient composition of spelt wheat', *J. Food Comp. Anal.*, 9, 81–84.

Read, A., Wright, A. and Abdel-Aal, E.-S. M. (2015), '*In vitro* bioaccessibility and monolayer uptake of lutein from wholegrain baked foods', *Food Chem.*,174, 263–269.

Ross, A. B., Shepherd, M. J., Schüpphaus, M., Sinclair, V., Alfaro, B., Kamal-Eldin, A. and Åman, P. (2003), 'Alkylresorcinols in cereals and cereal products', *J. Agric. Food Chem.*, 51, 4111–4118.

Ruiz-Carnicer, A., Comino, I., Segura, V., Ozuna, C. V., de Lourdes Moreno, M., López-Casado, M. Á., Torres, M. I., Barro, F. and Sousa, C. (2019), 'Celiac immunogenic potential of α-gliadin epitope variants from *Triticum* and *Aegilops* species', *Nutrients*, 11, 220.

Şahin, Y., Yıldırım, A. B., Yücesan, B., Zencirci, N., Erbayram, Ş. and Gürel, E. (2017), 'Phytochemical content and antioxidant activity of einkorn (*Triticum monococcum* ssp. *monococcum*), bread (*Triticum aestivum* L.), and durum (*Triticum durum* Desf.) wheat', *Progr. Nutr.*, 19, 450–459.

Schalk, K., Lang, C., Wieser, H., Koehler, P. and Scherf, K. A. (2017), 'Quantitation of the immunodominant 33-mer peptide from α-gliadin in wheat flours by liquid chromatography tandem mass spectrometry', *Sci. Rep.*, 7, 45092.

Schopf, M. and Scherf, K. A. (2018), 'Wheat cultivar and species influence variability of gluten ELISA analyses based on polyclonal and monoclonal antibodies R5 and G12', *J. Cereal Sci.*, 83, 32–41.

Serpen, A., Gökmen, V., Karagöz, A. and Köksel, H. (2008), 'Phytochemical quantification and total antioxidant capacities of emmer (*Triticum dicoccon* Schrank) and einkorn (*Triticum monococcum* L.) wheat landraces', *J. Agric. Food Chem.*, 56, 7285–7292.

Shevkani, K., Singh, N., Bajaj, R. and Kaur, A. (2017), 'Wheat starch production, structure, functionality and applications–a review', *Int. J. Food Sci. Technol.*, 52, 38–58.

Shewry, P. R. and Hey, S. (2015), 'Do "ancient" wheat species differ from modern common wheat in their contents of bioactive components?', *J. Cereal Sci.*, 65, 236–243.

Sievers, S., Rohrbach, A. and Beyer, K. (2020), 'Wheat-induced food allergy in childhood: ancient grains seem no way out', *Eur. J. Nutr.*, 59, 2693–2707.

Skrajda-Brdak, M., Konopka, I., Tańska, M. and Czaplicki, S. (2019), 'Changes in the content of free phenolic acids and antioxidative capacity of wholemeal bread in relation to cereal species and fermentation type', *Eur. Food Res. Technol.*, 245, 2247–2256.

Starzyńska-Janiszewska, A., Stodolak, B., Socha, R., Mickowska, B. and Wywrocka-Gurgul, A. (2019), 'Spelt wheat tempe as a value-added whole-grain food product', *LWT Food Sci. Technol.*, 113, 108250.

Suchowilska, E., Wiwart, M., Kandler, W. and Krska, R. (2012), 'A comparison of macro-and microelement concentrations in the whole grain of four *Triticum* species'. *Plant Soil Environ.*, 58, 141–147.

Szymanski, R. M. and Morris, C.F (2015), 'Internal structure of carbonized wheat (*Triticum* spp.) grains: relationships to kernel texture and ploidy', *Veget. Hist. Archaeobot.*, 24, 503–515.

Taddei, F., Gazza, L., Conti, S., Muccilli, V., Foti, S. and Pogna, N. E. (2009), 'Starch-bound 2S proteins and kernel texture in einkorn, *Triticum monococcum* ssp. *monococcum*', *Theor. Appl. Genet.*, 119, 1205–1212.

Tang, G. and Russell, R. M. (2009), 'Carotenoids as provitamin A', in G. Britton, H. Pfander, and S. Liaaen-Jensen (Eds.), *Carotenoids, Vol. 5: Nutrition and Health* (pp. 149-172), Basel: Birkhäuser.

Vaccino, P., Becker, H. A., Brandolini, A., Salamini, F. and Kilian, B. (2009), 'A catalogue of *Triticum monococcum* genes encoding toxic and immunogenic peptides for celiac disease patients', *Mol. Gen. Genom.*, 281, 289-300.

Van Hung, P., Maeda, T. and Morita, N. (2006), 'Waxy and high-amylose wheat starches and flours–characteristics, functionality and application', *Trend Food Sci. Technol.*, 17, 448-456.

Velu, G., Singh, R. P., Crespo-Herrera, L., Juliana, P., Dreisigacker, S., Valluru, R., Stangoulis, J., Sohu, V. S., Mavi, G. S., Mishra, V. K., Balasubramaniam, A., Chatrath, R., Gupta, V., Singh, G. P. and Joshi, A. K. (2018), 'Genetic dissection of grain zinc concentration in spring wheat for mainstreaming biofortification in CIMMYT wheat breeding', *Sci. Rep.*, 8, 13526.

Yilmaz, V. A., Brandolini, A. and Hidalgo, A. (2015), 'Phenolic acids and antioxidant activity of wild, feral and domesticated diploid wheats', *J. Cereal Sci.*, 64, 168-175.

Zanini, B., Basche, R., Ferraresi, A., Ricci, C., Lanzarotto, F., Marullo, M., Villanacci, V., Hidalgo, A. and Lanzini, A. (2015a), 'Randomised clinical study: gluten challenge induces symptom recurrence in only a minority of patients who meet clinical criteria for non-coeliac gluten sensitivity', *Aliment. Pharmacol. Ther.*, 42, 968-976.

Zanini, B., Villanacci, V., De Leo, L. and Lanzini, A. (2015b) '*Triticum monococcum* in patients with celiac disease: a phase II open study on safety of prolonged daily administration', *Eur. J. Nutr.*, 54, 1027-1029.

Zevallos, V. F., Raker, V., Tenzer, S., Jimenez-Calvente, C., Ashfaq-Khan, M., Rüssel, N., Pickert, G., Schild, H., Steinbrink, K. and Schuppan, D. (2017), 'Nutritional wheat amylase-trypsin inhibitors promote intestinal inflammation via activation of myeloid cells', *Gastroenterol.*, 152, 1100-1113.

Ziegler, J. U., Flockerzie, M., Longin, C. F. H., Würschum, T., Carle, R. and Schweiggert, R. M. (2016a), 'Development of lipophilic antioxidants and chloroplasts during the sprouting of diverse *Triticum* spp.', *J. Agric. Food Chem.*, 64, 913-922.

Ziegler, J. U., Steiner, D., Longin, C. F. H., Würschum, R. M., Schweiggert, R. M. and Carle, R. (2016b), 'Wheat and the irritable bowel syndrome–FODMAP levels of modern and ancient species and their retention during bread making', *J. Funct. Foods*, 25, 257-266.

Ziegler, J. U., Wahl, S., Würschum, T., Longin, C. F. H., Carle, R. and Schweiggert, R. M. (2015), 'Lutein and lutein esters in whole grain flours made from 75 genotypes of 5 *Triticum* species grown at multiple sites', *J. Agric. Food Chem.*, 63, 5061-5071.

Zrcková, M., Capouchová, I., Paznocht, L., Eliášová, M., Dvořák, P., Konvalina, P., Janovská, D., Orsák, M. and Bečková, L. (2019), 'Variation of the total content of polyphenols and phenolic acids in einkorn, emmer, spelt and common wheat grain as a function of genotype, wheat species and crop year', *Plant Soil Environ.*, 65, 260-266.

Chapter 10

Understanding the nutritional and nutraceutical properties of sorghum[1]

Sarah Cox, USDA-ARS, USA; Weiqun Wang, Kansas State University, USA; Seong-Ho Lee, University of Maryland, USA; and Dmitriy Smolensky, USDA-ARS, USA

1 Introduction

Sorghum bicolor is a drought-resistant crop grown as a dietary staple around the world, particularly in Africa and Asia [1–3]. However, it remains relatively little used as a food source in many developed countries. While sorghum is used primarily for cattle feed and biofuel production in the United States, increased interest is now being shown in its use as a food due to its potential health benefits. Sorghum has a wide variety of bioactive compounds. These include polyphenols, proteins, lipids and fiber, all of which have been demonstrated to show potential health benefits such as anti-cancer and anti-inflammatory effects [4]. Sorghum has also shown the potential to affect the microflora of the

1 **Disclaimer:** Names are necessary to report factually on available data; however, the U.S. Department of Agriculture neither guarantees nor warrants the standard of the product, and use of the name by the U.S. Department of Agriculture implies no approval of the product to the exclusion of others that may also be suitable. USDA is an equal opportunity provider and employer.

http://dx.doi.org/10.19103/AS.2021.0087.21
Published by Burleigh Dodds Science Publishing Limited, 2021.

gastrointestinal tract by selectively increasing or decreasing different types of bacteria [5].

Numerous studies have shown the potential anti-cancer effects of sorghum, particularly in preventing colorectal cancer [5–7]. Research has also explored the anti-inflammatory effects of sorghum polyphenols. Sorghum polyphenols have also shown benefits in treating obesity and diabetes [8]. While there are some anti-nutritional effects associated with sorghum polyphenols, research is being undertaken to minimize those impacts. This chapter discusses these topics in more detail.

2 Variation in the nutritional value of sorghum grain

Sorghum has a wide range of macronutrients, micronutrients and bioactive compounds that show promising health benefits. Sorghum has, for example, high levels of tannins and other polyphenols that have strong antioxidant properties [9]. These bioactive compounds and health benefits are listed in Table 1. The composition of these compounds can vary by the variety of sorghum and by the environment in which the sorghum is grown. In one study 10 wild genotypes of sorghum were gathered to analyze their micronutrient composition in comparison to domestic varieties. Several genotypes were found to have high protein and iron content and one wild genotype was found to be particularly high in zinc [45]. The total phenolic content, protein digestibility and mineral content of 16 varieties of raw sweet sorghum flour from Sudan have been compared. Almost no significant differences were observed between the samples regarding protein digestibility. Total phenolic content varied among the samples with a maximum of 12.98 mg/g and a minimum of 0.89 mg/g of total phenolic acids. Tannins were between 12.97 mg/g and 1.58 mg/g among the raw sorghum flour samples [46]. Thirty-five Ethiopian sorghum varieties, including white-, brown- and red-pigmented grain, were also examined for mineral and chemical content, nutritional and anti-nutritional effects. Significant variation was found among the samples in minerals including iron, which ranged from 2.262 to 14.08 mg/100 g, phosphorous ranged from 367.965 ppm to 112.554 mg/100 g and magnesium ranged from 207.53 to 62.06 mg/100 g. Tannin levels ranged from 1.36 to 3337.2 mg/100 g within the samples [47].

A wide diversity of nutrients was found in 390 accessions of sorghum with a range of 1.0–4.3% fat, 8.1–18.8% protein and 61.7–71.1% starch. Durra and bi-color durra sorghum from India and Ethiopia had the greatest protein and fat content while kafir-type sorghum from the United States, India and South Africa had the highest starch but the lowest protein content [48]. In a survey of sorghum varieties from around the world, 381 accessions were evaluated for their polyphenolic content. Around 55% of evaluated samples contained proanthocyanidins and 13% contained 3-deoxyanthocyanins, with

Published by Burleigh Dodds Science Publishing Limited, 2021.

6% of samples containing both [49]. When 50 different sorghum varieties from Burkina Faso were evaluated for their phenolic content, most varieties had less than 0.25% w/w content of tannins, with red varieties having the highest total phenolic content [50].

3 Polyphenols in sorghum

As the previous section suggests, sorghum grains have a wide range of polyphenols, which can differ by both composition and quantity in each sorghum variety. Polyphenolic compounds have strong anti-proliferative and apoptotic effects on cancer cells and reduce pro-inflammatory cytokines. Sorghum polyphenols also reduce fat storage, increase glucose tolerance, and modify levels of bacteria in the gut. Sorghum polyphenols can be used to reduce inflammation and to help treat HIV-positive individuals.

Polyphenols present in sorghum include the following groups:

- phenolic compounds;
- tannins; and
- flavonoids.

Sorghum possesses both bound and free forms of phenolic compounds. Bound phenols are esterified to cell walls and represent a significant portion of the phenols in sorghum [51]. Ferulic acid is the most common of the bound phenolic acids present in sorghum and is a derivative of cinnamic acid, whilst other commonly bound polyphenols in sorghum are caffeic, sinapic and p-coumaric acids [52].

Tannins are a second type of polyphenolic compounds in sorghum. Tannins fall into two distinct categories: hydrolysable or condensed tannins. In sorghum the majority of tannins present are condensed tannins [53]. Also known as proanthocyanidins, tannins are associated with some anti-nutritional effects such as inhibition of protein digestibility [54]. A third type of polyphenolic compound in sorghum is flavonoids. Subgroups of flavonoids include:

- anthocyanidins;
- flavones;
- flavanones; and
- flavanonols.

3-deoxyanthocyanidins are the most common type of anthocyanins in sorghum [55]. 3-deoxyanthocyanins are associated with darker pigmented sorghum grains, with black sorghum showing the highest levels [56]. 3-deoxyanthocyanin levels are also significantly higher in sorghum grown under direct sunlight.

Table 1 Sorghum studies

Use	Model	Sorghum used	Effect	Reference
Anti-cancer	SW480 cancer cells	Shawaya short black 1, IS11316, QL33,QL33/QL36, B923296, QT12	Anti-proliferative/Apoptosis	[10]
	HepG2, Caco2 cancer cells	PI574081 extract	Anti-proliferative/apoptosis/cell cycle arrest	[11]
	MCF-7 Cancer Cells	Red sorghum from Tamil Nadu, India	Anti-proliferative	[12]
	MDA-MB 231, MCF-7 Cancer cells	Hwanggeumchal	Tumor suppression, migration inhibition, anti-metastatic	[13]
	Mouse model	PI570481, SC84 and sumac bran extract	Tumor size and load reduction	[14]
	PC3M prostate cancer cells	Donganme extract	Anti-proliferative/Apoptotic /Reduced metastases	[15]
	HepG2, HCT-15 cancer cells	TX430, commercial sumac sorghum, PI570481	Anti-proliferative	[16]
	Colon cancer cells/ Wistar rats	Red and white sorghum	Anti-proliferative	[17]
	Porcine pancreatic alpha-amylase	Sumac sorghum, black sorghum	Alpha amylase and aromatase inhibition	[18]
	MCF-7 cancer cells, YAMC	ATX631 × RTX436, TX2911, TX430	Estrogenic effects/apoptosis	[19]
	YAMC		Estrogenic effects	[20]
	HEPG2 cells	Ultrasound assisted extracted red sorghum	Anti-proliferative	[21]
Anti-inflammation	THP-1 human macrophages	Kafirin extract from sorghum	Inflammation inhibition /reduced cytokine production/ reduction of ROS	[22]
	Clinical trial	Extruded sorghum	Reduced inflammation/oxidative stress	[23]
	Clinical trial	Red sorghum flour	Higher antioxidant activity	[24]
	Mouse and cancer cells	Black sorghum bran extract	Reduced inflammation/reduced production of TNF-a, IL-1B	[25]

	HUVEC cells	Shawaya short black 1	Reduced oxidative stress/reduced inflammation	[26]
	Wistar rat model	BRS 305, BRS 309, BRS 310	Reduced inflammation/oxidative stress	[27]
Anti-obesity	Wistar rat model	Extruded sorghum flour	Decreased lipogenesis	[28]
	Wistar rat model	Extruded or raw sorghum bran UDG110	Reduced fat storage/reduced glucose and insulin levels	[29]
	Rat model	Sorghum flour factions	Decreased liver fat, increased glucose tolerance and insulin resistance	[30]
	Wistar rat model	Extruded sorghum flour	Decreased fat/increased glucose tolerance	[31]
	Clinical trial	Extruded sorghum	Reduced body fat	[32]
Microbiome effects	Human fecal samples	Black or sumac sorghum extract	Selectively increased/modified bacterial levels	[33]
	Human fecal samples	Sorghum	Increased in select bacteria growth, increased short chain fatty acids	[34]
	Rat model	Extruded sorghum flour	Modified bacterial levels	[35]
	Rat model	Sumac and high tannin black sorghum	Increased bacterial richness, modified bacterial levels	[36]
Other uses	Human polymorphonuclear and mononuclear cells	Sorghum leaves	Increased natural killer cell activation, decreased ROS formation and cell migration	[37]
	Rat model	Sorghum leaves	Anti-inflammatory effect	[38]
	Clinical trial	Sorghum leaves	Improved CD4 counts, increased hemoglobin	[39]
	Human foreskin fibroblasts, human aortic endothelial cells	Sorghum leaves	Inhibited *T. gondii* growth	[40]
		Sorghum leaves	High polyphenolic content	[41]
	Agar medium with multiple microbe varieties	Algerian sorghum samples: AS12, AS20, I27, H40	Inhibited microbial growth	[42]
	Mouse macrophages	PI570481 bran extract	Reduced legionella replication, increased TNF and IL6 production	[43]
	Human blood samples	Shawaya short black 1 sorghum extract	Increased platelet aggregation and microparticles release	[44]

Luteolin and apigenin are flavones which are commonly found in sorghum [57]. A common flavanone present in sorghum is naringenin [56]. Taxifolin is also a common flavanonol found in sorghum [58].

Black and brown sorghum varieties were found to have the highest total phenolic and proanthocyanidin content. Sorghum varieties with high antioxidant activity are characterized by a wide variety of polyphenols rather than a specific phenol [59]. Red sorghum was found to have a significantly higher content of polyphenols than white sorghum, though a wide range of polyphenol content was reported for both types [60]. White sorghum was found to have three times higher levels of ferulic acid than red sorghum, with p-coumaric and protocatechuic acids two times higher; however, red sorghum had twice as much caffeic and gallic acids present. Flavonoid content was higher in white sorghum, but red sorghum had higher levels of naringenin, vitexin and apigenin [61].

Comparing different sorghum grains for their polyphenol and antioxidant content, a strong correlation was found between antioxidant activity and total polyphenol content. White sorghum has the lowest levels of polyphenols while red and brown brewing sorghum grains have the highest levels. Most free phenols are protocatechuic acid and taxifolin, with ferulic acid the most common bound phenol [62]. Raw pigmented sorghum flour is higher in kafirin and tannin than white sorghum and cooked pigmented sorghum has greater resistant starch than white sorghum [63]. In a sample set of 50 varieties of sorghum grain, most had less than 0.25% tannins with red sorghum having the highest phenolic content of the varieties studied [50]. When assessing the utilization of white sorghum in the making of tortilla chips by mixing cornflour with increasing amounts of sorghum flour, total phenolic content increases with the increased addition of sorghum flour. However antioxidant levels had no significant change between the mixtures, potentially due to production conditions [64]. The impact of germination on the total polyphenolic content of 50 sorghum varieties was assessed and showed that un-germinated sorghum varieties had the highest phenolic content while levels of proanthocyanidins, 3-deoxyanthocyanidins and flavan-4-ols decreased by germination [65].

While sorghum is a particularly rich source, other grains have a variety of polyphenols as well. A summary of sources can be found in Table 2. When comparing old and modern Tunisian durum wheat varieties, modern wheat was found to have higher total phenolic content, with a range of 177.48–272.62 mg gallic acid/100 g dry weight. Ferulic acid was the most common phenolic acid found in the wheat samples [66]. Ferulic acid is higher in wheat seed coats than the endosperm, with greater quantities being found in wheat bran than flour [75]. Twenty varieties of colored barley grown in southern Italy were evaluated for their phenolic content. Total phenolic content of the samples ranged from 1929 µg $^{g-1}$ to 2917 µg $^{g-1}$ with insoluble bound phenolic acids making up 88.3%

of total phenolic content. There were significant differences in antioxidant activity, among the samples, with a maximum of 13.4 μmol g^{-1} [80]. Evaluation of the polyphenolic content in six varieties of Thai rice, including white, red and purple rice, found that a red rice variety had the highest total phenol content and a purple rice variety had the highest anthocyanin content. Anthocyanins were too low to be measured in white rice, which also had the lowest total phenol content [68]. Commercial oats were found to have a total phenol content of up to 25.05 mg per 40 g dry weight. The most common polyphenol was ferulic acid [69].

The polyphenols of grains other than sorghum have also been found to have health benefits. When extracted and enriched, avenanthramides, a unique polyphenol found in oats prevented the release of pro-inflammatory cytokine interleukin-8 (IL-8) which affects inflammatory response [77]. Phenolic extracts from eight varieties of Australian pigmented rice were measured for their total phenol content and antioxidant ability prior to in vitro treatment of murine C3H10T1/2 multipotent cells to evaluate their effects in modulating adipogenicity (a factor in obesity). Peroxisome proliferator-activated receptor gamma (PPARγ), which regulates adipocyte differentiation in fat deposition, was also reduced by red, purple and brown rice [76]. Mice fed a high dose black barley diet had reduced total cholesterol and low-density lipoprotein cholesterol as well as increased high-density lipoprotein cholesterol in comparison to high-fat diet mice. Mice that were fed the black barley extract also had increased antioxidant gene expression [79].

4 Nutraceutical properties of sorghum: anti-cancer and anti-inflammatory effects

There has been a significant amount of research into the effects of sorghum polyphenolic compounds on cancer. Black and brown sorghum as well as pigmented rice were found to have significant anti-proliferative effects on SW480 colon cancer cells; black sorghum increased p53 activity as well as increasing effector caspases 3 and 7, indicating apoptotic effects [10]. In an in vitro study (using tissue culture) conducted on two cancer cell lines, HepG2 and CaCo2, polyphenols were extracted from sorghum from a panel of 15 genotypes. Extracts with a high polyphenol content induced apoptosis and cell cycle arrest, thus demonstrating anti-proliferative effects on cancer cells [11].

When breast cancer cell line MCF-7 was treated with anthocyanins extracted from red sorghum bran using a methanol-based solvent, the anthocyanin-rich extract induced morphological changes to the MCF-7 cells which inhibited cancer cell viability [12]. Hwanggeumchal sorghum has been found to slow tumor growth caused by breast cancer MCF-7 and MDA-MB 231 cells in mice. Additionally, cells treated in vitro with sorghum extract experienced a downregulation of STAT5b

Table 2 Benefits of grains other than sorghum

Grain	Polyphenols discussed	Effects	Reference
Tunisian durum wheat	Ferulic acid	Wide variety of polyphenols	[66]
Barley	p-hydroxybenzoic acid, gallic acid, vanillic acid	Significant differences of phytonutrients between outer layers and endosperm	[67]
Rice	Anthocyanins	Wide variety of polyphenols between samples	[68]
Oats	Ferulic acid	Unique phenol – avenanthramide; wide range of phenol quantities	[69]
Quinoa flour	Polyphenols	Significantly higher polyphenolic and antioxidant levels than wheat	[70]
Purple corn	Polyphenols	Natural source of pesticide	[71]
Purple corn	Polyphenols	Natural source of pesticide	[72]
wheat, sorghum, chickpeas and green gram	Total polyphenols, tannins	Sorghum polyphenols had lowest bioavailability	[73]
Wheat flour	Hydroxybenzoic acid glucoside, tryptophan, 6-C-glucosyl-8-C-arabinosyl-apigenin and diferulic acids	Cooking increased bound polyphenols	[74]
Wheat	Ferulic acid	Ferulic acid content is high in wheat seed coats	[75]
Pigmented Rice	Total phenolic, anthocyanidin, proanthocyanidins	Obesity prevention, lipid accumulation reduction	[76]
Oats	Avenanthramides, polyphenols	Anti-inflammatory	[77]
Oats	Avenanthramides	Inhibit NF-kB activation, decrease endothelial pro-inflammatory cytokine production	[78]
Black highland barley	Ferulic acid, p-coumaric acid	Strong antioxidant ability	[79]
Colored barley	Total phenolic acids	Significant variations in polyphenol levels among samples	[80]

and STAT3 signaling pathways, and induced cell cycle arrest. Compared to the results of similar treatments using wheat, millet and panicum, sorghum was shown to be the most effective in suppressing oncogenic proteins [13]. The anti-cancer effects of a sorghum ethyl-acetate extract were evaluated using PC3M prostate cancer cells, showing the extract as having an anti-proliferative effect dependent on time and dose of extract, as well as apoptotic effects. After seeding prostate cancer cells into mouse prostates and treating them for 5 weeks with the extract, tumors were significantly smaller in a dose-dependent manner. Metastases were also significantly reduced depending on dose [15].

Multiple variations of food-grade extraction solvents and temperatures have been evaluated to improve phenolic extraction from sorghum bran using a novel high polyphenol strain as well as other commercially available varieties. A 70% ethanol and 5% citric acid solvent was shown to be the most effective extraction solvent and had the highest anti-proliferative effect on colon cancer cells. Higher temperatures during the extraction process decreased the anti-proliferative properties of the extracts. When compared to green tea extract, the high polyphenol variety of sorghum was more effective at preventing cancer cell proliferation, despite lower total polyphenols [16]. While extruded red sorghum was found to have anti-proliferative effects on colon cancer, white sorghum had hyper-proliferative effects on colonic epithelial cells. Both red and white extruded sorghum also reduced body weight in the rats through increased satiety [17].

Sumac sorghum has high levels of proanthocyanins and moderate levels of flavonoids, while black sorghum lacks significant proanthocyanins but possesses high levels of 3-deoxyanthocyanins [18]. The proanthocyanins in sumac sorghum inhibit both amylase and aromatase in non-malignant mouse colonocytes. Amylase breaks down starch and its inhibition has the potential to lower blood sugar, while aromatase is an important enzyme in breast cancer because it converts androgens to estrogens, affecting the growth of cancerous cells. The same study found that black and white sorghums had higher quantities of flavonoids compared to red sorghum [19]. When the effect of sorghum polyphenols was evaluated against cells expressing alpha and beta estrogen receptors, red sorghum had no significant anti-cancer properties, while white and black sorghums were able to significantly reduce cancer cell count. Strong synergistic effects were found among the flavonoids apigenin and naringenin in activating estrogen receptors [20].

There have been many studies to optimize the process by which sorghum polyphenols are extracted. The use of ultrasound-assisted extraction was (UAE) was found to be more efficient than conventional solvent extraction in extracting polyphenols. Extracts that were obtained using UAE had higher levels of anti-proliferative activity against HEPG2 cells compared to those extracted using conventional solvent extraction [21].

The impact of phenolic compounds on inflammation is another area of interest. In an in vitro study using derived human macrophages, co-treated with LPS, kafirin protein extract from sorghum was shown to reduce multiple pro-inflammatory cytokines and reactive oxygen species [22]. In a 7-week-long randomized simple blind clinical study, patients with chronic kidney disease undergoing regular dialysis treatment were fed either a probiotic milk and extruded sorghum mixture, or a pasteurized milk and extruded corn mixture once every day. Patients who consumed the probiotic milk with extruded sorghum mixture showed a reduction in inflammation and oxidative stress as well as lower levels of malondialdehyde, which is a biomarker of oxidative stress, and increased levels of superoxide dismutase, an important antioxidant, potentially due to the high levels of zinc present in the sorghum as well as the phenolic compounds present [23]. In a clinical study, pasta made with red sorghum whole grain flour, a control non-sorghum pasta and white whole grain flour sorghum pasta were fed to human participants. Participants who ate the pasta containing red sorghum flour had higher antioxidant activity two hours after consumption than white sorghum and non-sorghum groups [24].

Both black and sumac sorghum bran extracts were found to reduce inflammation in mice [25]. When oxidative stress was induced in human epithelial cells via hydrogen peroxide treatment, a 2-hour exposure to higher concentrations of black sorghum extract exerted antioxidant and anti-inflammatory effects in epithelial cells [26]. To evaluate the impact of the addition of sorghum flour in high-fat diets, rats were fed three different types of sorghum flour as part of a high-fat diet for 35 days. The BRS 310 sorghum flour sample, a red tannin-free sorghum, caused a reduction in TNF – an expression in adipose tissue as well as in thiobarbituric acid reactive substances (TBARS) [27].

Another potential use for sorghum is utilizing the leaves and stalk of the plant, instead of using only the grain for health benefits. This would not only reduce environmental waste but also increase income for producers. Jobelyn™ is a brand of supplement made from sorghum leaf powder. This supplement has been evaluated for potential anti-inflammatory and immune system effects in human polymorphonuclear cells and mononuclear cells. The supplement was found to decrease the formation of reactive oxygen species and decrease cell migration due to inflammatory chemo-attractants as well as increasing natural killer cell activation in vitro in CD3-CD56+ NK cells [37].

The anti-inflammatory effects of Jobelyn have also been evaluated in vivo. Inflammation was induced in rats via intraplantar injection of carrageenan, a known inflammation activator, which causes swelling. When carrageenan treated rats were fed 50–200 mg/kg Jobelyn, there was a reduction of leukocytes and malondialdehyde at the site of inflammation, demonstrating a distinct anti-inflammatory effect [38]. A clinical human study conducted with Jobelyn, which evaluated its effects on CD4+ cells counts in HIV positive patients, demonstrated that, when combined with the use of antiretrovirals, Jobelyn could significantly and more rapidly improve CD4 cell counts and increase hemoglobin when compared to treatment with antiretrovirals alone [39]. In separate studies, polyphenols extracted from red sorghum leaves were also found to inhibit the parasite *Toxoplasma gondii* [40]. Studies of extraction methods showed that acetone, ethanol and methanol were found to extract similar amounts of polyphenols and flavonoids, while water extracted significantly less polyphenols than the 70% aqueous solvents [41].

5 Nutraceutical properties of sorghum: controlling obesity, antimicrobial effects and impact on gastrointestinal function

Another interesting health benefit which is currently being explored is the effect of sorghum in controlling and reducing obesity. When extruded sorghum was added to a high-fat diet and fed to male Wistar rats, markers of fatty liver disease and lipogenesis in the liver were decreased. Quantification of liver gene expression via RT-QCR showed that mRNA expression of PPAR-α, a protein which regulates lipid metabolism via fatty acid oxidation, increased in the liver of the sorghum-fed rats. Another mechanism of decreased fatty liver in the rats was reduced expression of sterol regulatory element binding protein 1c, which regulates lipid synthesis and development of fatty tissue [28]. When evaluating raw or extruded sorghum bran added to a high-fat diet in male Wistar rats, both raw and extruded sorghum bran addition to the diet reduced fat storage, insulin levels and interleukins 1B, 4 and 6 when compared to a high-fat diet without sorghum bran [29]. Obese rats fed fractions of brown sorghum whole grain flour, bran or decorticated flour, in addition to a high-fat diet for 12 weeks, had an increase in glucose tolerance and insulin resistance, as well as a decrease in liver fat [30]. The sorghum-fed rats had increased glucose tolerance and decreased adipose tissue and adiposity, accompanied by reduced expression of fatty acid synthase (a lipogenic enzyme) and increased expression of lipoprotein lipase (a triglyceride breaking enzyme) in adipose tissues. Feeding sorghum in high-fat diets also resulted in a decrease of pro-inflammatory TNF-α and increase of anti-inflammatory interleukin-10 [31].

There have also been a number of human clinical studies on the impact of sorghum phenolic compounds on obesity. In a single-blind, randomized crossover trial, overweight men consumed either extruded sorghum or extruded wheat for two 8-week periods. Subjects who consumed sorghum had a greater reduction in waist circumference, BMI and bodyweight when compared to subjects who consumed wheat. It is important to note that the sorghum diet contained higher levels of dietary fiber due to sorghum having higher fiber levels than wheat [32].

Antimicrobial effects are also associated with the grain portion of sorghum. When evaluating the antimicrobial properties of polyphenolic extracts of Algerian sorghum samples, total polyphenols ranged from 301.89 to 3214.46 mg GAE/100 g of sorghum flour. They were found to be moderately to highly effective at inhibiting microbial growth, particularly in gram positive bacteria [42]. The increase in pro-inflammatory cytokines seems to have inhibited the growth of *Legionella pneumophila* [43].

In vitro, sorghum was shown to have an anti-aggregation effect in human blood samples. The effect of black sorghum extract on the activation and aggregation of platelets as well as the release of platelet microparticles was investigated. Human blood samples were treated with various concentration of the extract and found that the extract reduced aggregation of platelets as well as the release of platelet microparticles but did not have a significant effect on platelet activation [44].

Sorghum polyphenols can also have an impact on the microbiome of the intestinal tract. The impact of sorghum polyphenols on the microbiome of normal weight and obese individuals was investigated using fecal samples. Samples treated with the sorghum extracts mixed with fructooligosaccharides (FOS) significantly increased the levels of *Bifidobacterium*, *Lactobacillus*, *Prevotella* and *Roseburia* – bacteria genera that are commonly found in the intestines [33]. In another study, sorghum and other cereals were used to treat infant fecal samples. The addition of cereals such as sorghum increased the growth of beneficial bacteria digesting plant polysaccharides as well as the production of short-chain fatty acids [34]. Rats fed extruded sorghum showed increased levels of *Bacterioides* bacteria and decreased *Firmicutes* bacteria, which have been correlated with intestinal dysbiosis [35]. Bacterial richness also increased significantly in rats fed sumac and high tannin black sorghum with a potential impact in reducing colitis [36].

6 Anti-nutritional effects of sorghum polyphenols

While polyphenols found in sorghum have many benefits, they can also have anti-nutritional effects. Among these negative effects is interference in the absorption of minerals. In rabbits fed low or high tannin sorghum diets, rabbits consuming high tannin sorghum had decreased body weight and inhibited lipase, trypsin and a-amylase activity along with a decrease in calcium absorption [81]. Comparing tannin, phytate and mineral levels in whole sorghum grain and in processed breakfast cereal, breakfast cereals with lower tannin and phytate content had higher availability of calcium, iron and zinc, suggesting that tannins reduce mineral availability [82]. Rats fed red and white sorghum had lower calcium absorbance than control diet rats [83]. Comparing the use of corn, wheat and sorghum-based diets for broiler chickens, it was found that sorghum-fed chickens had the lowest growth and energy utilization due to higher levels of phenolic compounds and kafirin. Chickens fed red sorghum had higher metabolizable energy (the energy left for functional use after excretion loss) and higher starch digestibility than white sorghum [84]. It has been found that when sorghum was fermented as gruel and had the enzymes wheat phytase and mushroom polyphenol oxidase added, polyphenolic content decreased while iron accessibility increased [85]. A diet mixing sorghum and rice found that sorghum reduced protein digestibility and energy metabolization compared to exclusively rice diets [86]. Chicks fed a high tannin sorghum diet had significantly reduced calcium, phosphorous, iron, potassium, magnesium and cobalt, as well as reduced weight gain [87].

Polyphenols can also have a negative impact on the digestibility of starch and protein. It has been found that sorghum with high tannins increased resistant starch levels and decreased starch digestibility [88]. When porridges were mixed with sorghum extracts it was found that high phenolic sorghum extracts increased resistant starch [89]. When replacing up to 50% of the starch weight with purified sorghum tannins, resistant starch levels increased, probably due to tannins binding amylose, while slowly digestible starch (SDS) levels decreased [90]. White and pigmented sorghum were cooked to measure the impact of polyphenols in pigmented sorghum flour on the digestibility of starch. Cooked white sorghum flour was found to have less resistant starch than cooked pigmented sorghum flour, which was higher in total phenols. The cooked pigmented sorghum flour also had lower rapidly digestible starch than the cooked white sorghum flour [63].

When assessing the effect of nixtamalization on the bioaccessibility of protein in red and white sorghum flour, during in vitro digestion, sorghum that underwent nixtamalization and cooking were found to have higher protein bioaccessibility due to a breakdown in tannin-protein complexes [91]. Pigs fed a sorghum diet had lower nitrogen digestion and excreted higher levels of nitrogen in their feces, indicating poor protein absorption [92]. Mixed sorghum and quinoa have been assessed for protein digestibility and antioxidant levels. Sorghum and quinoa flours showed no significant change in protein, lipid, fiber or antioxidant composition after heat treatment, and the sorghum quinoa flour mixture had the benefits of increased protein quality from the quinoa flour, and increased antioxidant capacity due to the sorghum flour [93].

7 Food processing effects on sorghum polyphenols

Preparation methods can affect the availability of micronutrients and bioactive compounds in sorghum. Three sorghum genotypes were cooked via either dry heat in a conventional oven or through extrusion. Analysis of phenolic content after the cooking process revealed that extrusion resulted in a less phenolic content compared to dry heat cooking [94]. However, extruded sorghum still had a higher phenolic content than unprocessed sorghum bran as well as greater antioxidant activity [95]. Another study found that extrusion processing increased total phenol content of sorghum bran by up to 14.1% under optimal conditions compared to non-extruded sorghum bran [96]. Catechin levels were found to be higher in pigs fed a sorghum-extruded mix whilst procyanindins were reduced, due to depolymerization caused by the extrusion process [97]. In contrast total polyphenolic content levels were found to decrease significantly via cooking, particularly when microwaved. However, total flavonoid content was only significantly effected in terms of bioavailability by microwave cooking [73]. Cooking and nixtamalization, a process where grain is soaked and cooked in an alkaline solution prior to being hulled, together were shown to increase the bioavailability of total phenols in sorghum, while nixtamalization decreased condensed tannin levels [98]. Total phenolic content was reduced via metabolization by bacteria during fermentation [99].

Sorghum tea is another way sorghum is consumed. The sorghum tea-making process was found to increase total polyphenol, flavonoid and condensed tannin content without reducing antioxidant activity [100]. Soaking, steaming and roasting steps each had different effects on the total polyphenolic content of sorghum tea. Steaming increased ferulic and p-coumaric acid levels, while roasting significantly increased total phenolic content, total flavonoid content and proanthocyanidin content compared to raw sorghum grain [101]. In fermented sorghum grain tea, total polyphenols were significantly reduced by soaking, steaming and fermenting. However, roasting the grain caused an increase in total phenolic content [102].

Fermentation is also used to produce sorghum foods. It has been found that fermentation and cooking sorghum increase total polyphenol content and antioxidant activity compared to unfermented sorghum grain [103]. When comparing fermented and unfermented red sorghum, total phenolic content was increased by fermentation [99]. Breakfast cereal made from fermented flour had increased protein content and minerals such as magnesium, calcium and iron, but decreased tannin and phenolic levels [104]. Malting was found to decrease phenolic content. While flavonoid content decreased during malting, anthocyanin content significantly increased [105]. Inhibitory abilities of the polyphenols against enzymes also decreased as roasting temperature increased [106]. When sorghum is mixed as part of a powdered chocolate beverage, it is possible to increase the bioactive compounds of the beverage without damaging the sensory aspects of the product [107].

8 Sorghum as a healthy gluten-free food

Sorghum is being considered as an alternative source of grain for those who are unable to consume gluten. Sorghum has been evaluated as a gluten-free alternative grain safe for consumption by celiac disease patients. Results indicated that gliadin-like peptides were not present in any of the sorghum samples, suggesting sorghum is safe for consumption for individuals with celiac disease [108]. When the cornstarch to sorghum flour ratio was increased while evaluating the use of white sorghum flour in making bread, the specific volume of the bread as well as the crumb grain score increased. Optimal ingredient ratios were found to be 0.55% cornstarch to sorghum flour, with 90% water and 3% hydroxypropyl methylcellulose added, which are used to increase viscosity of the dough [109]. An evaluation of the most beneficial bacterial cultures for use in producing sourdough bread made with sorghum flour revealed that of 103 lactic acid bacteria (LAB) strains and 20 yeast strains isolated from Nigerian sorghum flour, 3 specific LAB, *Pediococcus pentosaceus* SA8, *Weisella confusa* SD8 and *P Pediococcus pentosaceus* LD7, and 1 yeast, *Saccharomyces cerevisiae* YC1, met the requirements necessary to serve as starter cultures. Qualities necessary for yeast selection were high CO_2 production, a high tolerance for low pH and acetic acid, while LAB qualities included proteolytic activity, acidification and high exopolysaccharide production [110].

Pasta made with sorghum flour had high protein and dietary fiber content but had slower starch digestion and lower estimated glycemic index when compared to other gluten-free pasta samples. The bioavailability of the polyphenols in the sorghum flour pasta was significantly higher than other types of pasta [111]. When sorghum flour was added to durum wheat semolina pasta, resistant starch and total polyphenolic content increased. However, the cooking process reduced the total phenolic content and antioxidant activity overall, by reducing free phenolic acids while increasing bound phenolic acids [112]. The development of gluten-free cookies and savory pies using high tannin sorghum had significantly higher resistant and slowly digestible starches [113].

There have also been efforts to improve the health benefits of gluten-free food. The addition of phytase to sorghum flour increased the bioavailability of minerals such as iron, calcium and zinc by reducing *myo*-inositol phosphates, particularly the inositol hexaphosphate (IP6) fraction. The IP6 fraction and IP5 fraction of *myo*-inositol phosphates are known to inhibit mineral absorption [114]. Sorghum was prepared using ultrasonication to explore potential impacts on kafirin. The ultrasonication process modified the secondary structure of kafirin, which allowed hydrolysis by pepsin-pancreatin hydrolysis and increased antioxidant activity and digestibility [115].

9 Conclusion and future trends

Efforts are being made to breed varieties with potential health benefits. In evaluating the genetic diversity of local sorghum grown in Ghana, 41 varieties were sorted into seven genetically similar clusters. They found that there were three distinct genotypes that stood out that could be used to improve the breeding programs in Ghana [116]. Ideal environmental characteristics are also being investigated to improve crop yield and growth. A study of sorghum grown in Australia found that sorghum hybrids with multiple tillers make sorghum more sensitive to the environment but may increase yield stability [117]. Several varieties also showed chill tolerance and early maturity, which indicated they could be used to breed sorghum in more temperate climates [118]. Six sorghum varieties were grown under controlled conditions in two growth chambers at with high temperatures of 32°C or 38°C. Phenolic content and antioxidant content were significantly influenced by genotype while temperature had no significant impact in most cases [119]. Sixty sorghum varieties grown in India suggested that genotype had a stronger influence on total phenolic content [60].

Sorghum is an underrated food with a diverse variety of micronutrients and bioactive compounds. With polyphenols showing promise in potential anti-cancer and anti-inflammatory properties, sorghum is becoming an increasingly attractive food source. While it does possess some limitations regarding cooking and preparation methods as well as anti-nutritional properties, methods are being explored to overcome these issues. Efforts are also being made to breed new sorghum types to increase polyphenol content, as well as to improve other aspects of the crop. However, while studies suggest that there may be many health benefits in consuming sorghum, more research is necessary to identify and support these findings.

10 References

1. Taylor, J. and Duodu, K. G. *Sorghum and Millets: Chemistry, Technology, and Nutritional Attributes*: Elsevier; 2018.
2. Rooney, W. *Achieving Sustainable Cultivation of Sorghum Volume 2: Sorghum Utilization around the World*: Burleigh Dodds Science Publishing Limited; 2018.
3. Rooney, W. *Achieving Sustainable Cultivation of Sorghum Volume 1: Genetics, Breeding and Production Techniques*: Burleigh Dodds Science Publishing Limited; 2018.
4. Girard, A. L. and Awika, J. M. Sorghum polyphenols and other bioactive components as functional and health promoting food ingredients. *Journal of Cereal Science* 2018 84:112–124.
5. de Morais Cardoso, L., Pinheiro, S. S., Martino, H. S. D. and Pinheiro-Sant'Ana, H. M. Sorghum (Sorghum bicolor L.): nutrients, bioactive compounds, and potential impact on human health. *Critical Reviews in Food Science and Nutrition* 2017 57(2):372–390.
6. Awika, J. M. Chapter 3. Sorghum: its unique nutritional and health-promoting attributes. In: *Gluten-Free Ancient Grains* Taylor, J. R. N. and Awika, J. M. (Eds): Woodhead Publishing; 2017: 21–54.
7. Vanamala, J. K. P., Massey, A. R., Pinnamaneni, S. R., Reddivari, L. and Reardon, K. F. Grain and sweet sorghum (Sorghum bicolor L. Moench) serves as a novel source

of bioactive compounds for human health. *Critical Reviews in Food Science and Nutrition* 2018 58(17):2867–2881.
8. Xiong, Y., Zhang, P., Warner, R. D. and Fang, Z. Sorghum grain: from genotype, nutrition, and phenolic profile to its health benefits and food applications. *Comprehensive Reviews in Food Science and Food Safety* 2019 18(6):2025–2046.
9. Civáň, P. When bitter is better. *Nature Plants* 2019 5(12):1205–1206.
10. Rao, S., Chinkwo, K., Santhakumar, A., Johnson, S. and Blanchard, C. Apoptosis induction pathway in human colorectal cancer cell line SW480 exposed to cereal phenolic extracts. *Molecules* 2019 24(13):2465.
11. Smolensky, D., Rhodes, D., McVey, D. S., Fawver, Z., Perumal, R., Herald, T. and Noronha, L. High-polyphenol sorghum bran extract inhibits cancer cell growth Through ROS Induction, cell cycle arrest, and apoptosis. *Journal of Medicinal Food* 2018 21(10):990–998.
12. Devi, P. S., Kumar, M. S. and Das, S. M. Evaluation of antiproliferative activity of red sorghum bran anthocyanin on a human breast cancer cell line (mcf-7). *International Journal of Breast Cancer* 2011 2011:891481.
13. Park, J. H., Darvin, P., Lim, E. J., Joung, Y. H., Hong, D. Y., Park, E. U., Park, S. H., Choi, S. K., Moon, E. S., Cho, B. W., Park, K. D., Lee, H. K., Kim, M. J., Park, D. S., Chung, I. M. and Yang, Y. M. Hwanggeumchal sorghum induces cell cycle arrest, and suppresses tumor growth and metastasis through Jak2/STAT pathways in breast cancer xenografts. *PLOS ONE* 2012 7(7):e40531.
14. Lee, S.-H., Lee, J., Amarakoon, D., Lou, Z., Herald, T., Cox, S., Noronha, L., Perumal, R. and Smolensky, D. Tumor suppressive activity of high phenolic sorghum brans in colon cancer mouse model and proposed mechanisms. *Current Developments in Nutrition* 2020 4(Supplement_2):329–329.
15. Ryu, J.-M., Jang, G. Y., Woo, K. S., Kim, T. M., Jeong, H. S. and Kim, D. J. Effects of sorghum ethyl-acetate extract on PC3M prostate cancer cell tumorigenicity. *Journal of Functional Foods* 2017 37:449–459.
16. Cox, S., Noronha, L., Herald, T., Bean, S., Lee, S. H., Perumal, R., Wang, W. and Smolensky, D. Evaluation of ethanol-based extraction conditions of sorghum bran bioactive compounds with downstream anti-proliferative properties in human cancer cells. *Heliyon* 2019 5(5):e01589.
17. Llopart, E. E., Cian, R. E., López-Oliva, M. M. E., Zuleta, Á., Weisstaub, A. and Drago, S. R. Colonic and systemic effects of extruded whole-grain sorghum consumption in growing Wistar rats. *British Journal of Nutrition* 2017 118(8):589–597.
18. Hargrove, J. L., Greenspan, P., Hartle, D. K. and Dowd, C. Inhibition of aromatase and α-amylase by flavonoids and proanthocyanidins from Sorghum bicolor bran extracts. *Journal of Medicinal Food* 2011 14(7–8):799–807.
19. Yang, L., Allred, K. F., Geera, B., Allred, C. D. and Awika, J. M. Sorghum phenolics demonstrate estrogenic action and induce apoptosis in nonmalignant colonocytes. *Nutrition and Cancer* 2012 64(3):419–427.
20. Yang, L., Allred, K. F., Dykes, L., Allred, C. D. and Awika, J. M. Enhanced action of apigenin and naringenin combination on estrogen receptor activation in non-malignant colonocytes: implications on sorghum-derived phytoestrogens. *Food and Function* 2015 6(3):749–755.
21. Luo, X., Cui, J., Zhang, H., Duan, Y., Zhang, D., Cai, M. and Chen, G. Ultrasound assisted extraction of polyphenolic compounds from red sorghum (Sorghum bicolor L.) bran

and their biological activities and polyphenolic compositions. *Industrial Crops and Products* 2018 112:296–304.
22. Sullivan, A. C., Pangloli, P. and Dia, V. P. Kafirin from Sorghum bicolor inhibition of inflammation in THP-1 human macrophages is associated with reduction of intracellular reactive oxygen species. *Food and Chemical Toxicology: An International Journal Published for the British Industrial Biological Research Association* 2018 111:503–510.
23. Lopes, R. D. C. S. O., de Lima, S. L. S., da Silva, B. P., Toledo, R. C. L., Moreira, M. E. D. C., Anunciação, P. C., Walter, E. H. M., Carvalho, C. W. P., Queiroz, V. A. V., Ribeiro, A. Q., et al. Evaluation of the health benefits of consumption of extruded tannin sorghum with unfermented probiotic milk in individuals with chronic kidney disease. *Food Research International* 2018 107:629–638.
24. Khan, I., Yousif, A. M., Johnson, S. K. and Gamlath, S. Acute effect of sorghum flour-containing pasta on plasma total polyphenols, antioxidant capacity and oxidative stress markers in healthy subjects: a randomised controlled trial. *Clinical Nutrition* 2015 34(3):415–421.
25. Burdette, A., Garner, P. L., Mayer, E. P., Hargrove, J. L., Hartle, D. K. and Greenspan, P. Anti-inflammatory activity of select sorghum (Sorghum bicolor) brans. *Journal of Medicinal Food* 2010 13(4):879–887.
26. Francis, N., Rao, S., Blanchard, C. and Santhakumar, A. Black sorghum phenolic extract regulates expression of genes associated with oxidative stress and inflammation in human endothelial cells. *Molecules* 2019 24(18):3321.
27. Moraes, É. A., Natal, D. I. G., Queiroz, V. A. V., Schaffert, R. E., Cecon, P. R., de Paula, S. O., Benjamim, L. D. A., Ribeiro, S. M. R. and Martino, H. S. D. Sorghum genotype may reduce low-grade inflammatory response and oxidative stress and maintains jejunum morphology of rats fed a hyperlipidic diet. *Food Research International* 2012 49(1):553–559.
28. de Sousa, A. R., de Castro Moreira, M. E., Toledo, R. C. L., dos Anjos Benjamin, L., Queiroz, V. A. V., Veloso, M. P., de Souza Reis, K. and Martino, H. S. D. Extruded sorghum (Sorghum bicolor L.) reduces metabolic risk of hepatic steatosis in obese rats consuming a high fat diet. *Food Research International* 2018 112:48–55.
29. Salazar-López, N. J., González-Aguilar, G. A., Rouzaud-Sández, O., Loarca-Piña, G., Gorinstein, S. and Robles-Sánchez, M. Sorghum bran supplementation ameliorates dyslipidemia, glucose dysregulation, inflammation and stress oxidative induced by a high-fat diet in rats. *CyTA–Journal of Food* 2020 18(1):20–30.
30. Moraes, É. A., Marineli, RdS., Lenquiste, S. A., Queiroz, V. A. V., Camargo, R. L., Borck, P. C., Carneiro, E. M. and Maróstica Júnior, M. R. Whole sorghum flour improves glucose tolerance, insulin resistance and preserved pancreatic islets function in obesity diet-induced rats. *Journal of Functional Foods* 2018 45:530–540.
31. Arbex, P. M., Moreira, M. E. D. C., Toledo, R. C. L., de Morais Cardoso, L., Pinheiro-Sant'ana, H. M., Benjamin, L. D. A., Licursi, L., Carvalho, C. W. P., Queiroz, V. A. V. and Martino, H. S. D. Extruded sorghum flour (Sorghum bicolor L.) modulate adiposity and inflammation in high fat diet-induced obese rats. *Journal of Functional Foods* 2018 42:346–355.
32. Anunciação, P. C., Cardoso, L. D. M., Alfenas, R. D. C. G., Queiroz, V. A. V., Carvalho, C. W. P., Martino, H. S. D. and Pinheiro-Sant'Ana, H. M. Extruded sorghum consumption associated with a caloric restricted diet reduces body fat in overweight men: a randomized controlled trial. *Food Research International* 2019 119:693–700.

33. Ashley, D., Marasini, D., Brownmiller, C., Lee, J. A., Carbonero, F. and Lee, S. O. Impact of grain sorghum polyphenols on microbiota of normal weight and overweight/obese subjects during in vitro fecal fermentation. *Nutrients* 2019 11(2):217.
34. Gamage, H. K. A. H., Tetu, S. G., Chong, R. W. W., Ashton, J., Packer, N. H. and Paulsen, I. T. Cereal products derived from wheat, sorghum, rice and oats alter the infant gut microbiota in vitro. *Scientific Reports* 2017 7(1):14312.
35. de Sousa, A. R., de Castro Moreira, M. E., Grancieri, M., Toledo, R. C. L., de Oliveira Araújo, F., Mantovani, H. C., Queiroz, V. A. V. and Martino, H. S. D. Extruded sorghum (Sorghum bicolor L.) improves gut microbiota, reduces inflammation, and oxidative stress in obese rats fed a high-fat diet. *Journal of Functional Foods* 2019 58:282–291.
36. Ritchie, L. E., Sturino, J. M., Carroll, R. J., Rooney, L. W., Azcarate-Peril, M. A. and Turner, N. D. Polyphenol-rich sorghum brans alter colon microbiota and impact species diversity and species richness after multiple bouts of dextran sodium sulfate-induced colitis. *FEMS Microbiology Ecology* 2015 91(3).
37. Benson, K. F., Beaman, J. L., Ou, B., Okubena, A., Okubena, O. and Jensen, G. S. West African Sorghum bicolor leaf sheaths have anti-inflammatory and immune-modulating properties in vitro. *Journal of Medicinal Food* 2013 16(3):230–238.
38. Umukoro, S., Oluwole, O. G., Eduviere, A. T., Adrian, O. I. and Ajayi, A. M. Jobelyn® exhibited anti-inflammatory, antioxidant, and membrane-stabilizing activities in experimental models. *Journal of Basic and Clinical Physiology and Pharmacology* 2015 26(5):501–508.
39. Ayuba, G. I., Jensen, G. S., Benson, K. F., Okubena, A. M. and Okubena, O. Clinical efficacy of a West African Sorghum bicolor-based traditional herbal preparation Jobelyn shows increased hemoglobin and CD4+ T-lymphocyte counts in HIV-positive patients. *Journal of Alternative and Complementary Medicine* 2013 20(1):53–56.
40. Abugri, D. A , Witola, W. H., Jaynes, J. M. and Toufic, N. In vitro activity of Sorghum bicolor extracts, 3-deoxyanthocyanidins, against toxoplasma gondii. *Experimental Parasitology* 2016 164:12–19.
41. Tugli, L. S., Essuman, E. K., Kortei, N. K., Nsor-Atindana, J., Nartey, E. B. and Ofori-Amoah, J. Bioactive constituents of waakye; a local Ghanaian dish prepared with Sorghum bicolor (L.) Moench leaf sheaths. *Scientific African* 2019 3:e00049.
42. Farida, K., Messaoud, B., Bernard, W. and Salah, A. Antimicrobial activity of aqueous methanolic extracts of Algerian cultivars of sorghum (Sorghum bicolor (L.) Moench) *Acta Scientifica Naturalis* 2020 7(1):71 85.
43. Gilchrist, A. K., Smolensky, D., Ngwaga, T., Chauhan, D., Cox, S., Perumal, R., Noronha, L. E. and Shames, S. R. High-polyphenol extracts from Sorghum bicolor attenuate replication of Legionella pneumophila within RAW 264.7 macrophages. *FEMS Microbiology Letters* 2020 367(7).
44. Nignpense, B. E., Chinkwo, K. A., Blanchard, C. L. and Santhakumar, A. B. Black sorghum phenolic extract modulates platelet activation and platelet microparticle release. *Nutrients* 2020 12(6):1760.
45. Abdelhalim, T. S., Kamal, N. M. and Hassan, A. B. Nutritional potential of wild sorghum: grain quality of Sudanese wild sorghum genotypes (Sorghum bicolor L. Moench). *Food Science and Nutrition* 2019 7(4):1529–1539.
46. Elnasikh, M. H., Ahmed, M. I., Abuali, A. I., Ahmed, N. M., Ahmed, B. I., Babiker, A. E. and Hassan, A. B. Protein content, in-vitro protein digestibility (IVPD), anti-nutritional factor and mineral content and bioavailability of sixteen sweet sorghum (Sorghum

bicolor L Moench) grain genotypes grown in Sudan. *Journal of the Saudi Society for Food and Nutrition (JSSFN)* 2020 13(1):33–45.

47. Tasie, M. M. and Gebreyes, B. G. Characterization of nutritional, antinutritional, and mineral contents of thirty-five sorghum varieties grown in Ethiopia. *International Journal of Food Science* 2020 2020:8243617.
48. Rhodes, D. H., Hoffmann, L., Jr., Rooney, W. L., Herald, T. J., Bean, S., Boyles, R., Brenton, Z. W. and Kresovich, S. Genetic architecture of kernel composition in global sorghum germplasm. *BMC Genomics* 2017 18(1):15–15.
49. Rhodes, D. H., Hoffmann, L., Rooney, W. L., Ramu, P., Morris, G. P. and Kresovich, S. Genome-wide association study of grain polyphenol concentrations in global sorghum [Sorghum bicolor (L.) Moench] germplasm. *Journal of Agricultural and Food Chemistry* 2014 62(45):10916–10927.
50. Dicko, M. H., Hilhorst, R., Gruppen, H., Traore, A. S., Laane, C., van Berkel, W. J. H. and Voragen, A. G. J. Comparison of content in phenolic compounds, polyphenol oxidase, and peroxidase in grains of fifty sorghum varieties from Burkina Faso. *Journal of Agricultural and Food Chemistry* 2002 50(13):3780–3788.
51. Luthria, D. L. and Liu, K. Localization of phenolic acids and antioxidant activity in sorghum kernels. *Journal of Functional Foods* 2013 5(4):1751–1760.
52. Awika, J. M. and Rooney, L. W. Sorghum phytochemicals and their potential impact on human health. *Phytochemistry* 2004 65(9):1199–1221.
53. Butler, L. G., Riedl, D. J., Lebryk, D. G. and Blytt, H. J. Interaction of proteins with sorghum tannin: mechanism, specificity and significance. *Journal of the American Oil Chemists' Society* 1984 61(5):916–920.
54. Elmaki, H. B., Babiker, E. E. and El Tinay, A. H. Changes in chemical composition, grain malting, starch and tannin contents and protein digestibility during germination of sorghum cultivars. *Food Chemistry* 1999 64(3):331–336.
55. Awika, J. M., Rooney, L. W. and Waniska, R. D. Properties of 3-deoxyanthocyanins from sorghum. *Journal of Agricultural and Food Chemistry* 2004 52(14):4388–4394.
56. Dykes, L., Seitz, L. M., Rooney, W. L. and Rooney, L. W. Flavonoid composition of red sorghum genotypes. *Food Chemistry* 2009 116(1):313–317.
57. Bradwell, J., Hurd, M., Pangloli, P., McClure, A. and Dia, V. P. Storage stability of sorghum phenolic extracts' flavones luteolin and apigenin. *LWT* 2018 97:787–793.
58. Dykes, L. and Rooney, L. W. Sorghum and millet phenols and antioxidants. *Journal of Cereal Science* 2006 44(3):236–251.
59. Rao, S., Santhakumar, A. B., Chinkwo, K. A., Wu, G., Johnson, S. K. and Blanchard, C. L. Characterization of phenolic compounds and antioxidant activity in sorghum grains. *Journal of Cereal Science* 2018 84:103–111.
60. Aruna, C. R., Ratnavathi, C. V., Suguna, M., Ranga, B., Praveen Kumar, P., Annapurna, A., Bahadure, D. M. and Toapi, V. A. Genetic variability and GxE interactions for total polyphenol content and antioxidant activity in white and red sorghums (Sorghum bicolor). *Plant Breeding* 2020 139(1):119–130.
61. Przybylska-Balcerek, A., Frankowski, J. and Stuper-Szablewska, K. Bioactive compounds in sorghum. *European Food Research and Technology* 2019 245(5):1075–1080.
62. Shen, S., Huang, R., Li, C., Wu, W., Chen, H., Shi, J., Chen, S. and Ye, X. Phenolic compositions and antioxidant activities differ significantly among sorghum grains with different applications. *Molecules* 2018 23(5):1203.
63. Rocchetti, G., Giuberti, G., Busconi, M., Marocco, A., Trevisan, M. and Lucini, L. Pigmented sorghum polyphenols as potential inhibitors of starch digestibility: an

in vitro study combining starch digestion and untargeted metabolomics. *Food Chemistry* 2020 312:126077.
64. Gutiérrez-Salomón, A. L., Aguilar-Raymundo, V. G. and Barajas-Ramírez, J. A. Tortilla chips made with white sorghum and corn: comparison of sensory and physicochemical characteristics with corn-made commercial products. *Journal of Food Processing and Preservation* 2020 44(9):e14642.
65. Dicko, M. H., Gruppen, H., Traoré, A. S., van Berkel, W. J. and Voragen, A. G. Evaluation of the effect of germination on phenolic compounds and antioxidant activities in sorghum varieties. *Journal of Agricultural and Food Chemistry* 2005 53(7):2581–2588.
66. Boukid, F., Dall'Asta, M., Bresciani, L., Mena, P., Del Rio, D., Calani, L., Sayar, R., Seo, Y. W., Yacoubi, I. and Mejri, M. Phenolic profile and antioxidant capacity of landraces, old and modern tunisian durum wheat. *European Food Research and Technology* 2019 245(1):73–82.
67. Šimić, G., Horvat, D., Lalić, A., Koceva Komlenić, D., Abičić, I. and Zdunić, Z. Distribution of β-Glucan, phenolic acids, and proteins as functional phytonutrients of hull-less barley grain. *Foods* 2019 8(12):680.
68. Ratseewo, J., Warren, F. J. and Siriamornpun, S. The influence of starch structure and anthocyanin content on the digestibility of Thai pigmented rice. *Food Chemistry* 2019 298:124949.
69. Soycan, G., Schär, M. Y., Kristek, A., Boberska, J., Alsharif, S. N. S., Corona, G., Shewry, P. R. and Spencer, J. P. E. Composition and content of phenolic acids and avenanthramides in commercial oat products: are oats an important polyphenol source for consumers? *Food Chemistry: X* 2019 3:100047.
70. Ballester-Sánchez, J., Gil, J. V., Haros, C. M. and Fernández-Espinar, M. T. Effect of incorporating white, red or black quinoa flours on free and bound polyphenol content, antioxidant activity and colour of bread. *Plant Foods for Human Nutrition* 2019 74(2):185–191.
71. Tayal, M., Somavat, P., Rodriguez, I., Martinez, L. and Kariyat, R. Cascading effects of polyphenol-rich purple corn pericarp extract on pupal, adult, and offspring of tobacco hornworm (Manduca sexta L.). *Communicative and Integrative Biology* 2020 13(1):43–53.
72. Singh, S. and Kariyat, R. R. Exposure to polyphenol-rich purple corn pericarp extract restricts fall armyworm (Spodoptera frugiperda) growth. *Plant Signaling and Behavior* 2020 15(9):1784545.
73. Hithamani, G. and Srinivasan, K. Bioaccessibility of polyphenols from wheat (Triticum aestivum), sorghum (Sorghum bicolor), green gram (Vigna radiata), and chickpea (Cicer arietinum) as influenced by domestic food processing. *Journal of Agricultural and Food Chemistry* 2014 62(46):11170–11179.
74. Podio, N. S., Baroni, M. V., Pérez, G. T. and Wunderlin, D. A. Assessment of bioactive compounds and their in vitro bioaccessibility in whole-wheat flour pasta. *Food Chemistry* 2019 293:408–417.
75. Klepacka, J. and Fornal, Ł. Ferulic acid and its position among the phenolic compounds of wheat. *Critical Reviews in Food Science and Nutrition* 2006 46(8):639–647.
76. Callcott, E. T., Santhakumar, A. B., Strappe, P., Luo, J. and Blanchard, C. L. Polyphenols from Australian-grown pigmented red and purple rice inhibit adipocyte differentiation. *Journal of Cereal Science* 2018 81:140–146.

77. Sur, R., Nigam, A., Grote, D., Liebel, F. and Southall, M. D. Avenanthramides, polyphenols from oats, exhibit anti-inflammatory and anti-itch activity. *Archives of Dermatological Research* 2008 300(10):569-574.
78. Guo, W., Wise, M. L., Collins, F. W. and Meydani, M. Avenanthramides, polyphenols from oats, inhibit IL-1β-induced NF-κB activation in endothelial cells. *Free Radical Biology and Medicine* 2008 44(3):415-429.
79. Shen, Y., Zhang, H., Cheng, L., Wang, L., Qian, H. and Qi, X. In vitro and in vivo antioxidant activity of polyphenols extracted from black highland barley. *Food Chemistry* 2016 194:1003-1012.
80. Suriano, S., Iannucci, A., Codianni, P., Fares, C., Russo, M., Pecchioni, N., Marciello, U. and Savino, M. Phenolic acids profile, nutritional and phytochemical compounds, antioxidant properties in colored barley grown in southern Italy. *Food Research International* 2018 113:221-233.
81. Al-Mamary, M., Molham, A.-H., Abdulwali, A.-A. and Al-Obeidi, A. In vivo effects of dietary sorghum tannins on rabbit digestive enzymes and mineral absorption. *Nutrition Research* 2001 21(10):1393-1401.
82. Wu, G., Ashton, J., Simic, A., Fang, Z. and Johnson, S. K. Mineral availability is modified by tannin and phytate content in sorghum flaked breakfast cereals. *Food Research International* 2018 103:509-514.
83. Galán, M. G., Weisstaub, A., Zuleta, A. and Drago, S. R. Effects of extruded whole-grain sorghum (Sorghum bicolor (L.) Moench) based diets on calcium absorption and bone health of growing Wistar rats. *Food and Function* 2020 11(1): 508-513.
84. Moss, A. F., Khoddami, A., Chrystal, P. V., Sorbara, J.-O. B., Cowieson, A. J., Selle, P. H. and Liu, S. Y. Starch digestibility and energy utilisation of maize- and wheat-based diets is superior to sorghum-based diets in broiler chickens offered diets supplemented with phytase and xylanase. *Animal Feed Science and Technology* 2020 264:114475.
85. Towo, E., Matuschek, E. and Svanberg, U. Fermentation and enzyme treatment of tannin sorghum gruels: effects on phenolic compounds, phytate and in vitro accessible iron. *Food Chemistry* 2006 94(3):369-376.
86. Teixeira, L., Pinto, C. F. D., de Mello Kessler, A. M. and Trevizan, L. Effect of partial substitution of rice with sorghum and inclusion of hydrolyzable tannins on digestibility and postprandial glycemia in adult dogs. *PLOS ONE* 2019 14(5):e0208869.
87. Hassan, I. A. G., Elzubeir, E. A. and El Tinay, A. H. Growth and apparent absorption of minerals in broiler chicks fed diets with low or high tannin contents. *Tropical Animal Health and Production* 2003 35(2):189-196.
88. Barros, F., Awika, J. M. and Rooney, L. W. Interaction of tannins and other sorghum phenolic compounds with starch and effects on in vitro starch digestibility. *Journal of Agricultural and Food Chemistry* 2012 60(46):11609-11617.
89. Lemlioglu-Austin, D., Turner, N. D., McDonough, C. M. and Rooney, L. W. Effects of sorghum [Sorghum bicolor (L.) Moench] crude extracts on starch digestibility, estimated glycemic index (EGI), and resistant starch (RS) contents of porridges. *Molecules* 2012 17(9):11124-11138.
90. Mkandawire, N. L., Kaufman, R. C., Bean, S. R., Weller, C. L., Jackson, D. S. and Rose, D. J. Effects of Sorghum (Sorghum bicolor (L.) Moench) tannins on α-amylase activity and in vitro digestibility of starch in raw and processed flours. *Journal of Agricultural and Food Chemistry* 2013 61(18):4448-4454.

91. Cabrera-Ramírez, A. H., Luzardo-Ocampo, I., Ramírez-Jiménez, A. K., Morales-Sánchez, E., Campos-Vega, R. and Gaytán-Martínez, M. Effect of the nixtamalization process on the protein bioaccessibility of white and red sorghum flours during in vitro gastrointestinal digestion. *Food Research International* 2020 134:109234.
92. Pan, L. and An, D. Nitrogen utilization is lower for sorghum-based diets compared with corn-based diets in pigs. *Livestock Science* 2020 237:104066.
93. Medina Martinez, O. D., Lopes Toledo, R. C., Vieira Queiroz, V. A., Pirozi, M. R., Duarte Martino, H. S. and Ribeiro de Barros, F. A. Mixed sorghum and quinoa flour improves protein quality and increases antioxidant capacity in vivo. *LWT* 2020 129:109597.
94. Cardoso, LdM., Pinheiro, S. S., de Carvalho, C. W. P., Queiroz, V. A. V., de Menezes, C. B., Moreira, A. V. B., de Barros, F. A. R., Awika, J. M., Martino, H. S. D. and Pinheiro-Sant'Ana, H. M. Phenolic compounds profile in sorghum processed by extrusion cooking and dry heat in a conventional oven. *Journal of Cereal Science* 2015 65:220–226.
95. Salazar Lopez, N. J., Loarca-Piña, G., Campos-Vega, R., Gaytán Martínez, M., Morales Sánchez, E., Esquerra-Brauer, J. M., Gonzalez-Aguilar, G. A. and Robles Sánchez, M. The extrusion process as an alternative for improving the biological potential of sorghum bran: phenolic compounds and antiradical and anti-inflammatory capacity. *Evidence-Based Complementary and Alternative Medicine: eCAM* 2016 2016:8387975.
96. Ortiz-Cruz, R. A., Ramírez-Wong, B., Ledesma-Osuna, A. I., Torres-Chávez, P. I., Sánchez-Machado, D. I., Montaño-Leyva, B., López-Cervantes, J. and Gutiérrez-Dorado, R. Effect of extrusion processing conditions on the phenolic compound content and antioxidant capacity of sorghum (Sorghum bicolor (L.) Moench) bran. *Plant Foods for Human Nutrition* 2020 75(2):252–257.
97. Gu, L., House, S. E., Rooney, L. W. and Prior, R. L. Sorghum extrusion increases bioavailability of catechins in weanling pigs. *Journal of Agricultural and Food Chemistry* 2008 56(4):1283–1288.
98. Luzardo-Ocampo, I., Ramírez-Jiménez, A. K., Cabrera-Ramírez, Á. H., Rodríguez-Castillo, N., Campos-Vega, R., Loarca-Piña, G. and Gaytán-Martinez, M. Impact of cooking and nixtamalization on the bioaccessibility and antioxidant capacity of phenolic compounds from two sorghum varieties. *Food Chemistry* 2020 309:125684.
99. Svensson, L., Sekwati-Monang, B., Lutz, D. L., Schieber, A. and Ganzle, M. G. Phenolic acids and flavonoids in nonfermented and fermented red sorghum (Sorghum bicolor (L.) Moench). *Journal of Agricultural and Food Chemistry* 2010 58(16):9214-9220.
100. Xiong, Y., Zhang, P., Luo, J., Johnson, S. and Fang, Z. Effect of processing on the phenolic contents, antioxidant activity and volatile compounds of sorghum grain tea. *Journal of Cereal Science* 2019 85:6–14.
101. Wu, L., Huang, Z., Qin, P. and Ren, G. Effects of processing on phytochemical profiles and biological activities for production of sorghum tea. *Food Research International* 2013 53(2):678–685.
102. Sun, H., Wang, H., Zhang, P., Ajlouni, S. and Fang, Z. Changes in phenolic content, antioxidant activity, and volatile compounds during processing of fermented sorghum grain tea. *Cereal Chemistry* 2020 97(3):612–625.
103. Salih, S. A., Ahmed, K. E., Ezzdeen, L. T. and Hamza, A. A. Effect of fermentation and processing of Sorghum bicolor grains to produce traditional Sudanese Hulu-Mur on

phytochemicals and their biological activities. *GSC Biological and Pharmaceutical Sciences* 2020 10(3):76–84.

104. Kapoor, T., Abraham, A. R. and Jain, S. Development and evaluation of ready to eat breakfast cereal mix formulated using treated sorghum flour. *International Journal of Research in Agricultural Sciences* 2020 7(2) 2348–3997.
105. Khoddami, A., Mohammadrezaei, M. and Roberts, T. H. Effects of sorghum malting on colour, major classes of phenolics and individual anthocyanins. *Molecules* 2017 22(10):1713.
106. Irondi, E. A., Adegoke, B. M., Effion, E. S., Oyewo, S. O., Alamu, E. O. and Boligon, A. A. Enzymes inhibitory property, antioxidant activity and phenolics profile of raw and roasted red sorghum grains in vitro. *Food Science and Human Wellness* 2019 8(2):142–148.
107. Queiroz, V. A. V., Aguiar, A. S., de Menezes, C. B., de Carvalho, C. W. P., Paiva, C. L., Fonseca, P. C., da Conceição, R. R. P. A low calorie and nutritive sorghum powdered drink mix: influence of tannin on the sensorial and functional properties. *Journal of Cereal Science* 2018, 79:43–49.
108. Pontieri, P., Mamone, G., De Caro, S., Tuinstra, M. R., Roemer, E., Okot, J., De Vita, P., Ficco, D. B. M., Alifano, P., Pignone, D., Massardo, D. R. and Del Giudice, L. Sorghum, a healthy and gluten-free food for celiac patients as demonstrated by genome, biochemical, and immunochemical analyses. *Journal of Agricultural and Food Chemistry* 2013 61(10):2565–2571.
109. Velázquez, N., Sánchez, H., Osella, C. and Santiago, L. G. Using white sorghum flour for gluten-free breadmaking. *International Journal of Food Sciences and Nutrition* 2012 63(4):491–497.
110. Ogunsakin, A. O., Vanajakshi, V., Anu-Appaiah, K. A., Vijayendra, S. V. N., Walde, S. G., Banwo, K., Sanni, A. I. and Prabhasankar, P. Evaluation of functionally important lactic acid bacteria and yeasts from Nigerian sorghum as starter cultures for gluten-free sourdough preparation. *LWT–Food Science and Technology* 2017 82:326–334.
111. Palavecino, P. M., Ribotta, P. D., León, A. E. and Bustos, M. C. Gluten-free sorghum pasta: starch digestibility and antioxidant capacity compared with commercial products. *Journal of the Science of Food and Agriculture* 2019 99(3):1351–1357.
112. Khan, I., Yousif, A., Johnson, S. K. and Gamlath, S. Effect of sorghum flour addition on resistant starch content, phenolic profile and antioxidant capacity of durum wheat pasta. *Food Research International* 2013 54(1):578–586.
113. Soares, R. R. A., Vasconcelos, C. M., Oliveira, M. Vd, Minim, V. P. R., Queiroz, V. A. V. and Barros, F. Starch digestibility and sensory acceptance of gluten free foods prepared with tannin sorghum flour. *Pesquisa Agropecuária Brasileira* 2019 54.
114. Rebellato, A. P., Orlando, E. A., Thedoropoulos, V. C. T., Greiner, R. and Pallone, J. A. L. Effect of phytase treatment of sorghum flour, an alternative for gluten free foods and bioaccessibility of essential minerals. *Journal of Food Science and Technology* 2020 57(9):3474–3481.
115. Sullivan, A. C., Pangloli, P. and Dia, V. P. Impact of ultrasonication on the physicochemical properties of sorghum kafirin and in vitro pepsin-pancreatin digestibility of sorghum gluten-like flour. *Food Chemistry* 2018 240:1121–1130.
116. Danquah, A., Galyuon, I. K., Otwe, E. P. and Asante, D. K. Genetic diversity in some Ghanaian and Malian sorghum [*Sorghum bicolor* (L) Moench] accessions using SSR markers. *The African Journal of Biotechnology* 2019 18(27):591–602.

117. Clarke, S. J., McLean, J., George-Jaeggli, B., McLean, G., de Voil, P., Eyre, J. X. and Rodriguez, D. Understanding the diversity in yield potential and stability among commercial sorghum hybrids can inform crop designs. *Field Crops Research* 2019 230:84–97.
118. Schaffasz, A., Windpassinger, S., Friedt, W., Snowdon, R. and Wittkop, B. Sorghum as a novel crop for Central Europe: using a broad diversity set to dissect temperate-adaptation. *Agronomy* 2019 9(9):535.
119. Wu, G., Johnson, S. K., Bornman, J. F., Bennett, S., Singh, V. and Fang, Z. Effect of genotype and growth temperature on sorghum grain physical characteristics, polyphenol content, and antioxidant activity. *Cereal Chemistry Journal* 2016 93(4):419–425.

Chapter 11

Developing millet-based cereal products with enhanced nutritional properties

Kwaku G. Duodu, John Lubaale and Eugenie Kayitesi, University of Pretoria, South Africa

1 Introduction

Millets are regarded as the sixth most important food cereals in the world in terms of production (Serna-Saldivar and Espinosa-Ramirez, 2019). They are an important cereal staple in many parts of the world, particularly in Africa and Asia. The term 'millets' is the general name given to a range of different species including pearl, foxtail, proso, finger, barnyard and kodo millets and also fonio and teff. A few recent reviews discuss various aspects of the millets such as taxonomy, distribution, production (Taylor, 2019), postharvest technologies (Beta and Ndolo, 2019), grain structure and chemical composition (Serna-Saldivar and Espinosa-Ramirez, 2019) and food and beverage nutritional attributes (Taylor and Kruger, 2019), among others. Another recent review on the unique nutritional and health-promoting attributes of millets by Taylor (2017) is an extremely useful resource.

Millets are processed into a wide range of foods for human consumption. The wide range of traditional food products made from millets mainly in Africa and Asia include nonalcoholic beverages and gruels, alcoholic beverages, porridges (thin and thick), steam-cooked products, boiled rice-like products, various breads and related products and different types of snack foods (Taylor and Duodu, 2019). Apart from these traditional products, millets are also used in producing various modern convenient food products using technologies such as extrusion cooking (Alavi et al., 2019). Due to their wide consumption,

http://dx.doi.org/10.19103/AS.2021.0087.22

it is clear that millets are an important source of calories, protein and other nutrients for many inhabitants of Africa and Asia. The methods of processing used to transform millet grains into foods can enhance the nutritional properties of these foods and these will be discussed in this chapter.

2 Nutritional properties of millets

2.1 Macronutrients

Millets are an important source of nutrients and an essential food for people in developing countries (Vinoth and Ravindhran, 2017). In Africa and Asia, millets serve as a major food component in traditional foods and beverages, such as bread, porridges (thick and thin), beer and other value-added products. Millets are a rich source of proteins, vitamins and minerals. Carbohydrates (60–75%) is the major component of the millet grain, followed by dietary fibre (2–22%) and protein (8–13%), and lipids and ash (1–3%) (Table 1). Millet contains about 70% starch and contributes to the energy value of the grain. Millet grain starch is made of about 15–30% amylose and 70–85% amylopectin (Vinoth and Ravindhran, 2017). Millets also contain relatively high proportions of non-starch polysaccharides (Muthamilarasan et al., 2016). Millets are gluten-free (Taylor and Emmambux, 2008) and are therefore an excellent food choice for consumers with celiac disease.

Millet proteins are good sources of the sulphur-containing amino acids (methionine and cysteine). Like other cereals, millets are deficient in essential amino acids such as lysine and tryptophan; this is, however, cultivar dependent. Taylor and Taylor (2017) reported that prolamins were the main storage proteins for most millet varieties. The contribution of prolamins to total grain protein for finger, foxtail, pearl and proso millets was reported to be about 42%, 76%, 43% and 58%, respectively (Taylor and Taylor, 2017). The total lipid content in foxtail, proso and finger millets ranged from 1% to 4.3%, while it ranged from 2% to 5% in the barnyard, little and pearl millets (Table 1). Ramashia et al. (2019) reported that fatty acid profile of the millet lipid is mainly characterised by the presence of high levels of linoleic (24%), oleic (49.8%) and palmitic acid (21–25%).

2.2 Micronutrients

Minerals and vitamins in common millet species are presented in Table 2. Millets are rich in B-group vitamins and also in minerals such as potassium, phosphorus, iron, copper, magnesium, zinc and manganese (Saleh et al., 2013; Ramashia et al., 2019; Adebiyi et al., 2018). The mineral content in millets ranges from 0.5 to 4.4 g/100 g, which is higher than that reported in cereals such as wheat (1.5%) and rice (0.6%) (Kumar et al., 2018). The authors also indicated that calcium content of finger millet was about eight times higher

Table 1 Macronutrient compositions of common millet varieties

	Finger millet	*Proso* millet	Foxtail millet	Little millet	Kodo millet	Barnyard millet	Pearl millet
Proximate (g/100 g)							
Moisture	7-13	11.9	11.2	11.5	12.8	11.9	12.4
Fat	1-2	1.1	2.4-4.3	4.7	1.4	2.2	5.0
Minerals/Ash	2.7	1.9	0.5-3.3	1.5	2.6	4.4	2.3
Dietary fibre	11-22	-	2.4-8.5	2.53	2.47	2	11.3
Carbohydrates	72-75	70.4	60-67	67	65-67	65	67.5
Energy (kcal)	328	341	331	341	309	307	361
Protein	7-3	12.5	12.3	7.7	8.3-10	6.2	11.6
Essential amino acids (mg/gN)							
Arginine	300	290	220	250	270	-	300
Histidine	130	110	130	120	120	-	140
Lysine	220	190	140	110	150	-	190
Tryptophan	100	50	60	60	50	-	110
Phenyl alanine	310	310	420	330	430	-	290
Tyrosine	220	-	-	-	-	-	200
Methionine	210	160	180	180	180	-	150
Cystine	140	-	100	90	110	-	110
Threonine	240	150	190	190	200	-	240
Leucine	690	760	1040	760	650	-	750
Isoleucine	400	410	480	370	360	-	260
Valine	480	410	430	350	410	-	330

Source: Amadou et al. (2013), Saleh et al. (2013), Adebiyi et al. (2017), Adebiyi et al. (2018) and Ramashia et al. (2019).

Table 2 Micronutrient compositions of common millet varieties

	Finger millet	Proso millet	Foxtail millet	Little millet	Kodo millet	Barnyard millet	Pearl millet
Vitamins (per 100 g)							
Carotene (μg)	42	0	32	0	0	0	132
B1-Thiamine (mg)	0.42	0.20	0.59	0.30	0.33	0.33	0.33
B2-Riboflavin (mg)	0.19	0.18	0.11	0.09	0.09	0.10	0.245
B3-Niacin (mg)	1.1	2.3	3.2	3.2	2.0	4.2	2.3
B4-Choline (mg)	-	748	-	-	-	-	-
Minerals and trace elements (mg/100 g)							
Calcium	344	14	31	17	27	20	42
Phosphorus	283	206	290	220	188	280	296
Iron	3.9	0.8	2.8	9.3	0.5	5.0–17.47	8.0
Magnesium	137	153	81	133	147	82	137
Sodium	11	8.2	4.6	8.1	4.6	-	10.9
Potassium	408	113	250	129	144	-	307
Copper	0.47	1.60	1.40	1.00	1.60	0.60	1.06
Manganese	5.49	0.60	0.60	0.68	1.10	0.96	1.15
Molybdenum	0.102	-	0.70	0.016	-	-	0.069
Zinc	2.3	1.4	2.4-41	3.7	0.7	3.0	3.1
Chromium	0.028	0.020	0.030	0.180	0.020	0.090	0.023
Sulphur	160	157	171	149	136	-	147
Chlorine	44	19	37	13	11	-	39

Source: Amadou et al. (2013), Saleh et al. (2013), Shobana et al. (2013), Verma and Patel (2013), Muthamilarasan et al. (2016), Adebiyi et al. (2017), Adebiyi et al. (2018) and Ramashia et al. (2019).

than wheat (348 mg/100 g). Barnyard millet and pearl millet are rich sources of iron, and could contribute to iron requirements for expectant mothers. The iron content of barnyard millet was reported to be 17.47 mg/100 g (Kumar et al., 2018). In a study by Chandel et al. (2014), it was reported that foxtail millet contains highest content of zinc (4.1 mg/100 g) among all millets and is also a good source of iron (2.8 mg/100 g). Millets are a good source of β-carotene (pro-vitamin A) and water-soluble B vitamins particularly riboflavin, niacin and folic acid (Kumar et al., 2018). The thiamine and niacin content of millets is similar to that reported in rice and wheat (Kumar et al., 2018). The highest β-carotene content in millets (132 μg /100 g) was reported in pearl millet. Niacin has been predominantly reported in barnyard millet (4.20 mg/100 g), followed by foxtail and little millet (3.2 mg/100 g) and pearl millet (2.3 mg/100 g).

2.3 Anti-nutrients

Despite the nutritional value of millets, a major limitation to bioaccessibility and bioavailability of the minerals and protein is due to the presence of anti-nutrients. These anti-nutrients include phytate/phytic acid, tannins, oxalates and goitrogens (Pelembe, 2001; Taylor and Emmambux, 2008). Phytic acid (myo-inositol hexaphosphate) is mainly found in the cereal aleurone layer (Serna-Saldivar and Espinosa-Ramirez, 2019) and plays a vital role as the form in which phosphorus is stored in plants. Phytates form insoluble complexes with metal ions such as iron and zinc and can negatively affect their bioavailability (Liebman and Al-Wahsh, 2011). Oxalates particularly bind to calcium (Ca) to form complexes (calcium oxalate crystals) that prevent Ca from being absorbed by the body and the oxalates themselves can cause irritation of the gut lining (Mohan et al., 2016). While the phytates and oxalates decrease bioavailability of minerals, the goitrogenic compounds and their metabolites cause off-odours and flavours in pearl millet (Reddy et al., 1986; Pelembe, 2001; Lestienne et al., 2007). Finger millet has been reported to contain tannins (0.04–3.47%), phytate (0.48%) and oxalate (0.27%) (Kumar et al., 2016). It is estimated that the tannin content (0.04–3.47%) of finger millet is higher than that in any other millet (Kakade and Hathan, 2014). Tannin content in finger millet varies among different varieties. The brown or dark brown pigmented varieties of finger millet contain about 0.61% compared to the white variety reported to have 0.05% tannin (Kumar et al., 2016). Tannins have been reported to reduce protein digestibility, feed intake, growth rate and net metabolizable energy in experimental animals (Kumar et al., 2016). Phytic acid content of whole grain from foxtail millet flours ranged from 5.4 to 9.9 mg/g, while for proso millet wholegrain flour, phytic acid content was 7.2 mg/g (Devisetti et al., 2014).

3 Use of traditional and modern processing methods as strategies for developing millet-based foods with enhanced nutritional properties

Food processing is an important step in food production and supply chain required to improve palatability, functionality and nutritional value of food. Millet, like other cereals, is generally subjected to various types of processing treatment before consumption. These processing treatments may include traditional methods such as fermentation and malting as well as modern processing methods such as extrusion.

3.1 Fermentation

In the absence of sophisticated processing techniques in the developing world, fermentation remains an important unit operation utilised for the processing of cereal grains into various other products (Amadou et al., 2013; Adebiyi et al., 2018). Fermentation can be defined as a biochemical process of transforming substrates into different products with the action of microorganisms (bacteria and yeast). This leads to the modification of the substrate and production of a diverse range of metabolites. Use of lactic acid bacteria (LABs) for production of fermented foods has been well documented (Soro-Yao et al., 2014; Adebiyi et al., 2018; Adebo, 2020). LAB species of the genera *Lactobacillus, Leuconostoc, Streptococcus and Pediococcus*, as well as yeast species of the genera *Saccharomyces*, have been identified as the main microorganisms in the fermentation process of cereal-based foods (Soro-Yao et al., 2014). They are known to contribute to microbial safety, flavour, improved shelf life and sensory quality of the final products (Leroy and De Vuyst, 2004; LeBlanc and Todorov, 2011). Lactic acid bacteria are generally recognised as safe (GRAS) substances and are technologically interesting as they produce a wide variety of metabolites including ethanol, acetic acid, aromatic compounds, exopolysaccharides, several enzymes and bacteriocins (Leroy and De Vuyst, 2004). They also ensure food safety by decreasing pH through the production of organic acids which leads to subsequent inhibitory effect on the growth of other microorganisms, especially foodborne pathogens (Gautam and Sharma, 2009; Salvucci et al., 2016).

Lactic acid bacteria also break down proteins and peptides to generate diverse metabolites which contribute to flavour and structure of foods and also produce vitamins (Gänzle et al., 2008; LeBlanc et al., 2013). Many LABs also produce heteropolysaccharides (composed of D-glucose, D-galactose and L-rhamnose) and homopolysaccharides (composed of only D-glucose and D-fructose) with potential functional properties (Notararigo et al., 2013). These compounds also play an important role in the rheology, mouthfeel and beneficial health effects of fermented foods (Jay et al., 2005; Gänzle et al., 2008; Galle et al., 2011). Fermentation also ensures the safety of the food by suppressing the growth and survival of undesirable microflora in food (Ross et al., 2002). Most importantly, the metabolic processes of microorganisms during fermentation lead to decreases in inhibitors of digestive enzymes (such as trypsin and amylase), and other anti-nutrients such as phytic acid and tannins (Osman, 2004; Eltayeb et al., 2007). This then leads to improvement in the nutritional quality of millet by increasing mineral and protein bioaccessibility and bioavailability. Elyas et al. (2002) found that natural fermentation of pearl millet reduced phytic acid and total polyphenols. In another study, Murali and Kapoor (2003) reported that fermentation resulted in increased extractability of minerals, amino acids, water-soluble proteins and vitamins. This was confirmed by other authors who reported such increases and associated them with proteolysis (by bacterial proteases) and metabolic synthesis during fermentation (Murali and Kapoor, 2003; Saleh et al., 2013; Galati et al., 2014).

For the most part, fermented millet products in developing countries are produced through spontaneous fermentation (Fig. 1a). According to Galati et al. (2014), spontaneous fermentation results through the competitive activities of a pool of microorganisms, with the best adapted strains and higher growth rates eventually dominating the substrate (Galati et al., 2014). Spontaneous fermentation is generally achieved by back-slopping. This is a process whereby a portion of the previous fermentation batch is introduced to serve as a starter culture (by transferring fermenting microorganisms) for the next batch

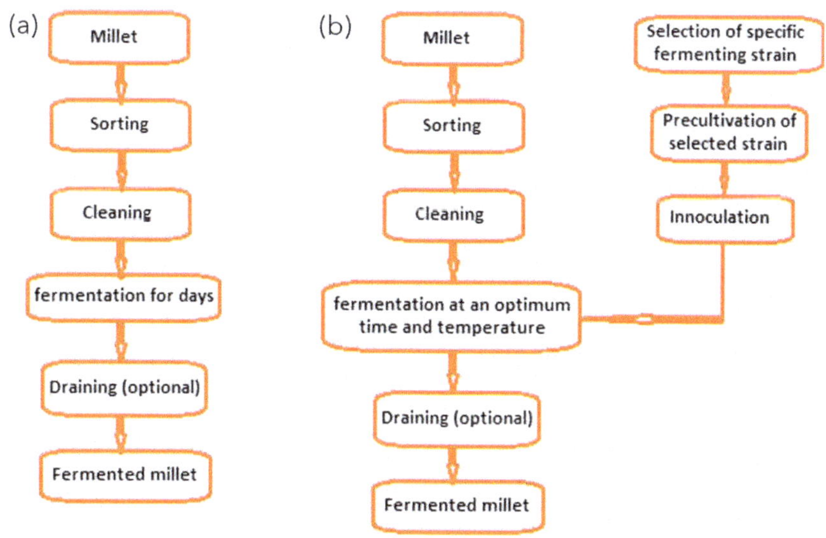

Figure 1 (a) Spontaneous (natural) fermentation processes for millet, (b) controlled fermentation process (adapted from Adebiyi et al., 2018).

(Aka et al., 2014). Challenges associated with spontaneous fermentation of millets have led to the need for the development, identification and selection of specific strains for a better fermentation process (Galati et al., 2014; Adebiyi et al., 2018). Fermentation has thus evolved to a more controlled technique, involving the use of starter cultures (Fig. 1b). Selection of such starter cultures are based on their ability to obtain products of desired characteristics and qualities in the intended final product. *L. fermentum*, *L. plantarum*, *L. salivarius*, *L. delbrueckii*, *L. amylolyticus*, *L. reuteri*, *L. paraplantarum*, *Lact. lactis*, *Leuc. mesenteroides*, *P. acidilactici*, *P. pentosaceus*, *Str. gallolyticus* and *W. confusa* species have been identified in efforts to optimise and standardise the fermentation of cereal-based foods (Soro-Yao et al., 2014).

Africa is home to a wide range of millet-based fermented products (Soro-Yao et al., 2014; Adebiyi et al., 2018; Adebo, 2020). Common fermented millet foods include *ogi*, a West-African porridge prepared from fermented millet, sorghum or maize (Adebiyi et al., 2018). Similar to *ogi* is *ben saalga*, a millet-based gruel popular in Burkina Faso, and *koko*, a Ghanaian millet porridge consumed as a meal or snack (Soro-Yao et al., 2014; Adebiyi et al., 2018). Fura (or furah) is a semi-solid dumpling prepared by steam-cooking of fermented pearl millet balls, followed by pounding in a mortar and rolling (Owusu-Kwarteng et al., 2010). It is consumed in Burkina Faso, Ghana and Nigeria (Jideani et al., 2002; Owusu-Kwarteng et al., 2010). Alcoholic and nonalcoholic beverages produced from fermented millet are widely consumed across the African continent. These include *pito*, a beverage consumed in Ghana and Nigeria (Ajiboye et al., 2014), and *katikalla*, a traditional drink in Ethiopia produced from finger millet (Bultosa and Taylor, 2004). Other important millet-based beverages include Sudanese *merrisa*, Namibian *oshikundu*, Ugandan and Rwandan *bushera* and *pombe* consumed in Mozambique (Mukisa et al., 2010; Adebiyi et al., 2018). Notable thin fermented breads made from millet are *injera* and *kisra* which are common staples in Eritrea, Ethiopia, Sudan and Somalia (Belton and Taylor, 2004). Other millet-based baked foods are usually composited with wheat flour in different proportions to obtain products such as cakes, biscuits or breads (Taylor et al., 2006; Saleh et al., 2013; Adebiyi et al., 2017).

3.2 Malting

In many parts of Africa and India, millets are malted or sprouted and the malts obtained are used in various food applications (Taylor, 2017; Adebiyi et al., 2018). Taylor and Dewar (2001) provide a definition of malting as the controlled germination of grain in moist air which promotes the development and mobilisation of endogenous hydrolytic enzymes that are otherwise inactive in the unsprouted grain. The enzymes mobilised include particularly amylases, proteases (Taylor, 2017; Adebiyi et al., 2018) and other enzymes such as phytases. The action of these enzymes modifies the structure of the grain and brings about hydrolysis of some of the starch and protein into sugars and amino acids.

An important consequence of these events during the malting process is that they generally lead to improvement of the nutritional quality of the grains mainly by reduction of anti-nutrients and rendering nutrients more bioavailable (Taylor, 2017). This improvement in nutritional quality of malted millet grains may then be carried through to food products made with the malted grain or in composite food products where the malted millet is included as a component. Some reported effects of malting also indicate an increase in levels of nutrients such as protein and minerals. There is a wide range of literature that reports on enhanced nutritional quality of millets after malting. A few will be summarised briefly here.

3.2.1 Effect on nutrient content

Traore et al. (2004) reported that malting increased protein content in pearl millet which was attributed to the decrease in carbohydrate compounds during malting. Malting increased protein content of pearl millet by 50% compared to unmalted grains. This was attributed to protein synthesis during malting (Akinola et al., 2017). Malting pearl millet with gibberellic acid and calcium chloride improved nutritional properties by bringing about an increase in levels of protein, ash, calcium, vitamin C and soluble solids (Balseiro et al., 2015). These could be related to the role of gibberellic acid in activating various enzymes such as proteases, phytases and amylases which break down their relevant substrates to produce an enhanced nutritional quality of the resulting malt. Obadina et al. (2017) reported that protein content of pearl millet grains increased with increasing malting period. This was attributed to the action of proteolytic enzymes that are activated during sprouting which break down polymeric proteins into amino acids and smaller peptides and also bring about accumulation of protein reserves.

Adebiyi et al. (2017) reported higher protein content of malted pearl millet flour compared to native flour which was attributed to accumulation of proteins and production of some additional amino acids during malting. A significant increase in essential amino acids was also observed and attributed to increased activities of hydrolytic enzymes and degradation of complex proteins to amino acids through proteolysis. The malted pearl millet also had higher mineral (Ca, Cu, Fe, Mg, Mn, P, Na and Zn) content which was attributed to improved extractability of minerals as a result of synthesis and cell wall solubilisation (possibly through the action of cell wall degrading enzymes mobilised during malting) (Adebiyi et al., 2017). These effects were carried through to biscuits made with the malted pearl millet which had higher protein content compared to biscuits made with the flour from native unmalted pearl millet. The malted pearl millet biscuits also had higher essential amino acid and mineral content (Adebiyi et al., 2017).

3.2.2 Effect on anti-nutrients

Makokha et al. (2002) observed that malting decreased phytic acid in finger millet by 23.9% and 45.3% after 72 and 96 hours, respectively. Malting also decreased phytate content possibly due to action of phytase enzymes activated by the malting process, and a significant increase in α-amylase activity was observed during malting (Traore et al., 2004). Malting decreased phytic acid, tannin and oxalate content by 45%, 46% and 29%, respectively (Hejazi and Orsat, 2016), in finger millet.

3.2.3 Effect on digestibility and bioaccessibility of nutrients

Archana et al. (2001) reported that malting significantly improved in vitro protein (by 14–26%) and starch (by 86–112%) digestibility of pearl millet grains. This was attributed to hydrolytic action of proteases and amylases as well as breakdown of anti-nutritional factors such as polyphenols and phytate. Platel et al. (2010) reported that malting increased bioaccessibility of iron by nearly fourfold in two varieties of finger millet. Malting also increased bioaccessibility of calcium. The authors attributed this to probable decrease in phytate content due to malting. Malting increased in vitro protein and starch digestibility of foxtail millet (Choudhury et al., 2011). Krishnan et al. (2012) reported that malting led to significant increases in bioaccessible Ca, Fe and Zn, and significant decreases in inhibitory factors, namely phytic acid, polyphenols and dietary fibre in finger millet. Malting of finger millet for 48 h at 30°C increased protein digestibility by about 23% relative to unmalted grain (Hejazi and Orsat, 2016).

The transfer of improved nutritional properties of malted millets to composite foods in which they are included as components is demonstrated by the work of Agrahar-Murugkar et al. (2015) and Syeunda et al. (2021). Agrahar-Murugkar et al. (2015) reported that biscuits made with composite flours containing malted finger millet had better nutritional quality in terms of improved protein and ash (mineral) content compared to biscuits made with refined flours. Syeunda et al. (2021) studied the effect of compositing malted finger millet with precooked cowpea in the formulation of a weaning food on anti-nutrients and nutritional quality. Malting decreased levels of anti-nutrients (polyphenols, tannins and phytate) in the finger millet. Composite flours and porridges containing malted finger millet had higher protein digestibility than composites containing unmalted finger millet.

3.3 Extrusion cooking

Extrusion is a processing technique that involves the application of high temperature under pressure combined with shear to transform food raw materials. The combined effect of high temperature, pressure and shear during extrusion cooks the food through modification of the biopolymers of the raw material by altering their physicochemical properties (Navale et al., 2015). Alterations such as starch gelatinisation, protein denaturation, degradation of vitamins and pigments, complexation of amylose and lipids, as well as inactivation of enzymes and anti-nutritional compounds occur during extrusion (Altan and Maskan, 2011). The ability of modern extruders to produce large volumes of product, their versatility and low cost of production per unit product has made extrusion cooking increasingly popular.

Several authors have indicated changes in the nutritional properties of cereal products following extrusion. The effects of extrusion cooking on the nutritional quality of millets are summarised in Table 3. Extrusion cooking increases the in vitro protein digestibility of millets (Malleshi et al., 1996; Onyango et al., 2004). Also, extrusion cooking increases in vitro soluble and insoluble protein digestibility (Onyango et al., 2005). The increase in in-vitro protein digestibility is attributed to the modification of protein structure (which exposes enzyme active sites), degradation of enzymatic inhibitors and decrease in starch–protein complexes (Cisse et al., 1998; Dahlin and Lorenz, 1993). In vitro availability of iron in pearl millet-based weaning foods was reported to be significantly higher in extruded products than in the roasted products by Cisse et al. (1998). This was proposed to be as a result of reduction of anti-nutritional compounds such as polyphenolic compounds. Thermal inactivation of protease inhibitors and anti-physiological factors also contributes to improved protein digestibility (Björck and Asp, 1983). Individual amino acids are reduced following extrusion cooking, and this is attributed to Maillard reaction (Onyango et al., 2005).

In-vitro starch digestibility is increased following extrusion cooking and this is attributed to the hydration, loss of structural integrity and partial solubilisation of starch molecules (Björck and Asp, 1983; García-Alonso et al., 1999; Urooj and Puttaraj, 1994). The shearing off of branches on amylopectin molecules during extrusion increases their accessibility to amylolytic enzymes thus improving starch digestibility (Camire, 2001). Similarly, water-soluble starch, a high-molecular weight polysaccharide is increased as a result of extrusion (Onyango et al., 2004). The disruption of the granular structure of starch and macromolecular chain splitting that occurs during extrusion accounts for the increase in the water-soluble starch (Valle et al., 1994).

The decrease in non-starch polysaccharides (NSP) following extrusion has been reported (Onyango et al., 2005). This is attributable to the disruption of

Table 3 Effects of extrusion on the nutritional quality of millets

Crop	Conditions	Outcomes	Reference
Finger millet, wheat	Refined wheat flour was substituted with extruded finger millet at 10%, 20% and 30%.	Substitution of refined wheat flour with extruded finger millet at 20% is optimal. It improves the total phenolic and antioxidant properties of composite bread while improving the polyphenolic content and antioxidant properties of the bread.	(Patil et al., 2016a)
Pearl millet, cowpea, peanut, peanut oil, skimmed milk powder and sugar	Pearl millet was processed by either roasting or extruding before milling into flour. Peanut and cowpea were roasted and milled before incorporating them into the pearl miller at ratios of pearl millet: peanut: cowpea: sugar: peanut oil: skimmed milk powder being 65:15:10:10:0:0 and 65:15:9:5:2:4.	• The iron content of the extruded formulations had a lower iron content when compared to the roasted one; however, the iron in the extruded formulation was more bioavailable in vitro. • Extruded products had a higher in-vitro protein digestibility.	(Cisse et al., 1998)
Finger millet, wheat	Finger millet flour was incorporated in refined whole wheat flour at 30%, 40% and 50%.	• With increasing finger millet percentage, the protein content of the noodles was lower since the finger millet has a low protein content. • Optimal substitution was 30%, and the resulting noodles exhibited a lower glycaemic response when compared to the reference noodles made with 100% refined wheat flour, indicating possible application in diabetic meal formulations. The reduced glycaemic response was attributed to the increased dietary fibre content with the addition of finger millet.	(Shukla and Srivastava, 2014)

Millet, soybean	The millet and soybean were malted before being mixed in proportions of 70:30 and 50:50, millet: soybean. The composites were then extruded.	Extruded malted millet-soybean composite at 70:30 showed enhanced protein content and protein quality expressed as amino acid content, and mineral content.	(Obatolu, 2002)
Pearl millet, soybean	Defatted soybean flour was incorporated in pearl millet at ratios of 10%, 20% and 30%. The flours were separated into two batches, and one was extruded and the second batch was fermented before extrusion. That is, fermented and extruded, and extruded without fermentation.	• Following extrusion, there was a significant increase in the in vitro protein digestibility and in vitro starch digestibility. • The increase in in-vitro protein and in-vitro starch digestibility corresponded with a substantial reduction in the tannin content, phytic acid content, saponin content and trypsin inhibitor. These reductions account for the increase in the protein and starch digestibility.	(Obaroakpo et al., 2016)
Finger millet, maize	Maize and finger millet flours were mixed in a ratio of 1:1. The mixed flours were then subject to either fermentation, fermentation and cooking, extrusion and fermentation or extrusion.	• Extrusion caused a reduction in the protein content, and this was attributed to Maillard's reaction. • The fat content was also reduced, and this attributed to the formation of starch–lipid complexes resistant to lipid extraction techniques. • Extrusion cooking also caused a significant reduction in the level of individual amino acids, and this was attributed to enzymatic browning. • In vitro protein digestibility was increased following extrusion of the flours. • In vitro starch digestibility was improved following extrusion.	(Onyango et al., 2004)

(Continued)

Table 3 (*Continued*)

Crop	Conditions	Outcomes	Reference
Finger millet, maize	Maize and finger millet flours were mixed in a ratio of 1:1. One batch was acidified by adjusting the moisture content of the blend to 19 g/100 g by pipetting appropriate amounts of 0.1 mol/L, 0.5 mol/L and 1.0 mol/L food-grade citric acid monohydrate or lactic acid. A second batch of the blend was subject to back slopped fermentation. The treated samples were then either cooked or extruded.	• In vitro starch, soluble and insoluble protein digestibility was increased following extrusion cooking. • There was a decrease in the amount of individual amino acids following extrusion. • There was a decrease in the NSP content of the product following extrusion. • There was a decrease in the insoluble dietary fibre content and a corresponding increase in the soluble dietary fibre content of the unfermented *uji* following extrusion. • Contrary to the unfermented *uji*, soluble dietary fractions decreased in the total NSP following extrusion of the fermented *uji* and acidified *uji*. • Resistant starch was formed in extruded unsoured *uji*, but fermented and acidified *uji* seemed to have lower resistant starch content.	(Onyango et al., 2005)
Pearl millet, soybean, Bengal gram and green gram	Decorticated pearl millet flour was composited with defatted soybean flour (15%) in one formulation and Bengal gram and green gram (20% and 10%, respectively) in a second formulation.	• Extruded products exhibited improved in vitro carbohydrate solubility when compared to the control (decorticated pearl millet). • The extrusion also enhanced the in vitro protein digestibility. • There was an increase in both soluble and insoluble dietary fibre following extrusion.	(Sumathi et al., 2007)

Finger millet, pearl millet, mung bean and skimmed milk powder	Pearl millet and finger millet flours (60% of each) were blended with toasted mung bean flour (30%) and skimmed milk powder (10%).	• There was a consistent increase in the total, soluble and insoluble dietary fibre following extrusion. • Amino acid lysine was most affected by extrusion cooking; other amino acids were not significantly affected. • In vitro protein digestibility was enhanced by extrusion cooking.	(Malleshi et al., 1996)
Finger millet, soybean	Decorticated soybean flour was incorporated in the finger millet flour at 10%, 15% and 20% ratios. Carrot juice was added through an inbuilt piston-type pump to the extruder barrel to maintain the dough moisture at 18% throughout the extrusion process.	• Extrusion temperature and screw speed did not significantly affect the crude protein, iron, calcium and total ash content. • Increasing the extrusion temperature decreased the fat content of the extrudates. • Iron and calcium were not affected by extrusion cooking, but the content of zinc seemed to decrease with increase in screw speed.	(Asefa and Melaku, 2017)
Pearl millet, pigeon pea	Pearl millet flour was blended with pigeon pea flour in proportions of 10%, 15%, 18%, 20% and 25% pigeon pea content	• There was a decrease in fat content following extrusion in some of the blended gruels, which was attributed to high screw speed. • Iron and zinc content seemed to increase following extrusion cooking.	(Nwezeh and Ndaliman, 2018)
Finger millet	Three finger millet varieties of varying amylose content (low, medium and high amylose content)	Increasing extrusion temperature and moisture during extrusion increased resistant starch formation while an increase in screw speed decreased the resistant starch formation.	(Nishani et al., 2017)
Pearl millet, cowpea	Varying composites of pearl millet and cowpea were subjected to varying extrusion conditions of extrusion screw speed and moisture content.	Increase in the screw speed, and moisture content slightly decreased the protein content of the composites.	(Filli et al., 2012)
Finger millet	Extrusion of finger millet flour at varying extrusion conditions.	Extrusion resulted in significant reduction in total phenolic content, total flavonoid content and antioxidant activity.	(Patil et al., 2016b)

glycosidic bonds and weak linkages between polysaccharide chains of fibre polysaccharides. However, high NSP has been reported following extrusion cooking, and this is dependent on the processing conditions, raw materials and the enzyme system employed in starch solubilisation (Björck et al., 1984). While amyloglucosidase hydrolyses α-(1–4) and α-(1–6) bonds of amylose and amylopectin, enzyme systems combining α-amylase and amyloglucosidase (Björck et al., 1984) or α-amylase and pullulanase (Onyango et al., 2005) are much more effective in hydrolysis of NSP.

Extrusion cooking can also lead to the formation of resistant starch in millets (Nishani et al., 2017; Onyango et al., 2005). Formation of resistant starch is associated with amylose retrogradation (Englyst et al., 1992; Shamai et al., 2003) resulting into formation of enzyme-resistant amylose–amylose linkages. However, following the classification of resistant starch of several raw and processed starch-rich foods, values ranging from less than 1% (i.e. negligible) to more than 15% (i.e. very high) have been classified as negligible and very high, respectively, by Goñi et al. (1997). Therefore, the resistant starch formed in millets following extrusion cooking is negligible (0.3–0.5%) (Onyango et al., 2005).

Increase in extrusion temperatures has been reported to decrease the fat content of millet (Asefa and Melaku, 2017). This decrease is attributed to the vaporisation of short-chain fatty acids as the extruded product exits the die. It has been reported that extrusion conditions (temperature, screw speed and die diameter) have no effect on the mineral content of extrudates (Asefa and Melaku, 2017; Nwezeh and Ndaliman, 2018), hence the increase in mineral content following extrusion could be as a result of contamination from the extrusion screws and barrel.

Increase in the soluble dietary fibre fraction after extrusion is associated with solubilisation of some insoluble dietary fibre fractions, disruption of lignocellulose links in the cell walls and disintegration of larger molecules of fibre during thermal treatment resulting in the formation of low-molecular weight, soluble constituent fragments such as arabinose, xylose, galactose and glucose (Björck et al., 1984; Fornal et al., 1987; Periago et al., 1996; Periago et al., 1997). Extrusion under acidic conditions facilitates the conversion of soluble dietary fibre to insoluble dietary fibre by polymerisation of the short-chain fibre fragments to form large insoluble complexes or Maillard compounds that are consequently analysed as lignin (Camire, 2001). Increase in soluble dietary fibre could also be attributed to the generation of highly reactive anhydro-compounds (1,6-anhydrosaccharides) which react with starch or fragmented starch through transglycosidation reactions to form new branched glucans (Theander and Westerlund, 1987; Vasanthan et al., 2002). These new branch glucans are resistant to amylase hydrolysis and could account for an increase in soluble dietary fibre content.

Anti-nutrients such as phenolic compounds in millets are adversely affected by extrusion cooking. Onyango et al. (2005) reported reduction in tannin content following extrusion, and Patil et al. (2016b) observed reduction in total phenolic content and total flavonoid content. Phenolic compounds could undergo thermal degradation, as well as changes in their chemical reactivity or the form of insoluble complexes that reduce their extractability (Alonso et al., 2000). Reduction in phytate content following extrusion of millets has also been reported (Obaroakpo et al., 2016). This reduction as a result of extrusion could be attributed to hydrolysis of inositol hexaphosphate to penta-, tetra- and triphosphates (Alonso et al., 2001).

3.4 Use of compositing millets with other nutrient-rich foodstuffs

Although millets are nutritionally comparable to cereals such as wheat, barley and rice in terms of carbohydrates, proteins and minerals (Devi et al., 2014), the utilisation of millet is still low and mostly confined to the lower economic class (Jaybhaye et al., 2014). Like most cereals, the protein quality of millets is low due to their relatively low levels of the limiting amino acid lysine (Mbithi-Mwikya et al., 2000) coupled with the presence of anti-nutritional compounds such as phytate, phenolic compounds and trypsin inhibitors (Devi et al., 2014) that confound nutrients and limit their bioavailability (Taylor, 2016). Due to reliance on cereal staples, countries within the sub-Saharan African region and several other developing countries are still grappling with protein-energy malnutrition (PEM) (Temba et al., 2016) and micronutrient deficiencies are a major problem in developing countries (van der Merwe et al., 2019).

To address the nutritional deficiencies of millets, several approaches including incorporation of legumes in millet flours have been exploited (Table 4). Legumes are rich in protein and essential amino acids, but also lacking in sulphur-containing amino acids (Temba et al., 2016). A combination of cereals with legumes would improve the protein and nutrient density of the subsequent food products (Temba et al., 2016). The reliance of most developing countries on cereals such as millets places these grains in a strategic position as vehicles for tackling mineral deficiencies in these countries, and food-to-food fortification has been explored as a method of improving mineral quality of millet (Table 4). Fortification using locally available, micronutrient-dense foods could ensure sustainable, and adequate micronutrient nutrition (Kruger et al., 2015). Due to the popularity of baked products such as bread and biscuits, they can serve as excellent sources of minerals, especially calcium, iron and zinc (Agrahar-Murugkar, 2020). However, the mineral balance in the human body cannot just be maintained by intake of a mineral-rich food source, but more importantly, by the amount of available minerals for absorption (Gaffney-Stomberg et al., 2010).

Table 4 Effect of compositing and food-to-food fortification on the nutritional quality of millets

Materials	Outcome	Product	Reference
Pearl millet, defatted soybean	Increase in in-vitro starch digestibility with increasing content of soybean flour.	Pearl millet-based composite flour	(Obaroakpo et al., 2016)
Finger millet, wheat	• Increasing the percentage of finger millet flour in the formulation increased the total phenolic content and antioxidant activity. • Extruded finger millet flour composites had higher total phenolic content and antioxidant activity than the native finger millet flour composites.	Finger millet-incorporated bread from composite flour	(Patil et al., 2016a)
Pearl millet, cowpea	• Increase in the amount of cowpea increased the protein content but decreased the fat and carbohydrate content. • Increase in cowpea content increases the lysine, tyrosine, phenylalanine, leucine, valine and leucine content of the composite.	Millet-cowpea-based *fura*	(Filli et al., 2012)
Finger millet, soybean	Increase in the soybean content increased the protein and fat content but did not significantly affect the iron, calcium and total ash content.	Nutritious complementary cereal-legume-vegetable food prepared by extrusion	(Asefa and Melaku, 2017)
Finger millet, wheat	• With increasing content of finger millet flour replacing the wheat flour, crude fibre, crude fat, total ash, carbohydrate, physiological energy, insoluble dietary fibre, soluble dietary fibre, total dietary fibre iron and calcium was increased. • The protein, amylose, amylopectin and starch content were, however, decreased. • Protein fractions of albumin, glutelin and prolamin were reduced with increasing finger millet content while the globulin content increased.	Nutritious finger millet flour-incorporated noodles with hypoglycaemic effect	(Shukla and Srivastava, 2014)

Pearl millet, pigeon pea	• Increasing the pigeon pea content of the flour blend increased the protein content while fat content was decreased. • Increase in pigeon pea content slightly but significantly reduced the calcium, magnesium and phosphorus content of the products.	An instant extruded gruel from a blend of pearl millet and pigeon pea	(Nwezeh and Ndaliman, 2018)
Pearl millet, maize, common bean, soybean and sugar	• The addition of soybean and common bean to the flours resulted in higher protein content and total ash content. • Total sugars, carbohydrates and starch content significantly decreased with the replacement of finger millet with the legumes. • Soybean addition to the flour blend caused an increase in fat content while common bean resulted in a decrease in fat content.	Pearl millet-based complementary food	(Kouakou et al., 2013)
Pearl millet, moringa seed, fenugreek seeds	• Addition of defatted moringa seed flour to pearl millet flour significantly increased the total ash, crude protein and crude fibre content but significantly decreased the crude fat, carbohydrate and total energy content. • Addition of defatted fenugreek seed flour to pearl millet flour significantly increased the total ash, crude protein, crude fibre, crude fat, carbohydrate and total energy content.	Nutrient-dense supplemented pearl millet flour	(Azhari et al., 2017)
Finger millet, wheat	• Incorporation of finger millet flour in refined wheat flour at a ratio of 1:1 and 2:1 (millet:wheat) produced bread with higher crude fibre, calcium, carbohydrate, B group vitamins, folic acid and total energy than bread from refined wheat flour. • Addition of finger millet slightly reduced the crude fat and protein content.	Finger millet-based composite bread	(Mythrayee and Pavithra, 2017)

(Continued)

Table 4 *(Continued)*

Materials	Outcome	Product	Reference
Pearl millet, wheat, horse gram, guar gum	• Refined wheat flour, pearl millet and horse gram were mixed in a ratio of 50:30:20, and varying proportions of guar gum (0.1–0.4%) were added to the cookie formulation. • Increase in the guar gum content increased the total ash, protein, fat and crude fat content but reduced the carbohydrate content of the cookies.	Pearl millet-based composite cookies	(Kukade et al., 2017)
Finger millet, wheat, amaranth seeds, soybean	• Gluten-free composite consisted of finger millet:sprouted amaranth:sprouted soy flours in a ratio of 85:25:10, while the composite flour ingredients were malted finger millet:whole wheat:sprouted amaranth:sprouted soy flours in a ratio of 50:40:20:10. • The protein and fat content of the composite cakes was lower than the control made from traditional cake formulation, while the phosphorous and total ash content was higher in the composite cakes.	Gluten- and egg-free finger millet-based cake	(Agrahar-Murugkar et al., 2018)
Pearl millet, peanuts	• Formulations of defatted peanut:pearl millet flours of 60:40, 40:60 and 50:50 were made, and a control of 100% refined wheat flour was also made. • There was no significant difference in carbohydrate content of the composite biscuits and the control, but increase in peanut flour caused a decrease in the carbohydrate content of the resulting biscuits. • Increasing peanut flour did not significantly affect the fat content but increased the protein and ash content.	Nutritious gluten-free biscuit from peanut-pearl millet composite flour	(Alhassan et al., 2019)
Pearl millet, soybean	Increasing the soybean flour content of the pearl millet blend increased the protein, crude fat, total ash and crude fibre content of the blends while decreasing the carbohydrate content	Millet-based complementary flours	(Ojokoh and Bello, 2014)

Finger millet, wheat	Incorporation of finger millet into wheat flour increased the total phenolic content, antioxidant activity and phytate content of the resulting chapattis	Finger millet-based flatbread (chapatti)	(Sharma et al., 2017)
Wheat, proso millet	Total ash content increased with increasing content of proso millet	Bread fortified with proso millet	(Gavurníková et al., 2011)
Finger millet, wheat	Total ash content was increased with increasing finger millet content in the muffin formulation while the dry gluten and protein content decreased.	Finger millet-based muffins	(Rajiv et al., 2011)
Pearl millet, defatted moringa seeds and fenugreek seeds	• Addition of defatted moringa seed flour and fenugreek seed flour to pearl millet increased the total phenolic content and phytate content of the millet blended flours. Addition of defatted moringa seed flour insignificantly decreased the tannin content of the pearl millet-blended flours while fenugreek seed flour did not affect the tannin content. • Incorporation of defatted moringa seed and fenugreek seed flours resulted in increased both protein content and in vitro protein digestibility.	Supplemented pearl millet flour	(Nour et al., 2015)
Finger millet, sorghum, wheat and oat	• High pressure-treated and native finger millet flour incorporated in high pressure-treated and native wheat flour, respectively, significantly increased the total ash, bioaccessible phenols and bound phenolic content and insignificantly increased the antioxidant activity, soluble and insoluble dietary fibre content. • The treatments also resulted in a significant decrease in free phenols and flavonoids and insignificant decrease in protein and digestible starch content. The β-glucan content was not affected.	Millet-based composite bread	(Angioloni and Collar, 2012)

(Continued)

Table 4 (*Continued*)

Materials	Outcome	Product	Reference
Finger millet, foxtail millet, wheat	• Incorporation of finger millet and foxtail millet in wheat flour for breadmaking increased the protein content, crude fibre and mineral content when compared to the control, with a decrease in carbohydrate and an increase in phytate and tannins. • Millet composite breads had low glycaemic response when compared to conventional wheat bread used as a standard due to their high fibre content.	Millet-based composite bread	(Chhavi and Sarita, 2012)
Pearl millet, moringa leaf, roselle calyx, baobab fruit pulp, mango, carrot, sunflower oil	• Fortification of pearl millet flour with moringa leaf powder, roselle calyces or baobab fruit pulp resulted in increased total and percentage bioaccessible iron and zinc content • Fortification also improved the total iron, zinc and calcium content. • Inclusion of mango, carrot and sunflower oil in pearl millet porridge as pro-vitamin A sources improved the percentage of iron and zinc bioaccessibility. • At high content (30:100 g), moringa leaf powder resulted in a significant drop in total and percentage of bioaccessible iron and zinc content. • Moringa leaf powder also increased phenolic acids and phytate content, while baobab fruit pulp increased phytate content, phenolic acid content and tannin content.	Pearl millet porridge fortified with plant foodstuff	(van der Merwe et al., 2019)
Finger-millet, skimmed milk powder, sesame seeds, cumin seeds, fresh moringa leaves	Several formulations were made to include sprouted finger millet, sprouted sesame seeds, moringa leaves and refined wheat flours.	Bread and biscuits fortified with plant foodstuff	(Agrahar-Murugkar, 2020)

Finger millet, wheat, soybean, mint leaf	The refined wheat flour was used as filler. Finger millet:soybean: mint leaf:refined wheat flour, 50:20:10:20, 60:20:10:10, 70:20:10:0. Two controls were formulated as shown below: Finger millet:refined wheat flour:Fe2SO3, 70:30:0.1337, 0:100:0.1468.	Iron-rich finger millet-based biscuits	(Singh and Srivastava, 2008)
Pearl millet, baobab fruit, moringa leaf	Fortification of pearl millet flour with moringa leave powder increased iron, magnesium, calcium and phosphorus content but did not affect the aluminium content. Fortification with baobab fruit pulp powder decreased the iron content but did not significantly affect the magnesium, calcium, zinc and aluminium content. Fortification with baobab fruit pulp powder increased both the amount and percentage of bioaccessible iron but did not affect the amount and percentage of bioaccessible zinc. Fortification with moringa increased the amount of bioaccessible iron but reduced the percentage bioaccessibility of zinc and iron as well as the amount of bioaccessible zinc	Pearl millet porridge fortified with plant foodstuff	(Adetola et al., 2019)
Finger millet, wheat	Fortification with finger millet increased the total ash, crude fibre and calcium content while decreasing the crude protein, energy and carbohydrate of the resulting bread.	White bread fortified with finger millet flour	(Devani et al., 2016)
Millet, wheat	• Substitution of wheat with millet at varying percentages significantly increased total ash and fat content but did not significantly affect the protein and carbohydrate content of the flours. • The biscuits contained substantially lower protein content than the control while the carbohydrate, total ash and fat content were not affected.	Millet-based cookies	(Chandi et al., 2015)
Finger millet, red kidney bean, wheat	Incorporation of malted finger millet did not significantly affect the total ash, protein and carbohydrate content but increased the dietary fibre and fat content considerably.	Nutritious bread made with composite flour-containing millet	(Bhol and Bosco, 2014)

(Continued)

Table 4 (*Continued*)

Materials	Outcome	Product	Reference
Finger millet, foxtail millet, little millet, proso millet, wheat	• Inclusion of finger millet caused a significant reduction in protein content while all other millets did not affect. • Crude fat, dietary fibre, crude fibre content were increased with the incorporation of millet. • All millets increased the phytic acid content except for barnyard millet that had no effect. • Inclusion of all millets resulted in lower damaged starch content. • Millet flours lowered the rapidly digestible starch and increased the slowly digestible and resistant starch.	Unleavened flatbread (chapatti) from composite flours	(Sharma and Gujral, 2019)
Finger millet, wheat	Incorporation of finger millet increased the protein, total ash, total fat and fibre content while reducing the carbohydrate content compared to the refined wheat cookies.	Cookies supplemented with oat and finger millet	(Giram et al., 2017)
Pearl millet, cashew nut, carrot	Incorporation of cashew nut and carrot powder into pearl millet flour increased the crude fibre, crude fat, total ash, protein and β-carotene content but decreased the carbohydrate content.	Pearl millet-based biscuits with cashew nut paste and carrot flour	(Omachi and Yusufu, 2017)
Millet grains, pumpkin seeds, carrots, cowpea leaves and skimmed milk powder	Incorporation of skimmed milk powder in the finger millet-based formulation increased the protein and energy content.	Nutrient-enhanced millet-based composite flour	(Tumwine et al., 2019)
Pearl millet, African breadfruit	Incorporation of African breadfruit flour in malted pearl millet at different percentages resulted in increased protein, total fat, total ash calcium, crude fibre, phosphorous, magnesium, iron and potassium content with a decrease in carbohydrate content	• Breads produced from millet, African breadfruit composite flours	(Okoye et al., 2019)

Kodo millet, soybean, wheat	• The protein, fibre, amylose/amylopectin ratio of dosa were found to be increased significantly with increase in millet flour blend level from 10% to 30%. Whereas, the total carbohydrate, total sugar and starch content were markedly decreased while increasing the level of millet flour blend. • Incorporation of kodo millet decreased the glycaemic index and glycaemic load.	Dosa (Indian rice pancake) from millet flour blend-incorporated composite flour	(Vijayakumar and Mohankumar, 2011)
Little millet, wheat	Inclusion of little millet in bread formulation resulted in an increase in the percentage of micronutrients such as iron, zinc, copper, phosphorus and fibre.	Little millet-incorporated functional bread	(Mannuramath et al., 2015)

Bioaccessibility has been defined as the fraction of a compound that is released from the food matrix into the gastrointestinal tract, making it available for intestinal absorption (Benito and Miller, 1998). It has been advocated that a combined approach of conventional food fortification and food-to-food fortification using locally available, micronutrient-dense foods is needed to ensure sustainable growth and development (Agrahar-Murugkar, 2020). Thus, tropical foodstuffs that are high in essential minerals, for example, moringa leaves (Glover-Amengor et al., 2017) and high in organic acids which are enhancers of mineral bioavailability (ascorbic acid and citric acid), for example, baobab fruit (Tembo et al., 2017) can be added to millet-based foods to improve their mineral nutritional quality. These organic acids chelate iron through binding with carboxyl and hydroxyl groups, and thus, increase the solubility and iron availability (Teucher et al., 2004). Various pro-vitamin A carotenoids have also been reported to positively impact iron absorption from cereal-based meals (García-Casal et al., 1998). Effects of incorporation of different crops on the mineral quality of millets have been explored (Table 4).

While there is a growing demand for wheat-based products in developing countries (Akhtar et al., 2011), there is an increasing demand by consumers for products with lower traditional bakery items such as eggs, refined wheat flour and butter (Agrahar-Murugkar et al., 2016). This has in turn increased the demand for more variety and novelty in cereal foods, in particular healthy foods (Schoenlechner et al., 2013). Studies have indicated the positive outcomes of consuming millets, such as increased insulin sensitivity and reduced triglyceride levels in diabetic mice fed on millet protein (Nishizawa et al., 2009). Moreover, due to increase in demand for gluten-free products by consumers as a result of health complications like celiac disease, there is a growing need to find alternatives for wheat flour in baked products (Agrahar-Murugkar et al., 2018; Alhassan et al., 2019; Chhabra et al., 2017). Composite flours of different millets are one possibility to increase the nutritional value of the final bread products as explored by different authors (Table 4). Other products such as pasta (Shukla and Srivastava, 2014), *kibbeh* (Middle East ground meat) (Brasil et al., 2015) and energy bars (Sobana, 2017), among others, have also been explored.

4 Conclusion and future trends

Millets are an important cereal staple for millions of people particularly in Africa and Asia and will increasingly be relied upon by people in these regions for daily supply of nutrients and to help ensure food security. Traditional methods of processing such as fermentation and malting are used to produce a wide range of traditional food products. These processing methods play an important role in enhancing the nutritional properties of millet-based foods. Modern food-processing technologies such as extrusion cooking are also finding increasing use in producing nutritionally enhanced millet-based food products. Given the current rise in urbanisation especially in sub-Saharan African countries and its effect on consumer demands for convenient food choices, the use of such modern processing technologies for production of various millet-based food products to meet this demand will continue to increase. It is important that the nutritional quality of these products is used to inform how the modern technologies will be applied. Compositing millets with other grains such as legumes for enhanced protein quality has been in practice for a long time. In relation to this, an emerging frontier is the application of the concept of food-to-food fortification where millets are fortified with other plant foodstuffs that are rich in micronutrients as well as micronutrient bioavailability enhancers. This presents an affordable and efficient strategy to enhance the nutritional quality of millet-based food products to combat micronutrient deficiencies.

5 Where to look for further information

As mentioned earlier, the concept and practice of food-to-food fortification (combining micronutrient-rich plant foodstuffs with staple cereal crops like millets for enhanced nutritional quality) is gaining traction and increased research interest. In this regard, the following recent reviews are very informative:

- Chadare, F. J., Idohou, R., Nago, E., Affonfere, M., Agossadou, J., Fassinou, T. K., Kénou, C., Honfo, S., Azokpota, P., Linnemann, A. R. and Hounhouigan, D. J. (2019). Conventional and food-to-food fortification: An appraisal of past practices and lessons learned. *Food Science and Nutrition*, 7, 2781-2795. DOI: 10.1002/fsn3.1133.
- Teye, E. Deha, C. I., Dadzie, R. and MacArthur, R. L. (2020). Delivering the nutritional needs by food to food fortification of staples using underutilised plant species in Africa. *International Journal of Food Science*, 2020, 8826693. DOI: 10.1155/2020/8826693.
- Kruger, J., Taylor, J. R. N., Ferruzzi, M.G. and Debelo, H. (2020). What is food-to-food fortification? A working definition and framework for evaluation of efficiency and implementation of best practices. *Comprehensive Reviews in Food Science and Food Safety*, 19, 3618–3658, DOI: 10.1111/1541-4337.12624.

Other useful review articles which deal with the effect of processing on nutritional value of cereals include:

- Oghbaei, M. and Prakash, J. (2016). Effect of primary processing of cereals and legumes on its nutritional quality: A comprehensive review. *Cogent Food & Agriculture*, 2(1), 1136015. DOI: 10.1080/23311932.2015.1136015.

- Moreno, C. R., Fernández, P. C. R., Rodríguez, E. O. C., Carrillo, J. M. and Rochín, S. M. (2018). Changes in nutritional properties and bioactive compounds in cereals during extrusion cooking. *Extrusion of Metals, Polymers and Food Products*, pp. 104–124. DOI: 10.5772/intechopen.68753.

6 References

Adebiyi, J. A., Obadina, A. O., Adebo, O. A. and Kayitesi, E. (2017). Comparison of nutritional quality and sensory acceptability of biscuits obtained from native, fermented, and malted pearl millet (Pennisetum glaucum) flour, *Food Chem.* 232, 210–217.

Adebiyi, J. A., Obadina, A. O., Adebo, O. A. and Kayitesi, E. (2018). Fermented and malted millet products in Africa: expedition from traditional/ethnic foods to industrial value-added products, *Crit. Rev. Food Sci. Nutr.* 58(3), 463–474.

Adebo, O. A. (2020). African sorghum-based fermented foods: past, current and future prospects, *Nutrients* 12(4), 1111. doi: 10.3390/nu12041111.

Adetola, O. Y., Kruger, J., White, Z. and Taylor, J. R. N. (2019). Comparison between food-to-food fortification of pearl millet porridge with Moringa leaves and baobab fruit and with adding ascorbic and citric acid on iron, zinc and other mineral bioaccessibility, *LWT* 106, 92–97.

Agrahar-Murugkar, D. (2020). Food to food fortification of breads and biscuits with herbs, spices, millets and oilseeds on bio-accessibility of calcium, iron and zinc and impact of proteins, fat and phenolics, *LWT* 130, 109703.

Agrahar-Murugkar, D., Gulati, P., Kotwaliwale, N. and Gupta, C. (2015). Evaluation of nutritional, textural and particle size characteristics of dough and biscuits made from composite flours containing sprouted and malted ingredients, *J. Food Sci. Technol.* 52(8), 5129–5137.

Agrahar-Murugkar, D., Zaidi, A. and Dwivedi, S. (2018). Development of gluten free eggless cake using gluten free composite flours made from sprouted and malted ingredients and its physical, nutritional, textural, rheological and sensory properties evaluation, *J. Food Sci. Technol.* 55(7), 2621–2630.

Agrahar-Murugkar, D., Zaidi, A., Kotwaliwale, N. and Gupta, C. (2016). Effect of egg-replacer and composite flour on physical properties, color, texture and rheology, nutritional and sensory profile of cakes, *J. Food Qual.* 39(5), 425–435.

Ajiboye, T. O., Iliasu, G. A., Adeleye, A. O., Abdussalam, F. A., Akinpelu, S. A., Ogunbode, S. M., Jimoh, S. O. and Oloyede, O. B. (2014). Nutritional and antioxidant dispositions of sorghum/millet-based beverages indigenous to Nigeria, *Food Sci. Nutr.* 2(5), 597–604.

Aka, S., Konan, G., Fokou, G., Dje, K. M. and Bassirou, B. (2014). Review on African traditional cereal beverages, *Am. J. Res. Comm.* 2(5), 103–153.

Akhtar, S., Anjum, F. M. and Anjum, M. A. (2011). Micronutrient fortification of wheat flour: recent development and strategies, *Food Res. Int.* 44(3), 652–659.

Akinola, S. A., Badejo, A. A., Osundahunsi, O. F. and Edema, M. O. (2017). Effect of preprocessing techniques on pearl millet flour and changes in technological properties, *Int. J. Food Sci. Technol.* 52(4), 992–999.

Alavi, S., Mazumdar, S. D. and Taylor, J. R. N. (2019). Modern convenient sorghum and millet food, beverage and animal feed products and their technologies. In: Taylor,

J. R. N. and Duodu, K. G. (Eds), *Sorghum and Millets: Chemistry, Technology and Nutritional Attributes*, London, UK: AACC International and Elsevier, pp. 293-329.

Alhassan, M. W., Ojangba, T. and Amagloh, F. K. (2019). Development of gluten-free biscuit from peanut-pearl millet composite flour, *Am. J. Food Sci. Technol.* 7(2), 40–44.

Alonso, R., Aguirre, A. and Marzo, F. (2000). Effects of extrusion and traditional processing methods on antinutrients and *in vitro* digestibility of protein and starch in faba and kidney beans, *Food Chem.* 68(2), 159–165.

Alonso, R., Rubio, L. A., Muzquiz, M. and Marzo, F. (2001). The effect of extrusion cooking on mineral bioavailability in pea and kidney bean seed meals, *Anim. Feed Sci. Technol.* 94(1–2), 1–13.

Altan, A. and Maskan, M. (2011). Development of extruded foods by utilizing food industry by-products. In: Altan, A. and Maskan, M. (Eds), *Advances in Food Extrusion Technology*, USA: CRC Press LLC Press, pp. 121–160.

Amadou, O., Mahamadou, E. G. and Le, G. W. (2013). Millets, nutritional composition, some health benefits and processing–review, *Food Sci. Technol. (Campinas)* 25(7), 501–508.

Angioloni, A. and Collar, C. (2012). Effects of pressure treatment of hydrated oat, finger millet and sorghum flours on the quality and nutritional properties of composite wheat breads, *J. Cereal Sci.* 56(3), 713–719.

Archana, Sehgal, S. and Kawatra, A. (2001). *In vitro* protein and starch digestibility of pearl millet (Pennisetum glaucum L.) as affected by processing techniques, *Nahrung* 45(1), 25–27.

Asefa, Boru and Melaku, E. T. (2017). Evaluation of impact of some extrusion process variables on chemical, functional and sensory properties of complimentary food from blends of finger millet, soybean and carrot, *Afr. J. Food Sci.* 11(9), 302–309.

Azhari, A., Adiamo, O., Awad, R. and Babiker, E. (2017). Changes in chemical composition and total energy as affected by fermentation and/or cooking of pearl millet flour supplemented with Moringa or fenugreek seeds flour, *Int. Food Res. J.* 24(4), 1562.

Balseiro, G., Taron, A. and Garcia-Zapateiro, L. A. (2015). Nutritional properties of different composite flours from maize (*Zea mays*) variety ica v109 and pearl millet (*Pennisetum glaucum*) malted with calcium chloride and gibberellic acid, *Eur. Food Res. Technol.* 240(3), 471–475.

Belton, P. S. and Taylor, J. R. N. (2004). Sorghum and millets: protein sources for Africa, *Trends Food Sci. Technol.* 15, 94–98.

Benito, P. and Miller, D. (1998). Iron absorption and bioavailability: an updated review, *Nutr. Res.* 18(3), 581–603.

Beta, T. and Ndolo, V. U. (2019). Postharvest technologies. In: Taylor, J. R. N. and Duodu, K. G. (Eds), *Sorghum and Millets: Chemistry, Technology and Nutritional Attributes*, London, UK: AACC International and Elsevier, pp. 69–84.

Bhol, S. and Bosco, S. J. D. (2014). Influence of malted finger millet and red kidney bean flour on quality characteristics of developed bread, *LWT* 55(1), 294–300.

Björck, I. and Asp, N.-G. (1983). The effects of extrusion cooking on nutritional value–a literature review, *J. Food Eng.* 2(4), 281–308.

Björck, I., Asp, N.-G., Birkhed, D. and Lundquist, I. (1984). Effects of processing on availability of starch for digestion *in vitro* and *in vivo*; I Extrusion cooking of wheat flours and starch, *J. Cereal Sci.* 2(2), 91–103.

Brasil, T. A., Capitani, C. D., Takeuchi, K. P. and Ferreira, T. A. P. d. C. (2015). Physical, chemical and sensory properties of gluten-free kibbeh formulated with millet flour (Pennisetum glaucum (L.) R. Br.), *Food Sci. Technol.* 35(2), 361–367.

Bultosa, G. and Taylor, J. R. N. (2004). Teff. In: Wrigley, C., Corke, H. and Walker, C. E. (Eds), *Encyclopedia of Grain Science*, Amsterdam: Elsevier, pp. 281–290.

Camire, M. (2001). Extrusion and nutritional quality. In: Guy, R. (Ed), *Extrusion Cooking: Technologies and Application*, USA: Woodhead Publishing Limited and CRC Press LLC, pp.108–122.

Chandel, G., Kumar, M., Dubey, M. and Kumar, M. (2014). Nutritional properties of minor millets: neglected cereals with potentials to combat malnutrition, *Curr. Sci.* 107, 1109–1111.

Chandi, G. K., Lok, C. W., Jie, N. Y. and Seetharaman, K. (2015). Functionality of kamut and Millet flours in macro wire cut cookie systems, *J. Food Sci. Technol.* 52(1), 556–561.

Chhabra, N., Kaur, A. and Kaur, S. (2017). Development of composite tortilla chips: an approach with improved quality, *Pharma. Innov.* 6(9), 514.

Chhavi, A. and Sarita, S. (2012). Evaluation of composite millet breads for sensory and nutritional qualities and glycemic response, *Malays. J. Nutr.* 18(1), 89–101.

Choudhury, M., Das, P. and Baroova, B. (2011). Nutritional evaluation of popped and malted indigenous millet of Assam, *J. Food Sci. Technol.* 48(6), 706–711.

Cisse, D., Guiro, A. T., Diaham, B., Souane, M., Doumbouya, N. T. and Wade, S. (1998). Effect of food processing on iron availability of African pearl millet weaning foods, *Int. J. Food Sci. Nutr.* 49(5), 375–381.

Dahlin, K. and Lorenz, K. (1993). Protein digestibility of extruded cereal grains, *Food Chem.* 48(1), 13–18.

Devani, B. M., Jani, B. L., Kapopara, M. B., Vyas, D. M. and Ningthoujam, M. D. (2016). Study on quality of white bread enriched with finger millet flour, *Int. J. Agr. Environ. Biotechnol.* 9(5), 903–907.

Devi, P. B., Vijayabharathi, R., Sathyabama, S., Malleshi, N. G. and Priyadarisini, V. B. (2014). Health benefits of finger millet (Eleusine coracana L.) polyphenols and dietary fiber: a review, *J. Food Sci. Technol.* 51(6), 1021–1040.

Devisetti, R., Yadahally, S. N. and Bhattacharya, S. (2014). Nutrients and antinutrients in foxtail and proso millet milled fractions: evaluation of their flour functionality, *LWT* 59(2), 889–895.

Eltayeb, M. M., Hassan, A. B., Sulieman, M. A. and Babiker, E. E. (2007). Effect of processing followed by fermentation on antinutritional factors content of pearl millet (*Pennisetum glaucum* L.) cultivars, *Pak. J. Nutr.* 6, 463–467.

Elyas, S. H. A., El Tinay, A. H., Yousif, N. E. and Elsheikh, E. A. E. (2002). Effect of natural fermentation on nutritive value and *in vitro* protein digestibility of pearl millet, *Food Chem.* 78(1), 75–79.

Englyst, H. N., Kingman, S. M. and Cummings, J. H. (1992). Classification and measurement of nutritionally important starch fractions, *Eur. J. Clin. Nutr.* 46 (Suppl. 2), S33–S50.

Filli, K., Nkama, I., Jideani, V. A. and Abubakar, U. (2012). The effect of extrusion conditions on the physicochemical properties and sensory characteristics of millet-cowpea based fura, *Eur. J. Nutr. Food Saf.* 2(1), 1–23.

Fornal, Ł, Soral-Smietana, M., Smietana, Z. and Szpendowski, J. (1987). Chemical characteristics and physico-chemical properties of the extruded mixtures of cereal starches, *Starch/Stärke* 39(3), 75–78.

Gaffney-Stomberg, E., Sun, B. H., Cucchi, C. E., Simpson, C. A., Gundberg, C., Kerstetter, J. E. and Insogna, K. L. (2010). The effect of dietary protein on intestinal calcium absorption in rats, *Endocrinology* 151(3), 1071–1078.

Galati, A., Oguntotinbo, F. A., Moschetti, G., Crescimanno, M. and Settanni, L. (2014). The cereal market and the role of fermentation in cereal-based food production in Africa, *Food Rev. Int.* 30(4), 317–337.

Galle, S., Schwab, C., Arendt, E. K. and Gänzle, M. G. (2011). Structural and rheological characterisation of heteropolysaccharides produced by lactic acid bacteria in wheat and sorghum sourdough, *Food Microbiol.* 28(3), 547–553.

Gänzle, M. G., Haase, G. and Jelen, P. (2008). Lactose–crystallisation, hydrolysis and value-added derivatives, *Int. Dairy J.* 18(7), 685–694.

García-Alonso, A., Jiménez-Escrig, A., Martín-Carrón, N., Bravo, L. and Saura-Calixto, F. (1999). Assessment of some parameters involved in the gelatinization and retrogradation of starch, *Food Chem.* 66(2), 181–187.

García-Casal, M. N., Layrisse, M., Solano, L., Barón, M. A., Arguello, F., Llovera, D., Ramírez, J., Leets, I. and Tropper, E. (1998). Vitamin A and β-carotene can improve nonheme iron absorption from rice, wheat and corn by humans, *J. Nutr.* 128(3), 646–650.

Gautam, N. and Sharma, N. (2009). Purification and characterization of bacteriocin produced by strain of Lactobacillus brevis MTCC 7539, *Indian J. Biochem. Biophy.* 46(4), 337–341.

Gavurníková, S., Havrlentová, M., Mendel, Ľ., Čičová, I., Bieliková, M. and Kraic, J. (2011). Parameters of wheat flour, dough, and bread fortified by buckwheat and millet flours, *Polnohospodárstvo* 57(4), 144–153.

Giram, S. J., Agrawal, R. and Shere, D. (2017). Organoleptic and nutritional evaluation of cookies supplemented with oat and finger millet, *Int. J. Pure App. Biosci.* 5(5), 360–365.

Glover-Amengor, M., Aryeetey, R., Afari, E. and Nyarko, A. (2017). Micronutrient composition and acceptability of Moringa oleifera leaf-fortified dishes by children in Ada-East district, Ghana, *Food Sci. Nutr.* 5(2), 317–323.

Goñi, I., Garcia-Alonso, A. and Saura-Calixto, F. (1997). A starch hydrolysis procedure to estimate glycemic index, *Nutr. Res.* 17(3), 427–437.

Hejazi, S. N. and Orsat, V. (2016). Malting process optimization for protein digestibility enhancement in finger millet grain, *J. Food Sci. Technol.* 53(4), 1929–1938.

Jay, J. M., Loessner, M. J. and Golden, D. A. (2005). *Modern Food Microbiology*, 7th ed., New York, USA: Springer, Science and Business Media Inc.

Jaybhaye, R., Pardeshi, I., Vengaiah, P. and Srivastav, P. (2014). Processing and technology for millet based food products: a review, *J. Ready Eat. Food* 1(2), 32–48.

Jideani, V. A., Nkama, I., Agbo, E. B. and Jideani, I. A. (2002). Modelling of visual shelf life of millet (Pennisetum glaucum) dough (fura) as affected by time-temperature and product packaging parameters, *Cereal Chem.* 79(5), 710–714.

Kakade, S. B. and Hathan, B. S. (2014). Finger millet processing–review, *Int. J. Agric. Innov. Res.* 3(4), 1003–1008.

Kouakou, B., N'da-Kouassi, A. M., Halbin, K. J., Tagro, G., N'Guessan, K. F. and Dago, G. (2013). Biochemical characterisation and functional properties of weaning food made from cereals (millet, maize) and legumes (beans, soybeans), *Food Chem. J. Nutr.* 1(1), 22–32.

Krishnan, R., Dhamaraj, U. and Malleshi, N. G. (2012). Influence of decortication, popping and malting on bioaccessibility of calcium, iron and zinc in finger millet, *LWT* 48(2), 169–174.

Kruger, J., Mongwaketse, T., Faber, M., van der Hoeven, M. and Smuts, C. M. (2015). Potential contribution of African green leafy vegetables and maize porridge composite meals to iron and zinc nutrition, *Nutrition* 31(9), 1117–1123.

Kukade, A., Pawar, V., Syed, H. and Sharma, D. (2017). Development and evaluation of organoleptic and microbial quality of guar gum incorporated composite flour cookies, *J. Pharmacog. Phytochem.* 6, 716–720.

Kumar, A., Tomer, V., Kaur, A., Kumar, V. and Gupta, K. (2018). Millets: a solution to agrarian and nutritional challenges, *Agric. Food Secur.* 7, 31–46.

Kumar, S. I., Babu, C. G., Reddy, V. C. and Swathi, B. (2016). Anti-nutritional factors in finger millet, *J. Nutr. Food Sci.* 6(3). doi: 10.4172/2155-9600.1000491.

LeBlanc, J. G., Milani, C., Savoy de Giori, G., van Sinderen, D. and Ventura, M. (2013). Bacteria as vitamin suppliers to their host: a gut microbiota perspective, *Curr. Opin. Biotechnol.* 24, 160–168.

LeBlanc, J. G. and Todorov, S. D. (2011). Bacteriocin producing lactic acid bacteria isolated from Boza, a traditional fermented beverage from Balkan Peninsula–from isolation to application. In: *Science Against Microbial Pathogens: Communicating Current Research and Technological Advances*, Norristown, PA, USA: Formatex Research Centre, Vol. 3, pp. 1311-1320.

Leroy, F. and De Vuyst, L. (2004). Lactic acid bacteria as functional starter cultures for the food fermentation industry, *Trends Food Sci. Technol.* 15(2), 67–78.

Lestienne, I., Buisson, M., Lullien-Pellerin, V., Piq, C. and Treche, S. (2007). Losses of nutrients and anti-nutritional factors during abrasive decortication of two pearl millet cultivars (Pennisetum glaucum), *Food Chem.* 100(4), 1316 -1323.

Liebman, M. and Al-Wahsh, I. A. (2011). Probiotics and other key determinants of dietary oxalate absorption, *Adv. Nutr.* 2(3), 254-260.

Makokha, A. O., Oniang'o, R. K., Njoroge, S. M. and Kamar, O. K. (2002). Effect of traditional fermentation and malting on phytic acid and mineral availability from sorghum (Sorghum bicolor) and finger millet (Eleusine coracana) grain varieties grown in Kenya, *Food Nutr. Bull.* 23(3) (Suppl.), 241-245.

Malleshi, N. G., Hadimani, N. A., Chinnaswamy, R. and Klopfenstein, C. F. (1996). Physical and nutritional qualities of extruded weaning foods containing sorghum, pearl millet, or finger millet blended with mung beans and nonfat dried milk, *Plant Foods Hum. Nutr.* 49(3), 181-189.

Mannuramath, M., Yenagi, N. and Orsat, V. (2015). Quality evaluation of little millet (Panicum miliare) incorporated functional bread, *J. Food Sci. Technol.* 52(12), 8357–8363.

Mbithi-Mwikya, S., Ooghe, W., Van Camp, J., Ngundi, D. and Huyghebaert, A. (2000). Amino acid profiles after sprouting, autoclaving, and lactic acid fermentation of finger millet (Eleusine coracana) and kidney beans (Phaseolus vulgaris L.), *J. Agric. Food Chem.* 48(8), 3081-3085.

Mohan, V. R., Tresina, P. S. and Daffodil, E. D. (2016). Antinutritional factors in legume seeds: characteristics and determination. In: Caballero, B., Finglas, P. M. and Toldra, F. (Eds), *Encyclopedia of Food and Health*, Oxford: Academic Press (Elsevier), pp. 211-220.

Mukisa, I. M., Nsiimire, D. G., Byaruhanga, Y. B., Muyanja, C. M. B. K., Langsrud, T. and Narvhus, J. A. (2010). Obushera: descriptive sensory profiling and consumer acceptability, *J. Sens. Stud.* 25, 190–214.

Murali, A. and Kapoor, R. (2003). Effect of natural and pure culture fermentation of finger millet on zinc availability as predicted from HCL-extractability and molar ratios, *J. Food Sci. Technol.* 40(1), 112–114.

Muthamilarasan, M., Dhaka, A., Yadav, R. and Prasad, M. (2016). Exploration of millet models for developing nutrients rich graminaceous crops, *Plant Sci.* 242, 89–97.

Mythrayee, R. and Pavithra, A. (2017). Comparative study on nutritive content of finger millet-wheat composite bread fermented with lactic acid bacilli and yeast, *IOSR J. Biotechnol. Biochem.* 3(3), 15–21.

Navale, S., Swami, S. B. and Thakor, N. (2015). Extrusion cooking technology for foods: a review, *J. Ready Eat. Food* 2(3), 66–80.

Nishani, S., Rudra, S. G. and Verghese, E. (2017). 'Effect of extrusion processing variables on finger millet flours with respect to their functional properties using response surface methodology, *Current J. Appl. Sci. Technol.* 23(2), 1-9. doi: 10.9734/CJAST/2017/34495.

Nishizawa, N., Togawa, T., Park, K. O., Sato, D., Miyakoshi, Y., Inagaki, K., Ohmori, N., Ito, Y. and Nagasawa, T. (2009). Dietary Japanese millet protein ameliorates plasma levels of adiponectin, glucose, and lipids in type 2 diabetic mice, *Biosci. Biotechnol. Biochem.* 73(2), 351–360.

Notararigo, S., Nacher-Vazquez, M., Ibarburu, I., Werning, M. L., de Palencia, P. F., Duenas, M. T., Aznar, R., Lopez, P. and Prieto, A. (2013). Comparative analysis of production and purification of homo- and hetero-polysaccharides produced by lactic acid bacteria, *Carbohydr. Polym.* 93(1), 57–64.

Nour, A. A. M., Ahmed, I. A. M., Babiker, E. E., Ibrahim, M. and Ahmed, M. (2015). Effect of supplementation and cooking on *in vitro* protein digestibility and antinutrients of pearl millet flour, *Am. J. Food Sci. Health* 1(3), 69–75.

Nwezeh, G. and Ndaliman, M. (2018). Effects of extrusion components on the compositional characteristics of an instant gruel produced from millet-pigeon pea flour blend, *Asian Food Sci. J.* 3(1), 1–11. doi: 10.9734/afsj/2018/v3i1791.

Obadina, A. O., Arogbokun, C. A., Soares, A. O., Piler de Carvalho, C. W. P., Barboza, H. T. and Adekoya, I. O. (2017). Changes in nutritional and physico-chemical properties of pearl millet (Pennisetum glaucum) Ex-Borno variety flour as a result of malting, *J. Food Sci. Technol.* 54(13), 4442–4451.

Obaroakpo, J., Iwanegbe, I. and Ojokoh, A. (2016). The effect of fermentation and extrusion on the anti-nutritional composition and digestibility of millet and soybean flour blends, *Asian J. Biol.* 1(2), 1–10. doi: 10.9734/AJOB/2016/31174.

Obatolu, V. A. (2002). Nutrient and sensory qualities of extruded malted or unmalted millet/soybean mixture, *Food Chem.* 76(2), 129–133.

Ojokoh, A. and Bello, B. (2014). Effect of fermentation on nutrient and anti-nutrient composition of millet (Pennisetum glaucum) and soyabean (Glycine max) blend flours, *J. Life Sci.* 8(27), 3566–3570.

Okoye, J., Ani, E. and Anih, A. (2019). Quality characteristics of breads produced from millet-African bread fruit composite flours, *Am. J. Food Sci. Technol.* 7(6), 169–174.

Omachi, D. O. and Yusufu, P. A. (2017). Physico-chemical, sensory and microbiological assessment of millet based biscuits improved with cashew nuts (Anarcadium Occidentale), carrot flour (Daucus carota), *Am. J. Food Nutr.* 7(1), 13–22.

Onyango, C., Noetzold, H., Bley, T. and Henle, T. (2004). Proximate composition and digestibility of fermented and extruded uji from maize–finger millet blend, *LWT* 37(8), 827–832.

Onyango, C., Noetzold, H., Ziems, A., Hofmann, T., Bley, T. and Henle, T. (2005). Digestibility and antinutrient properties of acidified and extruded maize–finger millet blend in the production of uji, *LWT* 38(7), 697–707.

Osman, M. A. (2004). Changes in sorghum enzyme inhibitors, phytic acid, tannins, and *in vitro* protein digestibility occurring during Khamir (local bread) fermentation, *Food Chem.* 88(1), 129–134.

Owusu-Kwarteng, J., Tano-Debrah, K., Glover, R. L. K. and Akabanda, F. (2010). Process characteristics and microbiology of fura produced in Ghana, *Nat. Sci.* 8(8), 41–51.

Patil, S. S., Rudra, S. G., Varghese, E. and Kaur, C. (2016a). Effect of extruded finger millet (Eleusine coracan L.) on textural properties and sensory acceptability of composite bread, *Food Biosci.* 14, 62–69.

Patil, S. S., Varghese, E., Rudra, S. G. and Kaur, C. (2016b). Effect of extrusion processing on phenolics, flavonoids and antioxidant activity of millets, *Int. J. Food Ferm. Technol.* 6(1), 177.

Pelembe, L. A. M. (2001). Pearl millet malting: factors affecting product quality, PhD Thesis, South Africa: University of Pretoria.

Periago, M. J., Englyst, H. N. and Hudson, G. J. (1996). The influence of thermal processing on the non-starch polysaccharide (NSP) content and *in vitro* digestibility of starch in peas (Pisum sativum, L.), *LWT* 29(1–2), 33–40.

Periago, M. J., Ros, G. and Casas, J. L. (1997). Non-starch polysaccharides and *in vitro* starch digestibility of raw and cooked chick peas, *J. Food Sci.* 62(1), 93–96.

Platel, K., Eipeson, S. W. and Srinivasan, K. (2010). Bioaccessible mineral content of malted finger millet (Eleusine coracana), wheat (Triticum aestivum), and barley (Hordeum vulgare), *J. Agric. Food Chem.* 58(13), 8100–8103.

Rajiv, J., Soumya, C., Indrani, D. and Venkateswara Rao, G. (2011). Effect of replacement of wheat flour with finger millet flour (Eleusine corcana) on the batter microscopy, rheology and quality characteristics of muffins, *J. Texture Stud.* 42(6), 478–489.

Ramashia, S. E., Anyasi, T. A., Gwata, E. T., Meddows-Taylor, S. and Jideani, A. I. O. (2019). Processing, nutritional composition and health benefits of finger millet in sub-Saharan Africa, *Food Sci. Technol* 39(2), 253–266.

Reddy, V. P., Faubin, J. M. and Hoseney, R. C. (1986). Odour generation in ground, stored pearl millet, *Cereal Chem.* 63, 383–406.

Ross, R. P., Morgan, S. and Hill, C. (2002). Preservation and fermentation: past, present and future, *Int. J. Food Microbiol.* 79(1–2), 3–16.

Saleh, A. S. M., Zhang, Q., Chen, J. and Shen, Q. (2013). Millet grains: nutritional quality, processing, and potential health benefits, *Compr. Rev. Food Sci. Food Saf.* 12(3), 281–295.

Salvucci, E., LeBlanc, J. G. and Perez, G. (2016). Technological properties of lactic acid bacteria isolated from raw cereal material, *LWT* 70, 185–191.

Schoenlechner, R., Szatmari, M., Bagdi, A. and Tömösközi, S. (2013). Optimisation of bread quality produced from wheat and proso millet (Panicum miliaceum L.) by adding emulsifiers, transglutaminase and xylanase, *LWT* 51(1), 361–366.

Serna-Saldivar, S. O. and Espinosa-Ramirez, J. (2019). Grain structure and grain chemical composition. In: Taylor, J. R. N. and Duodu, K. G. (Eds), *Sorghum and Millets:*

Chemistry, Technology and Nutritional Attributes, London, UK: AACC International and Elsevier, pp. 85–129.
Shamai, K., Bianco-Peled, H. and Shimoni, E. (2003). Polymorphism of resistant starch type III, *Carbohydr. Polym.* 54(3), 363–369.
Sharma, B. and Gujral, H. S. (2019). Modulation in quality attributes of dough and starch digestibility of unleavened flat bread on replacing wheat flour with different minor millet flours, *Int. J. Biolog. Macrom.* 141, 117–124.
Sharma, B., Gujral, H. S. and Solah, V. (2017). Effect of incorporating finger millet in wheat flour on Mixolab behavior, chapatti quality and starch digestibility, *Food Chem.* 231, 156–164.
Shobana, S., Krishnaswamy, K., Sudha, V., Malleshi, N. G., Anjana, R. M., Palaniappan, L. and Mohan, V. (2013). Finger millet (*Ragi, Eleusine coracana L.*). Review of its nutritional properties, processing and plausible health benefits, *Adv. Food Nutr. Res.* 69, 1–39.
Shukla, K. and Srivastava, S. (2014). Evaluation of finger millet incorporated noodles for nutritive value and glycemic index, *J. Food Sci. Technol.* 51(3), 527–534.
Singh, P. and Srivastava, S. (2008). Development and quality evaluation of iron rich biscuit mixes using finger millet, *J. Commun. Mobiliz Sustain. Dev.* 3(2), 89–94.
Sobana, R. (2017). Quality evaluation of millet based composite sports bar, *Int. J. Food Sci. Nutr.* 2(4), 65–68.
Soro-Yao, A. A., Brou, K., Amani, G., Thonart, P. and Djè, K. M. (2014). The use of lactic acid bacteria starter cultures during the processing of fermented cereal-based foods in West Africa: a review, *Trop. Life Sci. Res.* 25(2), 81–100.
Sumathi, A., Ushakumari, S. R. and Malleshi, N. G. (2007). Physico-chemical characteristics, nutritional quality and shelf-life of pearl millet based extrusion cooked supplementary foods, *Int. J. Food Sci. Nutr.* 58(5), 350–362.
Syeunda, C. O., Anyango, J. O., Faraj, A. K. and Kimurto, P. K. (2021). *In vitro* protein digestibility of finger millet complementary porridge as affected by compositing precooked cowpea with improved malted finger millet, *J. Food Sci. Technol.* 58(2), 571–580. doi: 10.1007/s13197-020-04569-1.
Taylor, J. R. N. (2016). Millet pearl: overview. In: Wrigley, C., Corke, H., Seetharaman, K. and Faubion, J. (Eds), *Encyclopedia of Food Grains*, Oxford: Academic Press, pp. 190–198.
Taylor, J. R. N. (2017). Millets: their unique nutritional and health-promoting attributes. In: Taylor, J. R. N. and Awika, J. M. (Eds), *Gluten-Free Ancient Grains*, UK: Woodhead Publishing, and Elsevier, pp. 55–103.
Taylor, J. R. N. (2019). Sorghum and millets: taxonomy, history, distribution and production. In: Taylor, J. R. N. and Duodu, K. G. (Eds), *Sorghum and Millets: Chemistry, Technology and Nutritional Attributes*, London, UK: AACC International and Elsevier, pp. 1–21.
Taylor, J. R. N. and Dewar, J. (2001). Developments in sorghum food technologies, *Adv. Food Nutr. Res.* 43, 217–264.
Taylor, J. R. N. and Duodu, K. G. (2019). Traditional sorghum and millet food and beverage products and their technologies. In: Taylor, J. R. N. and Duodu, K. G. (Eds), *Sorghum and Millets: Chemistry, Technology and Nutritional Attributes*, London, UK: AACC International and Elsevier, pp. 259–292.
Taylor, J. R. N. and Emmambux, M. N. (2008). Gluten-free cereal products and beverages. In: Arendt, E. K. and Bello, F. D. (Eds), *Gluten-Free Foods and Beverages from*, Amsterdam: Millets, Elsevier, pp. 119–148.

Taylor, J. R. N. and Kruger, J. (2019). Sorghum and millets: food and beverage nutritional attributes. In: Taylor, J. R. N. and Duodu, K. G. (Eds), *Sorghum and Millets: Chemistry, Technology and Nutritional Attributes*, London, UK: AACC International and Elsevier, pp. 171–224.

Taylor, J. R. N., Schober, T. J. and Bean, S. R. (2006). Novel food and non-food uses for sorghum and millets, *J. Cereal Sci.* 44(3), 252–271.

Taylor, J. R. N. and Taylor, J. (2017). Proteins from sorghum and millets. In: Nadathur, S. R., Wanasundara, J. P. D. and Scanlin, L. (Eds), *Sustainable Protein Sources*, San Diego: Academic Press, pp. 79–104.

Temba, M. C., Njobeh, P. B., Adebo, O. A., Olugbile, A. O. and Kayitesi, E. (2016). The role of compositing cereals with legumes to alleviate protein energy malnutrition in Africa, *Int. J. Food Sci. Technol.* 51(3), 543–554.

Tembo, D. T., Holmes, M. J. and Marshall, L. J. (2017). Effect of thermal treatment and storage on bioactive compounds, organic acids and antioxidant activity of baobab fruit (Adansonia digitata) pulp from Malawi, *J. Food Compos. Anal.* 58, 40–51.

Teucher, B., Olivares, M. and Cori, H. (2004). Enhancers of iron absorption: ascorbic acid and other organic acids, *Int. J. Vitam. Nutr. Res.* 74 (6), 403–419.

Theander, O. and Westerlund, E. (1987). Studies on chemical modifications in heat-processed starch and wheat flour, *Starch/Stärke* 39(3), 88–93.

Traore, T., Mouquet, C., Icard-Verniere, C., Traore, A. S. and Treche, S. (2004). Changes in nutrient composition, phytate and cyanide contents and α-amylase activity during cereal malting in small production units in Ouagadougou (Burkina Faso), *Food Chem.* 88(1), 105–114.

Tumwine, G., Atukwase, A., Tumuhimbise, G. A., Tucungwirwe, F. and Linnemann, A. (2019). Production of nutrient-enhanced millet-based composite flour using skimmed milk powder and vegetables, *Food Sci. Nutr.* 7(1), 22–34.

Urooj, A. and Puttaraj, S. (1994). Effect of processing on starch digestibility in some legumes–an *in vitro* study, *Nahrung* 38(1), 38–46.

Valle, G. D., Quillien, L. and Gueguen, J. (1994). Relationships between processing conditions and starch and protein modifications during extrusion-cooking of pea flour, *J. Sci. Food Agric.* 64(4), 509–517.

Van der Merwe, R., Kruger, J., Ferruzzi, M. G., Duodu, K. G. and Taylor, J. R. N. (2019). Improving iron and zinc bioaccessibility through food-to-food fortification of pearl millet with tropical plant foodstuffs (Moringa leaf powder, roselle calyces and baobab fruit pulp), *J. Food Sci. Technol.* 56(4), 2244–2256.

Vasanthan, T., Gaosong, J., Yeung, J. and Li, J. (2002). Dietary fiber profile of barley flour as affected by extrusion cooking, *Food Chem.* 77(1), 35–40.

Verma, V. and Patel, S. (2013). Value added products from nutri-cereals, finger millet (Eleusine coracana), *Emirates J. Food Agric.* 25(3), 169–176.

Vijayakumar, T. P. and Mohankumar, J. B. (2011). Quality evaluation of dosa from millet flour blend incorporated composite flour, *Elixir Food Sci.* 34, 2624–2629.

Vinoth, A. and Ravindhran, R. (2017). Biofortification in millets: a sustainable approach for nutritional security, *Front. Plant Sci.* 8(29), 29. doi: 10.3389/fpls.2017.00029.

Index